北京市园林绿化局 组织编写
Directed by Beijing Municipal Forestry and Parks Bureau

北京鸟类图谱

主编 赵欣如 朱 雷
副主编 黄瀚晨 关雪燕 关翔宇 李兆楠 张 琴

EDITOR-IN-CHIEF ZHAO Xinru ZHU Lei
DEPUTY EDITOR-IN-CHIEF HUANG Hanchen GUAN Xueyan
GUAN Xiangyu LI Zhaonan ZHANG Shen

中国林业出版社
·北 京·

CHINA FORESTRY PUBLISHING HOUSE
BEIJING

内容简介

本书对北京地区分布的 22 目 74 科 243 属的 508 种鸟，以中英文对照的形式按陆禽、游禽、涉禽、攀禽、猛禽、鸣禽六大生态类群依次呈现，每个鸟种精选 1~3 幅彩色照片，反映其形态特征和栖息生境，并简要记述鸟种的形态特征、分布与生境、习性与生态等。

本书适合从事野生动物研究与保护、生物多样性保护、科学普及、自然教育的科技人员以及鸟类爱好者、教师、学生阅读和使用。

Introduction

A total of 508 species of birds of Beijing, in 22 orders, 74 families and 243 genera, are described in order based on the six functional groups: landfowls, natatores, wading birds, scansorial birds, raptors and passerines. One to three photographs are used to present the characteristics of each bird species, followed with an introduction to morphology, distribution, habitat, behaviours and ecology of the species in both Chinese and English.

The following groups of people would be the most helped or entertained by this book: academic researchers, staffs, students and amateurs in the fields of wildlife research, biodiversity conservation, science communication and environmental education.

图书在版编目（CIP）数据

北京鸟类图谱 / 北京市园林绿化局组织编写；赵欣如，朱雷主编. —北京：中国林业出版社，2021.11（2022.5重印）
ISBN 978-7-5219-1384-2

Ⅰ. ①北… Ⅱ. ①北… ②赵… ③朱… Ⅲ. ①鸟类-北京-图谱 Ⅳ. ①Q959.708-64

中国版本图书馆 CIP 数据核字（2021）第 205156 号

审图号：京 S（2021）070 号

出版发行　中国林业出版社（100009　北京市西城区刘海胡同7号）
印　　刷　北京雅昌艺术印刷有限公司
版　　次　2021年11月第1版
印　　次　2022年5月第2次
开　　本　710mm × 1000mm　1/16
字　　数　1273千字
印　　张　38.5
定　　价　368.00元

编辑委员会
EDITORIAL BOARD

编写组
CONTRIBUTORS

策　划
王小平　吕　植

Project Planners
WANG Xiaoping　LU Zhi

主　编
赵欣如　朱　雷

Editor-in-chief
ZHAO Xinru　ZHU Lei

副主编
黄瀚晨　关雪燕　关翔宇　李兆楠
张　莘

Deputy Editor-in-chief
HUANG Hanchen　GUAN Xueyan　GUAN Xiangyu
LI Zhaonan　ZHANG Shen

中文撰写
朱　雷　李兆楠　黄瀚晨　赵　敏
关翔宇　赵欣如

Chinese Authors
ZHU Lei　LI Zhaonan　HUANG Hanchen
ZHAO Min　GUAN Xiangyu　ZHAO Xinru

中文审校
林　红　朱　雷　张　莘　黄瀚晨

Chinese Copy Editors
LIN Hong　ZHU Lei　ZHANG Shen　HUANG Hanchen

英文翻译
朱　雷　叶　红　刘　明　黄瀚晨
赵　敏　曾　晴

English Translators
ZHU Lei　YE Hong　LIU Ming
HUANG Hanchen　ZHAO Min　ZENG Qing

英文审校
张　莘　朱　雷　唐　瑞

English Copy Editors
ZHANG Shen　ZHU Lei　Terry TOWNSHEND

鸟种图片审核
关翔宇　关雪燕

Bird Photographs Reviewers
GUAN Xiangyu　GUAN Xueyan

绘　图
崔　月　张　莘　郑秋旸　朱　雷

Illustrators
CUI Yue　ZHANG Shen　ZHENG Qiuyang　ZHU Lei

排　版
柴鉴云

Layout
CHAI Jianyun

责任编辑
刘香瑞

Executive Editor
LIU Xiangrui

摄 影
PHOTOGRAPHERS

文 辉	关翔宇	沈 岩	关雪燕	郭亮/IBE	张 龙	宋炜东	刘爱华
杨 军	娄方洲	孙 琰	朱 雷	郑永胜	张玉平	叶思伦	郑秋旸
赵云天	张 永	刘 璐	黄长锁	牛 震	罗义华	刘 武	张小玲
严志文	徐永春	邱小熙	刘 强	李兆楠	崔 月	宋 健	甘礼清
董江天	蔡欣然	叶 腾	慕 童	刘建国	林长洛	Gerald GARDEBLED	
J.M.GARG/IBE		Dûrzan CÎRANO/IBE		李 甦	巩慧君	朱 晖	任旭清
刘一红	刘曙海	许明岗	李东明	李 强	杨 虹	张连喜	张 武
武 钢	林琦玮	林 曦	钟宏英	钟悦陶	钱 锋	曹晓荣	宋 旭
颜小勤	潘文星	魏子晨					

WEN Hui	GUAN Xiangyu	SHEN Yan	GUAN Xueyan	GUO Liang/IBE
ZHANG Long	SONG Weidong	LIU Aihua	YANG Jun	LOU Fangzhou
SUN Yan	ZHU Lei	ZHENG Yongsheng	ZHANG Yuping	YE Silun
ZHENG Qiuyang	ZHAO Yuntian	ZHANG Yong	LIU Lu	HUANG Changsuo
NIU Zhen	LUO Yihua	LIU Wu	ZHANG Xiaoling	YAN Zhiwen
XU Yongchun	QIU Xiaoxi	LIU Qiang	LI Zhaonan	CUI Yue
SONG Jian	GAN Liqing	DONG Jiangtian	CAI Xinran	YE Teng
MU Tong	LIU Jianguo	LIN Changluo	Gerald GARDEBLED	J.M.GARG/IBE
Dûrzan CÎRANO/IBE		LI Su	GONG Huijun	ZHU Hui
REN Xuqing	LIU Yihong	LIU Shuhai	XU Minggang	LI Dongming
LI Qiang	YANG Hong	ZHANG Lianxi	ZHANG Wu	WU Gang
LIN Qiwei	LIN Xi	ZHONG Hongying	ZHONG Yuetao	QIAN Feng
CAO Xiaorong	SONG Xu	YAN Xiaoqin	PAN Wenxing	WEI Zichen

序一

北京不仅是我国的首都，也是一座飞速发展的国际化大都市。由于其地理上的优势与特点，该地区动植物资源相对丰富，仅就鸟类而言，目前的记录已超过 500 种，是世界上鸟类种数最为丰富的首都城市之一。

北京地区的鸟类科学研究，可追溯到 19 世纪末，David (1877)、Sowerby (1914) 等外国研究者开始对北京地区的鸟类进行零星研究，我国学者寿振黄（1936）在相关研究中也涉及北京地区的鸟类。新中国成立以后，北京师范大学、首都师范大学、北京林业大学、北京自然博物馆、中国科学院动物研究所等高校、博物馆及科研机构对北京地区鸟类分布、鸟类居留情况、鸟类区系与生态分布及鸟种新记录等开展了一些有价值的研究，对北京地区的鸟类分类及生态研究积累了不少基础数据和信息。我从事鸟类学的研究是从北京地区开始的，早年曾在教学之余，专门对北京及其附近地区的冬季鸟类（1962）、夏季鸟类（1984）的生态分布进行过较为系统的研究，不仅报道了北京鸟类在冬、夏两季的组成，也关注鸟类在北京地区与景观和生态的联系，探讨一些鸟类与生物多样性的基本关系。

如今，人们已经认识到，鸟类是对环境变化最为敏感的类群，许多鸟种在一个地区的有或无及多与少，往往可以用来评价一个地区的环境质量和生态状况。因此，关注北京地区鸟类的分布及生存状况显得非常重要。

北京地区在动物区系上，虽然属古北界，但在鸟类群落上反映出的情况复杂多样，鸟类的组成以古北界鸟种为多，也有不少东洋界鸟种的渗透，还有一些广布种出现。因此，该地区鸟类区系成分复杂，鸟种数量繁多。

北京地区仅占我国国土面积的 0.17%，而鸟类种数却占全国鸟种数的 35%，由此反映出该地区鸟类物种的丰富程度。随着环境学与生态学的发展，世界各国都越来越重视生物多样性的保护，我国也在大力推进生态保护的发展战略。对于北京地区而言，鸟类及相关生物类群的保护尤为重要。

《北京鸟类图谱》的出版，将会对北京地区鸟类及相关生物类群的保护起到积极作用，也会对相关研究与科学教育产生良好影响。该书是基于作者近年来对北京地区鸟类最新的观察与记录的系统梳理，编写的内容系统性强，种类的增加与更新远超出同类书籍。书中对每个鸟种都以中英文对照方式简要介绍，并配有表现鸟种形态特征、栖息生境的照片，使中外读者都能直观、快捷地获取相关鸟种信息，对鸟的形态、行为、分布，特别是在北京地区的栖息地及居留类型等内容有清晰的了解。这些信息的获取，对北京地区的自然保护机构、园林绿化部门、鸟类监测站点等专业人员开展日常保护与监测工作很有帮助，也给热爱鸟类的朋友提供了一本科学性、实用性很强的好书。该书的另一个特点是，在符合鸟类的分类体系前提下，全书按陆禽、游禽、涉禽、攀禽、猛禽、鸣禽等六大生态类群呈现，将生态习性相似的鸟种集中编排，方便更多读者的查阅使用。

我深信，依靠学科研究的力量，发挥职能机构的作用，鼓励民间公民科学的积极参与，就能有效保护和建设北京的生态，使北京的自然和社会健康而永续发展。

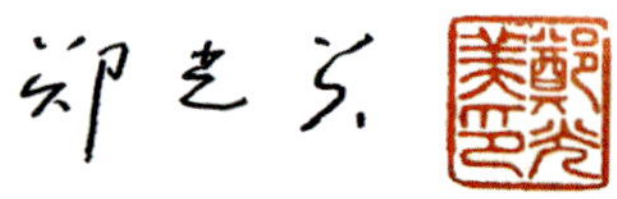

中国科学院院士
北京师范大学教授
2021 年 6 月 20 日

Foreward by ZHENG Guangmei

Beijing is the capital of China, as well as an international city. Because of its geographical position and characteristics, it is quite rich in wildlife resources. More than 500 species of birds have been recorded in Beijing, which means Beijing is amongst the capitals in the world with the highest richness of bird species.

The academic studies on birds of Beijing can be traced back to the end of the 19th century when a few studies were taken by foreign researchers (David, 1877; Sowerby, 1914), as well as the Chinese scientist Shou Zhenhuang (1936). Since the foundation of the People's Republic of China, many valuable studies have been carried out on distribution, habitat, species and new records by Beijing Normal University, Capital Normal University, Beijing Forestry University, Beijing Natural History Museum, Institute of Zoology, Chinese Academy of Sciences, etc., all of which contribute to knowledge on taxonomy, distribution and ecology of birds in Beijing. In my earlier years, I also conducted a few ornithological research focused on the ecology and distribution of wintering (1962) and summering birds (1984) in Beijing. My studies documented the bird communities in Beijing during winter and summer, and also revealed the links between birds, the landscape and environment, and explored the relationship between birds and the rest of biodiversity.

It has been widely realised that birds are among the most sensitive species to environmental changes. The presence and abundance of some species in an area could be considered as indicators of environmental quality and ecological status.

Beijing is located in the Palaearctic Realm which has a complex and diverse bird community. While most Beijing birds are Palaearctic species, there are also some Oriental species and some cosmopolitan species represented. Therefore, the composition of the bird flora in the Beijing area is complicated and with rich diversity.

Although Beijing occupies only 0.17% of China's land area, the species of birds numbers 35% of the total bird species of China. With the development of environmental science and ecology, countries all over the world are paying more and more attention to biodiversity conservation, and China is also vigorously promoting the strategy of ecological civilisation. The conservation of biodiversity in Beijing is, therefore, vital.

The publication of this "Birds of Beijing" will play an important role in the conservation of birds and related species in Beijing, and also contributes to related research, science and education. This book is a systematic review of the latest observations and records of birds in Beijing. The content is very comprehensive with fully updated information. In the book, each bird species was introduced in both Chinese and English, with photos showing its morphological characteristics as well as habitats. The reader can thus intuitively gain clear knowledge on morphology, behaviour, distribution, and especially habitats of birds in Beijing. The information is very helpful to the professional staffs of nature conservation agencies, landscaping departments, bird monitoring stations, etc., for daily monitoring work and conservation. It is also a scientific and practical book for the amateurs who are interested in birds. I would like to highlight that, based on the taxonomic classification system, the bird species described in the book are presented in six functional groups (landfowls, natatores, wading birds, scansorial birds, raptors and passerines). The compiling of similar ecological characteristics will help readers to use the book quickly and easily.

I believe, with the strength of scientific research, the role of functional institutions and the involvement of citizen science, the environment of Beijing can be effectively improved, and nature and human society of Beijing can both develop in a healthy and sustainable way.

ZHENG Guangmei

Academician, Chinese Academy of Sciences

Beijing Normal University

20th June 2021

序二

生物多样性及其提供的生态系统功能和服务支持着地球上所有形式的生命，是我们人类和地球的健康与福祉、经济增长和可持续发展的基础。近年来，国际社会对生物多样性保护的呼声愈发高涨，我国更是对此高度重视。在建设生态文明，共建地球生命共同体的理念指导下，中国的生态文明思想和成功实践正在快速走向世界，尊重自然、顺应自然、保护自然深入人心，有力地推动了生物多样性保护事业的发展。

北京是中国的首都，为了探索适宜北京生物多样性保护的策略和方法，北京市园林绿化局连续三年实施了“北京城市生物多样性恢复与公众自然教育”项目，产出了一系列基于北京自然禀赋和特色的标准、指南和报告。欣闻项目即将出版《北京鸟类图谱》一书，甚为高兴。北京处于东亚—澳大利西亚全球候鸟迁徙路线上。按照国际鸟盟的标准，北京市范围内划定了密云水库、官厅水库、东灵山、百花山自然保护区和松山自然保护区等 5 个重点鸟区，鸟类资源丰富、物种多样性高。

《北京鸟类图谱》由北京市园林绿化局组织专家团队和长期观鸟、懂鸟、爱鸟的志愿者们共同编写完成。书中收录北京地区有分布的 22 目 74 科 243 属共 508 种鸟类，以中英文对照的形式，图文并茂地按陆禽、游禽、涉禽、攀禽、猛禽、鸣禽六大生态类群依次呈现。图谱所采用的每一张照片都来之不易，从上万张素材中以严谨、认真的态度逐一选出，力求采编的图片能够准确反应鸟类的性状或生态。对于鸟类爱好者和中小学生而言，这是一本难得的科普读物；而对于专业的鸟类保护人员来讲，则是一本非常实用的基础性工具书。

联合国《生物多样性公约》第十五次缔约方大会第一阶段会议于 10 月 11~15 日在中国昆明成功召开，第二阶段会议将于 2022 年举办。大会将全面总结国际社会在生物多样性保护方面的经验，审议通过未来十年全球生物多样性治理的蓝图。在此期间，系统总结我国生物多样性保护的成效及经验尤为重要。北京作为中国的首善之都，是全世界瞩目的焦点，《北京鸟类图谱》的出版恰逢其时。我相信，图谱的出版，必将有助于北京市生物多样性保护工作，特别是为自然教育提供了不可或缺的参考书，也将成为大都市生物多样性社会认知实践的范例。

中国科学院植物研究所研究员
中国科学院生物多样性委员会副主任兼秘书长
世界自然保护联盟理事兼亚洲区会员委员会主席
2021 年 10 月

Foreward by MA Keping

Biodiversity and the ecosystem functions and services it provides support all forms of life on earth. Biodiversity is fundamental to the health and well-being, economic growth and sustainable development of our people and planet. In recent years, the international community is increasingly calling for biodiversity conservation, and China attaches great importance to it. Guided by the concept of building an ecological civilisation and building a global community of life, China's ideas and successful practices of ecological civilisation are rapidly spreading to the world. The principles of respecting and protecting nature and obeying nature's laws have gained popular support, giving a strong boost to conservation of biodiversity.

Beijing is the capital of China. In order to explore strategies and methods suitable for biodiversity conservation in Beijing, Beijing Municipal Forestry and Parks Bureau has carried out the "Beijing Urban Biodiversity Restoration and Public Nature Education" project for three consecutive years, and produced a series of standards, guidelines and reports based on Beijing's natural endowment and characteristics. I am delighted to learn that the project is about to publish a book entitled *Birds of Beijing*. Beijing is located on the East Asian-Australasian flyway. In line with BirdLife's standards, five Important Bird and Biodiversity Areas have been designated in Beijing: Miyun Reservoir, Guanting reservoir, Dongling Mountain, Baihuashan Nature Reserve and Songshan Nature Reserve, with rich bird resources and high species diversity.

Organised by Beijing Municipal Forestry and Parks Bureau, *Birds of Beijing* was compiled by a team of experts and volunteers who have been watching, understanding and loving birds for a long time. 508 species of birds, belonging to 22 orders, 74 families and 243 genera, are included in the book, which is illustrated in English and Chinese by six ecological groups, namely, land birds, wandering birds, wading birds, climbing birds, raptors and songbirds. Each photo used in the atlas is hard won. From tens of thousands of materials, we selected them one by one in a rigorous and serious manner, striving to accurately reflect the characteristics or ecology of birds. For bird lovers and primary and secondary school students, this is a rare popular science reading; for professional bird protection personnel, it is a very practical basic reference book.

The first phase of the 15th meeting of the Conference of the Parties to the

United Nations Convention on Biological Diversity, aka COP 15, was successfully held in Kunming, China from October 11 to 15, and the second phase will be held in 2022. The Conference will comprehensively review the experience of the international community in biodiversity conservation and consider and adopt a blueprint for global biodiversity governance in the coming decade. During this period, it is particularly important to systematically summarize the effectiveness and experience of biodiversity conservation in China. As the capital of China, Beijing is the focus of the world's attention, and the publication of *Birds of Beijing* is timely. I believe that the publication of the book will certainly contribute to the conservation of biodiversity in Beijing, especially provide an indispensable reference book for nature education, and serve as an example for the practice of social cognition of biodiversity in metropolises.

MA Keping

Professor, Institute of Botany, Chinese Academy of Sciences

Deputy Director and Secretary General of Biodiversity Commission, Chinese Academy of Sciences

Member of the International Union for Conservation of Nature and Chairman of the Asian Member Committee

October 2021

序三

北京，中华人民共和国的首都，拥有极为多样的鸟类，然而这些鸟类在很大程度上却不被大部分居民和游客所知。北京市有观测记录的鸟类达500多种，在世界首都中居前列，跟我的祖国瑞典全国所观测记录到的鸟类总数差不多!

北京鸟类极其丰富有几个方面的原因。北京及其周边地区自然栖息地种类丰富，如不同类型的森林、湿地、草原、平原及山区，最高海拔达2300多米。同时，北京位于“东亚—澳大利西亚迁飞区”上，在这条迁徙路线上每年有数百万迁徙鸟往返于东北亚的繁殖地和南方越冬地。

全球仅繁殖于中国的鸟类中，超过10种在北京有繁殖记载。这其中有褐头鸫、绿背姬鹟、白腹暗蓝鹟，它们在北京地区有最大的种群，广受观鸟者青睐。

北京全年都可以看到鸟类，但是在春秋的迁徙季节可以看到的鸟类种类和个体数量最多。在这期间，城市公园和大学校园成为鸟类最佳中途停留地，这些地方的部分区域“留野”的话就更好。很多在北京繁殖或者迁徙途径的鸟类都是环球旅行家。例如，科学家追踪到北京雨燕、大杜鹃、红脚隼一路飞至非洲——这样超长距离的旅程一年有两次。

我于1984年首次访问北京，之后有幸在北京呆过很长时间，其中有两年半在中国科学院任访问教授。在这段时间里，北京在很多方面发生了巨大的变化。像世界上很多首都一样，城市发展给自然环境带来很大的压力。然而，很显然当北京正致力于建设成21世纪清洁、绿色、宜居的现代化都市的同时，越来越多的人认识到北京丰富的生物多样性、认识到这种自然遗产是巨大的财富。维持和培育这种生物多样性需要保护和修复森林、草原、湿地在内的所有核心栖息地。

决策者和特别是年轻群体在内的公众相关知识和参与程度的提高，是打造适合鸟类、其他生物及人类更好生存环境的第一步。

本书是一本集大成的书籍，将为增加对北京鸟类的了解做出重要贡献。我相信本书将激发人们对首都北京鸟类的兴趣、激发人们去保护鸟类本身及其赖以生存的栖息地。

Per Alström

瑞典乌普萨拉大学鸟类学教授

2021年10月

Foreward by Per ALSTRÖM

Beijing, the capital of the People's Republic of China, has an enormously rich birdlife, which is largely unknown to most human inhabitants and visitors. More than 500 species of birds have been observed in the municipality, amongst the highest figures for any capital in the world, and similar to the total number of birds observed in my home country, Sweden!

There are several reasons for the tremendous diversity of birds in Beijing. The city and its surroundings hold a great variety of natural habitats, including different types of forests, wetlands and grasslands, and plains as well as hilly and mountainous areas, with the highest peak rising to more than 2,300 meters. Beijing is also located on the "East Asian-Australasian flyway", where many millions of birds migrate biannually between their breeding grounds in northeast Asia and winter quarters further south.

More than ten of the bird species that breed in the Beijing municipality breed only in China in the world. Some of these, e.g. the Grey-sided Thrush, Green-backed Flycatcher and Zappey's Flycatcher, have some of their strongest populations in the Beijing area, and are much sought-after by visiting birdwatchers.

Birds can be found throughout the year in Beijing, although the greatest diversity of species and largest numbers of individuals can be encountered during the migration periods, in spring and autumn. During this time, city parks and university campuses can be excellent stopover spots for birds, especially if part of these areas are left "wild". Many of the species that breed in or migrate through Beijing are real globetrotters. For example, the Beijing Swift, Common Cuckoo and Amur Falcons have been tracked by scientists all the way to Africa – an enormous distance travelled twice a year.

I have had the privilege to spend much time in Beijing since my first visit in 1984, and including nearly 2.5 years as a Visiting Professor at the Chinese Academy of Sciences. During this extended period, Beijing has gone through an enormous transformation in multiple aspects. As in many capital cities around the world, development has put massive pressure on the natural environment. However, it is clear that there is a growing awareness of Beijing's rich biodiversity and that this natural heritage is a major asset as the city works to become a clean, green and desirable modern city for the 21st century. To maintain and nurture this biodiversity will require the protection and restoration of all key habitats, including forests, grassland and wetlands.

Increased knowledge and engagement, both from decision-makers and the public, particularly from young people, is a first step towards creating an environment where birds, other organisms and humans can prosper.

This book is a significant achievement and an extremely important contribution towards our knowledge of the avifauna of Beijing. I am confident that it will stimulate a greater interest in the capital's birds and the protection of the birds themselves and of the habitats on which they depend.

Professor of Ornithology
Uppsala University
Sweden
October 2021

前言

北京地处华北平原西北端，西、北和东北部群山环绕，由西部的太行山脉和北部的燕山山脉形成环绕形山体，东南部为缓慢向渤海延伸的洪积平原，又称“北京湾”。平原海拔 20~60 米，山地一般海拔 1000~1500 米，东灵山为北京市最高峰，海拔 2303 米。由于地势、气候、海拔、景观、植被的多样与变化，使得鸟类及其他动植物种类十分丰富，鸟类的分布体现出垂直变化和季节性变化特征。同时，北京位于东亚—澳大利西亚的候鸟迁徙路线上，在每年的春、秋两季，北京的旅鸟种类异常丰富。有记录以来，北京地区的鸟类种数超过 500 种，隶属于 22 目 74 科 243 属。从鸟类的居留类型看，北京地区有常年留居的留鸟、夏季繁殖的夏候鸟、冬季越冬的冬候鸟、春秋季节迁徙途径北京的旅鸟，还有少量不曾在北京分布，却偶然出现被视为迷鸟的零星种类。

1949 年之前，对北京地区鸟类的研究甚少，仅有一些零星的文献报道，1877 年谭卫道（David）、1914 年苏柯仁（Sowerby）、1914—1921 年拉图许（La Touche）、1924—1938 年万卓志和胡本德（Wilder and Hubbard）、1936 年寿振黄、1946—1951 年赫明森（Hemmingsen）等有相关著作或报告。他们的研究往往是在调查我国东北、河北地区的鸟类时涉及北京地区，采集的标本也多都流入各国博物馆。

新中国成立以后，北京地区鸟类资源的调查与研究逐步展开。除了常见鸟种的调查之外，还对鸟类的生态开始关注，其中有北京师范大学郑光美先生（1962）发表的《北京及其附近地区冬季鸟类的生态分布》，郑光美先生（1984）发表的《北京及其附近地区夏季鸟类的生态分布》，对该地区鸟类区系的两个重要季节阐述研究结果；北京自然博物馆蔡其侃先生（1987）出版的《北京鸟类志》，系统整理北京地区鸟类分布的种类为 344 种；首都师范大学高武先生（1994）等出版的《北京脊椎动物检索表》整理的鸟种为 375 种；北京师范大学赵欣如先生（2014）出版的《北京鸟类图鉴》（第 2 版），整理的鸟种为 448 种，其中包括在北京地区生活的少量逃逸鸟种。随着北京地区鸟类研究的深入以及民间观鸟活动的开展，近年在北京地区发现和记录的鸟种仍在不断增加。黄瀚晨、赵欣如先生等对每种鸟逐一查询、考证、确认出处，最终统计的鸟种数已达 508 种之多。

《北京鸟类图谱》是“北京城市生物多样性恢复与公众自然教育”项目重要的产出，是对前人研究的继承和发扬，兼顾了科研与科普双重功能。“北京城市生物多样性恢复与公众自然教育”项目以习近平生态文明思想为指导，以生动践行“两山论”，探索全市生物多样性保护与恢复的新路径、新模式为目标，以解决林地绿地“绿而不活、有林没鸟”为突破，以构建“自然北京”为美好愿景。在该项目的支持下，我们根据最新研究编写了此书，依据最新的鸟类名录分种进行描述呈现。《北京鸟类图谱》的出版，将对北京地区鸟类及相关生物类群的保护起到积极作用，并且有助于鸟类科学研究和自然教育的开展。

《北京鸟类图谱》鸟种的学名和中文名称，采用郑光美（2017）《中国鸟类分类与分布名录》（第三版），请读者在阅读各类工具书和使用鸟种名称时要注意名称的来源，以减少近年来由于分类体系不同带来学术交流上的困扰。

为便于科研监测专业人员和更广泛的保护管理人员及社会热爱鸟类的人士使用本书，在郑光美（2017）《中国鸟类分类与分布名录》（第三版）名录的框架下，按鸟类的生态类群（陆禽、游禽、涉禽、攀禽、猛禽、鸣禽）按种排序呈现，便于各类读者阅读使用。

特别感谢张雁云教授对该书出版的指导与帮助！

由于编者专业能力和水平有限，书中难免出现错漏，敬请读者批评指正。

编著者

2021 年 6 月 18 日

Preface

Beijing is located in the north western part of the North China Plain, surrounded by mountains of the Taihang Mountains in the west and the Yan Mountains in the north. A floodplain extends to the Bohai Sea in the southeast, which is also known as the "Beijing Bay". The plain has an elevation of 20-60 m while the mountains are generally 1,000-1,500 m, with Dongling Mountain the highest peak in Beijing reaching an elevation of 2,303 m. Due to the diversity of terrain, climate, altitude, landscape, and vegetation, the species of birds and other wildlife are quite rich; distribution of birds varies with elevation and season. Meanwhile, located in the East Asian-Australasian Flyway, migrant species are extremely rich during spring and autumn in Beijing. Over 500 bird species have been recorded in Beijing, which belong to 22 orders, 74 families, and 243 genera. These birds include all-year residents, summer breeders, winter visitors, passage migrants and a few vagrants.

Before 1949, there was little research on birds in Beijing, with only a few reports, including David (1877), Sowerby (1914), La Touche (1914-1921), Wilder and Hubbard (1924-1938), Shaw (1936), Hemmingsen (1946-1951), etc. Typically, Beijing was included in the survey to Northeast China or Hebei; most of the specimens were collected for foreign museums.

The survey and research of bird resources in Beijing launched after the foundation of the People's Republic of China. In addition to the survey of common bird species, the ecology of birds started to be a focus. For example, Mr. Zheng Guangmei (1962, 1984) from Beijing Normal University published "*The Ecological Distribution of Wintering Birds in Beijing and Surrounding Areas*" and "*The Ecological Distribution of Summering Birds in Beijing and Surrounding Areas*", revealing the avian distribution patterns and community structures of two important seasons in Beijing. Also, *Bird of Beijing* by Mr. Cai Qikan (1987) from Beijing Natural History Museum systematically compiled 344 species of birds. *A Key for Identification of the Vertebrates of Beijing* by Mr. Gao Wu et al. (1994) from Capital Normal University listed 375 species of birds. *Atlas of Beijing Birds: Second Edition* published by Mr. Zhao Xinru from Beijing Normal University (2014) illustrated 448 bird species, including a few escaped species. With the development of studies and bird watching activities, an increasing number of bird species has been recorded in Beijing. Mr. Huang Hanchen and Zhao Xinru (2021, submitted for publication) listed a total of 508 species in *An Update to the Checklist of the Birds of Beijing* with carefully verification and confirmation.

This *Birds of Beijing* is an important outcome of the "Beijing Urban Biodiversity Restoration and Public Environment Education" project, which is an inheritance and development of previous research, and has a multiple function of scientific research and science popularisation. Guided by Xi Jinping's thought of ecological civilisation, the project aims to vividly put into practice the "Two Mountains Theory" and explore new approaches and modes of biodiversity conservation and restoration in the Beijing. The project also aims to solve the problem of "green but not alive; no birds in the forest", and to build a "natural Beijing" as its vision. With the support of the project, this book was written based on the latest research, presenting descriptions based on the latest bird taxonomy. The publication of the *Birds of Beijing* will have a positive impact on the conservation of birds and related taxa in Beijing and will contribute to scientific research and nature education on birds.

The scientific and Chinese names in this book were consistent with A Checklist on the *Classification and Distribution of the birds of China: Third Edition* (Zheng, 2017). Readers are advised to pay attention to the origin of the bird names when using various tool books and using the names of bird species, in order to reduce the obstacles in academic communication in recent years due to differences in taxonomic systems.

For the convenience of use by readers, and consistent with A Checklist on the *Classification and Distribution of the birds of China: Third Edition* (Zheng, 2017), this book is presented in categories of functional groups (landfowls, natatores, wading birds, scansorial birds, raptors and passerines). We would like to express our special thanks to Prof. Zhang Yanyun for his guidance and assistance in the publication of this book! Due to the limited professional competence and level of the editor, errors and inaccuracies are inevitable in this book. Your reviews, suggestions and comments are, therefore, welcomed.

Authors

18th June 2021

本书使用说明

本书使用的鸟类分类系统及中文、英文和学名主要依据《中国鸟类分类与分布名录》第三版（郑光美，2017）。在此基础上，我们亦根据世界鸟类名录（版本 11.2，国际鸟类学委员会，2021）等被广泛使用的国际性鸟类名录对个别被收录鸟种的英文名进行了调整。各鸟种保护级别，北京市级的按照北京市园林绿化局 2007 年第 1 号公告《北京市一级保护野生动物名录》、第 2 号公告《北京市二级保护野生动物名录》标注；国家级的按照国家林业和草原局、农业农村部 2021 年 2 月发布的《国家重点保护野生动物名录》标注；我们亦标注了各鸟种在国际自然保护联盟（IUCN）红色名录的濒危等级。

读者可通过目录或索引查找每一个鸟种所在的页码。鸟种页面描述与呈现的信息包括鸟种名称、辨识特征、分布与生境、习性与生态、保护级别和鸟种照片。内页阅读指南如下图所示。

HOW TO USE THIS BOOK

The classification, scientific names, English names and Chinese names of birds in this book generally follow those used in *A Checklist on the Classification and Distribution of the birds of China: Third Edition* (Zheng, 2017). We changed a few English names to more appropriate versions based on *IOC World Bird List* (Gill, F., et al.) and other widely used English bird checklists. For each bird species, municipality-level protected species are noted according to the Beijing Municipal Forestry and Parks Bureau's first and second announcements of 2007, *List of First-class Protected Wildlife in Beijing* and *List of Second-class Protected Wildlife in Beijing*; national-level protected species are noted according to *List of State Key Protected Wild Animals* issued by the National Forestry and Grassland Administration and the Ministry of Agriculture and Rural Affairs in February 2021; we have also noted the conservation status of each bird species based on the International Union for Conservation of Nature (IUCN) Red List of Threatened Species.

The page number of each bird species is given in both contents and index. Each species account includes species names, identification, distribution and habitat, behaviour and ecology, conservation status and photographs. The key features of species accounts and the list of terms used in bird topography are illustrated in the following figures.

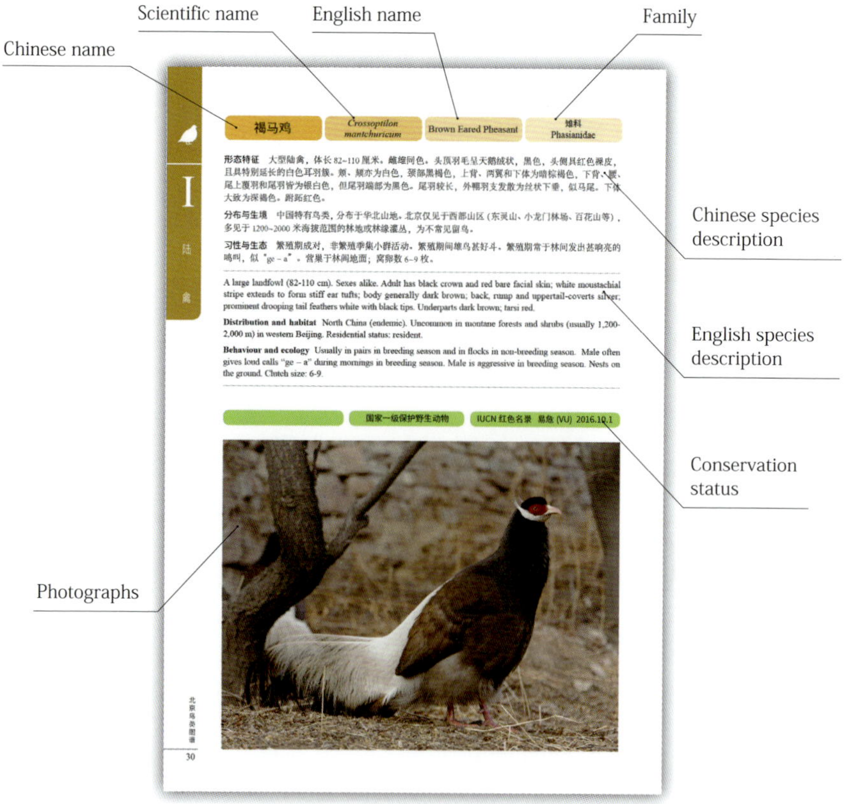

鸟类外部形态结构术语
BIRD TOPOGRAPHY

全身（小型鸣禽） Topography of a small passerine

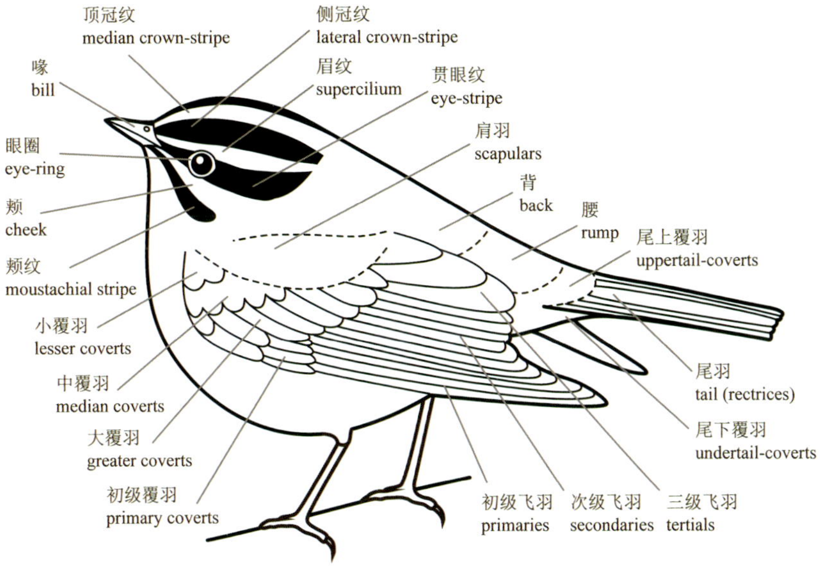

翅 Wing

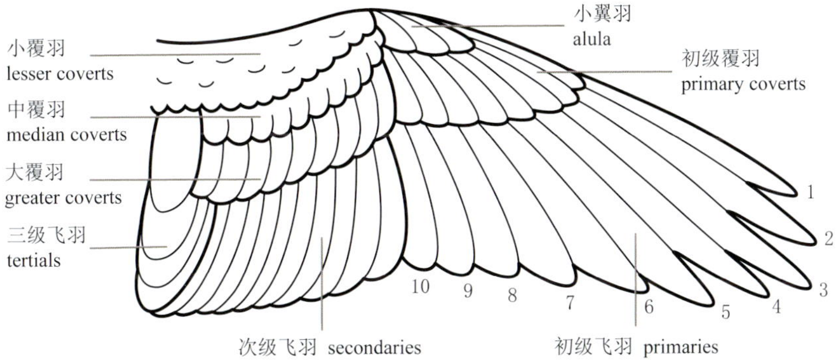

全身（鸥类） Topography of a gull

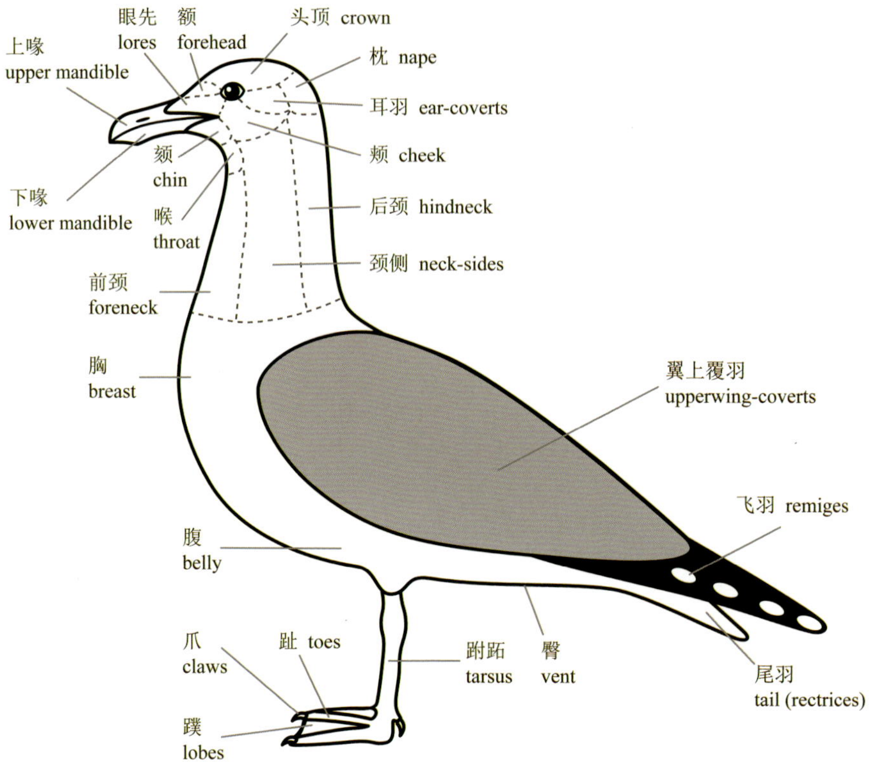

目 录

*注：为控制目录篇幅并方便读者使用，我们在目录中列出了非雀形目各目及雀形目各科对应的页码，并将相应类群所属鸟种按顺序列在目录中。

Contents

第一章 北京自然地理概况

北京位于华北平原的西北端，太行山脉余脉的西山和燕山山脉的军都山在南口关沟相交，形成一个向东南展开的半圆形大山弯，它所围绕的小平原东南部为缓慢向渤海延伸的扇形平原，又称“北京湾”。北京与天津相邻，并与天津一起被河北省环绕。地理坐标为北纬 39°38′~41°05′、东经 115°25′~117°30′。总面积 16807.8 平方公里，其中平原约占 38%（6390.3 平方公里），山区约占 62%（10417.5 平方公里），最高峰为门头沟区的东灵山，海拔 2303 米。2020 年，北京行政区划调整为 16 个区：东城区、西城区、朝阳区、丰台区、石景山区、海淀区、门头沟区、房山区、通州区、顺义区、昌平区、大兴区、怀柔区、平谷区、密云区、延庆区。

北京的气候为典型的北温带半湿润大陆性季风气候，夏季高温多雨，冬季寒冷干燥，春、秋短促。年平均气温在 8~12 摄氏度。年均降水量 600 毫米，降水季节分配很不均匀，全年降水的 70%~80% 集中在夏季 6~8 月。

北京行政区划图

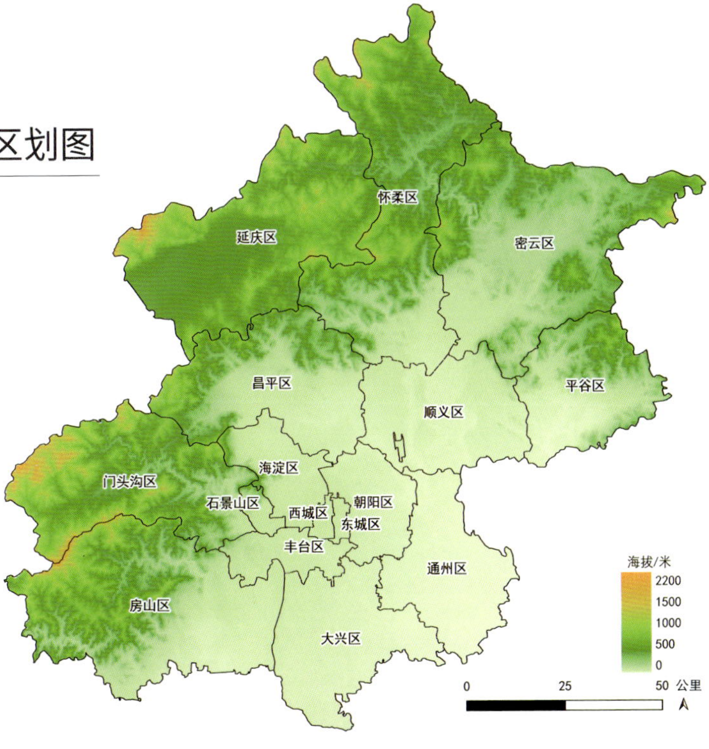

Chapter 1

Introduction of Nature and Geography of Beijing

Beijing is located in the northwest end of the North China Plain, where the western mountains of the remaining Taihang Mountains and the Jundu Mountains of the Yanshan Mountains meet at Nankouguangou, forming a large semicircular mountain bend that unfolds to the southeast. The small plain Nankouguangou surrounds a fan-shaped plain extending towards the Bohai Sea, which is known as "Beijing Bay". Beijing is adjacent to Tianjin and is surrounded by Hebei Province and Tianjin. It is located at 39°38′-41°05′ N, 115°25′-117°30′ E. The total area is 16,807.8 km^2, of which the plain accounts for about 38% (6,390.3 km^2) and the mountains account for about 62% (10,417.5 km^2), with the highest peak being Dongling Mountain in Mentougou District at an altitude of 2,303 m. In 2020, Beijing's administrative divisions comprised 16 districts: Dongcheng District, Xicheng District, Chaoyang District, Fengtai District, Shijingshan District, Haidian District, Mentougou District, Fangshan District, Tongzhou District, Shunyi District, Changping District, Daxing District, Huairou District, Pinggu District, Miyun District, and Yanqing District.

Beijing's climate is a typical semi-humid continental monsoon climate in the northern temperate zone, with hot and rainy summers, cold and dry winters, and short springs and autumns. The average annual temperature is 8-12℃ . The average annual precipitation is 600 mm, seasonally very uneven, with 70%-80% of the annual precipitation concentrated between June and August.

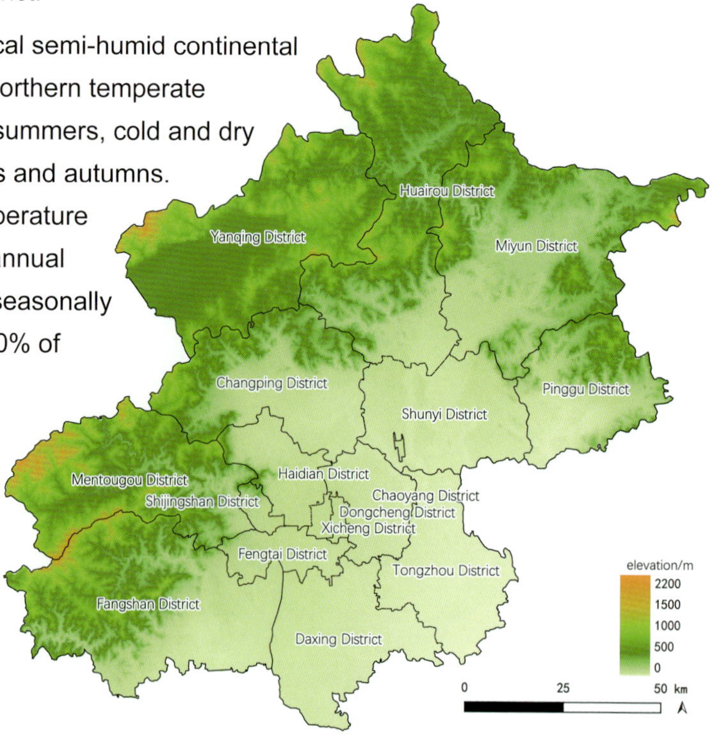

Administrative division of Beijing

第二章 北京鸟类区系特点

北京地区鸟类达 508 种，隶属于 22 目 74 科 243 属。鸟类组成非常复杂，虽然以古北界种类为主，但也有不少东洋界种类和广布种类，表现出我国古北界与东洋界物种的交错与混合，鸟类的分布型呈现多样，体现出两界及其亚界之间的渗透与过渡性。根据张荣祖（2011）《中国动物地理》，按照动物地理区划分级，北京属古北界东北亚界华北区（内含黄土高原亚区和黄淮平原亚区）。

北京地区分布的鸟种主要是北方种类：

古北型的大鸨（*Otis tarda*）、灰鹤（*Grus grus*）、普通雨燕（*Apus apus*）、大杜鹃（*Cuculus canorus*）、普通翠鸟（*Alcedo atthis*）、普通鵟（*Buteo japonicus*）、黑头䴓（*Sitta villosa*）、麻雀（*Passer montanus*）、灰喜鹊（*Cyanopica cyanus*）、喜鹊（*Pica pica*）、秃鼻乌鸦（*Corvus frugilegus*）等；

东北型的白枕鹤（*Grus vipio*）、白头鹤（*Grus monacha*）、白眉姬鹟（*Ficedula zanthopygia*）、金翅雀（*Chloris sinica*）、三道眉草鹀（*Emberiza cioides*）等；

华北型的褐马鸡（*Crossoptilon mantchuricum*）、山噪鹛（*Garrulax davidi*）等；

东北型中的广布种鸭、雁和鹭类以及灰斑鸠（*Streptopelia decaocto*）、珠颈斑鸠（*Streptopelia chinensis*）、戴胜（*Upupa epops*）、北领角鸮（*Otus semitorques*）、银喉长尾山雀（*Aegithalos glaucogularis*）等；

东北一华北型的牛头伯劳（*Lanius bucephalus*）、红尾伯劳（*Lanius cristatus*）、灰椋鸟（*Spodiopsar cineraceus*）等。

另外还有少量中亚亚界的蒙新区和青藏区的种类出现，如中亚型的石鸡（*Alectoris chukar*）、斑翅山鹑（*Perdix dauurica*）、毛腿沙鸡（*Syrrhaptes paradoxus*）、大鵟（*Buteo hemilasius*）、凤头百灵（*Galerida cristata*）、山鹛（*Rhopophilus pekinensis*）等；

喜马拉雅一横断山区型的棕腹啄木鸟（*Dendrocopos hyperythrus*）、中华朱雀（*Carpodacus davidianus*）、蓝鹀

（*Emberiza siemsseni*）等；

高地型的斑头雁（*Anser indicus*）、棕头鸥（*Chroicocephalus brunnicephalus*）、红腹红尾鸲（*Phoenicurus erythrogastrus*）、褐岩鹨（*Prunella fulvescens*）、粉红胸鹨（*Anthus roseatus*）等。

北京地区分布的鸟种也有一些南方种类：

东洋型的牛背鹭（*Bubulcus ibis*）、红翅凤头鹃（*Clamator coromandus*）、噪鹃（*Eudynamys scolopaceus*）、大鹰鹃（*Hierococcyx sparverioides*）、四声杜鹃（*Cuculus micropterus*）、黑翅鸢（*Elanus caeruleus*）、赤腹鹰（*Accipiter soloensis*）、红角鸮（*Otus sunia*）、黑枕黄鹂（*Oriolus chinensis*）、发冠卷尾（*Dicrurus hottentottus*）等；

南中国型的勺鸡（*Pucrasia macrolopha*）、黄腹山雀（*Pardaliparus venustulus*）、白头鹎（*Pycnonotus sinensis*）等。

由于北京地区在候鸟迁徙的通道上，每年春、秋两季，都会有种类繁多的雁鸭、鸥、鸻鹬、猛禽以及燕科、柳莺科、鹟科、鹟科、鹡鸰科、燕雀科、鹀科等雀形目鸟类出现在北京，使得北京鸟类的种类非常丰富。

Chapter 2

Characteristics of Beijing Bird Fauna

There are 508 species of Birds in Beijing, belonging to 22 orders, 74 families, and 243 genera. The community structure is complicated: most birds are Palaearctic species, with some Oriental species and cosmopolitan species. It shows the weaving and mixing of species from the Palearctic and Oriental realms, and the diversity of the distribution types, reflecting the interaction and transition between the two realms and their sub-realms. Based on *China Zoological Geography* (Zhang Rongzu, 2011), Beijing is in the Palaearctic realm, Northeast sub-realm, and North China region (including the subregions of the Loess Plateau and the Huanghuai Plain).

A majority of bird species of Beijing are northern species:

Palearctic type: Great Bustard (*Otis tarda*), Common Crane (*Grus grus*), Common Swift (*Apus apus*), Common Cuckoo (*Cuculus canorus*), Common Kingfisher (*Alcedo atthis*), Eastern Buzzard (*Buteo japonicus*), Chinese Nuthatch (*Sitta villosa*), Eurasian Tree Sparrow (*Passer montanus*), Azure-winged Magpie (*Cyanopica cyanus*), Common Magpie (*Pica pica*), Rook (*Corvus frugilegus*), etc.

Northeast type: White-naped Crane (*Grus vipio*), Hooded Crane (*Grus monachal*)、Yellow-rumped Flycatcher (*Ficedula zanthopygia*), Grey-capped Greenfinch (*Chloris sinica*), Meadow Bunting (*Emberiza cioides*), etc.

North type: Brown Eared Pheasant (*Crossoptilon mantchuricum*)、Plain Laughingthrush (*Garrulax davidi*), etc.

Widespread species in Northeast type: Ducks, geese, egrets and herons, Eurasian Collared Dove (*Streptopelia decaocto*), Spotted Dove (*Streptopelia chinensis*), Common Hoopoe (*Upupa epops*), Japanese Scops Owl (*Otus semitorques*), Silver-throated Bushtit (*Aegithalos glaucogularis*), etc.

Northeast-North type: Bull-headed Shrike (*Lanius bucephalus*), Brown Shrike (*Lanius cristatus*), White-cheeked Starling (*Spodiopsar cineraceus*),

In addition, there are a small number of species from the Mongolian-Xinjiang region and the Qinghai-Tibet region of Central Asia sub-realm, e.g. Central Asia type: Chukar Partridge (*Alectoris chukar*), Daurian Partridge (*Perdix dauurica*), Pallas's Sandgrouse (*Syrrhaptes paradoxus*), Upland Buzzard (*Buteo hemilasius*),

Crested Lark (*Galerida cristata*), Chinese Hill Babbler (*Rhopophilus pekinensis*), etc. Himalayas-Hengduan Mountains type: Rufous-bellied Woodpecker (*Dendrocopos hyperythrus*), Chinese Beautiful Rosefinch (*Carpodacus davidianus*), Slaty Bunting (*Emberiza siemsseni*), etc. Highland type: Bar-headed Goose (*Anser indicus*), Brown-headed Gull (*Chroicocephalus brunnicephalus*), White-winged Redstart (*Phoenicurus erythrogastrus*), Brown Accentor (*Prunella fulvescens*), Rosy Pipit (*Anthus roseatus*), etc.

There are also some southern species:

Oriental type: Cattle Egret (*Bubulcus ibis*), Chestnut-winged Cuckoo (*Clamator coromandus*), Common Koel (*Eudynamys scolopaceus*), Large Hawk-Cuckoo (*Hierococcyx sparverioides*), Indian Cuckoo (*Cuculus micropterus*), Black-winged Kite (*Elanus caeruleus*), Chinese Sparrowhawk (*Accipiter soloensis*), Oriental Scops Owl (*Otus sunia*), Black-naped Oriole (*Oriolus chinensis*), Hair-crested Drongo (*Dicrurus hottentottus*), etc.

Southern China type: Koklass Pheasant (*Pucrasia macrolopha*), Yellow-bellied Tit (*Pardaliparus venustulus*), Light-vented Bulbul (*Pycnonotus sinensis*), etc.

Beijing is located in the East Asian-Australasian flyway. Therefore, a wide variety of migratory birds migrate through Beijing, e.g. ducks, geese, gulls, shorebirds, raptors, as well as Hirundinidae, Phylloscopidae, Turdidae, Muscicapidae, Motacillidae, Fringillidae, etc.

第三章 北京鸟类的生态类群及分布

Chapter 3 The Functional Groups and Distribution of Birds of Beijing

北京的鸟类包括陆禽、游禽、涉禽、攀禽、猛禽、鸣禽等六大生态类群。在山区和平原，以及城市中心的各种生境中，各生态类群都有相同或不同代表种类的分布。由于鸟类善于飞行，它们选择栖息地的能力很强，分布往往随季节、食物等因素的变化而变化。

There are six functional groups of birds of Beijing: landfowls, natatores, wading birds, scansorial birds, raptors and passerines. In mountainous areas, plains, as well as urban, each functional group has a similar or different distribution of representative species. Due to their flight skill, most birds have the capability to choose different habitats, and so their distributions can change with seasons, food availability, etc.

1. 陆 禽

陆禽的后肢强壮，适于地面行走，翅短圆，喙强壮且多为“弓”形，适于啄食。代表种类有环颈雉 (*Phasianus colchicus*)、鹌鹑 (*Coturnix japonica*) 等。斑鸠和鸽虽然善飞翔，但取食主要在地面，因此也被归于陆禽。

北京地区的 14 种陆禽分布广泛，见于各区。褐马鸡 (*Crossoptilon mantchuricum*) 仅见于门头沟区和房山区（小龙门、东灵山、百花山、京西林场）；勺鸡 (*Pucrasia macrolopha*) 见于房山区、门头沟区、昌平区、延庆区、怀柔区、密云区、平谷区海拔 600 米以上的山地；环颈雉、珠颈斑鸠（*Streptopelia chinensis*）、山斑鸠 (*Streptopelia orientalis*) 见于城市公园、平原和山地；鹌鹑、毛腿沙鸡（*Syrrhaptes paradoxus*）在迁徙季节见于平原草地。

1. Landfowls

The legs and feet of landfowls are strong and suitable for walking on the ground; the wings are short and rounded; the beak is strong and mostly arched for pecking. Representative species are Common Pheasant (*Phasianus colchicus*), Japanese Quail (*Coturnix japonica*), etc. Given that they mainly feed on the

ground, doves and pigeons are regarded as landfowls, although they are good at flying.

Landfowls 14 species in Beijing are widely distributed in Beijing and occur in all districts. Brown Eared Pheasant (*Crossoptilon mantchuricum*) is only seen at Mentougou and Fangshan Districts (Xiaolongmen, Dongling Mountain, Baihua Mountain, West Beijing Forest Farm); Koklass Pheasant (*Pucrasia macrolopha*) can be found in mountain areas over 600 m (Fangshan District, Mentougou District, Changping District, Yanqing District, Huairou District, Miyun District, Pinggu District); Common Pheasant, Spotted Dove (*Streptopelia chinensis*), and Oriental Turtle Dove (*Streptopelia orientalis*), occur in urban parks, plains and mountain areas; Japanese Quail and Pallas's Sandgrouse (*Syrrhaptes paradoxus*) are seen at grasslands in plains during migration.

2. 游　禽

游禽的脚趾间具蹼（蹼有多种），擅游泳。尾脂腺发达，能分泌大量油脂涂抹于全身羽毛，以保护羽衣不被水浸湿。嘴形或扁或尖，适于在水中滤食或啄鱼。代表种类有绿头鸭 (*Anas platyrhynchos*)、䴙䴘、潜鸟、鸬鹚等。

北京地区的 77 种游禽，主要分布于平原和山区的各类湿地。尤以延庆区（野鸭湖）、密云区（不老屯）、昌平区（沙河水库）的迁徙季节，会出现种类繁多的游禽，包括天鹅、雁、河鸭、潜鸭、秋沙鸭、麻鸭、鸬鹚、鸥等，䴙䴘除了在上述水域分布外，还见于城市各区有水域的公园繁殖。近年鸳鸯 (*Aix galericulata*) 在怀柔区（怀沙河、怀九河）、门头沟区（百花山）、城市公园（北海、圆明园、颐和园、北京动物园、紫竹院、翠湖）均有分布或繁殖记录。

2.Natatores

Natatores have webbed feet. They are good at swimming. The uropygial glands are well-developed, and the secreted oil is used to keep plumage waterproof. The beak is flat or pointed, and suitable for filtering or catching fish in the water. Representative species are Mallard (*Anas platyrhynchos*), grebes, loons, and cormorants, etc.

Natatores 77 species in Beijing occur at wetlands in plains and mountains, especially Wild Duck Lake in Yanqing District, Bulaotun in Miyun District, Shahe

Reservoir in Changping District during migration, including swans, geese, ducks, pochards, shelducks, cormorants, gulls, etc. Grebes breeds in wetlands at urban parks. In recent years, Mandarin Duck (*Aix galericulata*) has been seen breeding in Huairou District (Huaisha River, Huaijiu River), Mentougou District (Baihua Mountain), and urban parks (Beihai Park, Yuanmingyuan Relic Park, Summer Palace, Beijing Zoo, Zizhuyuan Park, Cuihu Park).

3. 涉 禽

涉禽的外形具有“三长”特征，即喙长、颈长、后肢（腿和脚）长，适于涉水生活，因为后肢长可以在较深水处捕食和活动。它们趾间的蹼膜往往退化，因此不擅游水。典型的代表种类是鹤和鹭。还有体形较小但种类繁多的鸻类和鹬类都属于典型的涉禽。

北京地区的 96 种涉禽主要分布于城郊各区的湿地及有水域的公园，黄斑苇鳽 (*Ixobrychus sinensis*) 在挺水植物茂密处繁殖，苍鹭 (*Ardea cinerea*)、夜鹭

(*Nycticorax nycticorax*)、池鹭 (*Ardeola bacchus*)、白鹭 (*Egretta garzetta*) 等中大型鹭鸟也在城区周边及远郊地区的湿地和养鱼塘觅食，并在其附近林地的林冠层筑巢繁殖。灰鹤 (*Grus grus*) 喜到延庆区（野鸭湖、官厅水库）、密云区（不老屯）等地区越冬，白枕鹤 (*Grus vipio*) 在春季迁徙时，曾集群出现于密云区（陈各庄）、延庆区（野鸭湖）、平谷区（金海湖）。

3.Wading birds

Wading birds have long beaks, necks, and legs which make them skilled in wading, and feeding in deep water. The webbed membrane between toes is vestigial. Therefore, they are not good at swimming. Typical representative species are cranes, egrets and herons, as well as small while diverse plovers and sandpipers.

Wading birds 96 species in Beijing occur at suburban wetlands and urban

parks with open water. Yellow Bittern (*Ixobrychus sinensis*) breeds in dense emergent aquatic plants. Grey Heron (*Ardea cinerea*), Black-crowned Night Heron (*Nycticorax nycticorax*), Chinese Pond Heron (*Ardeola bacchus*), and Little Egret (*Egretta garzetta*) also feed in the wetlands and fishponds around the city and in the outer suburbs, and they build nests in the forest canopy in nearby forests. Common Crane (*Grus grus*) winters in Yanqing District (Wild Duck Lake, Guangting Reservoir), and Miyun District (Bulaotun), etc. Large groups of White-naped Crane (*Grus vipio*) were recorded in Miyun District (Chengezhuang), Yanqing District (Wild Duck Lake) and Pinggu District (Jinhai Lake).

4. 攀 禽

攀禽的足（脚）趾类型发生多种变化，适于在岩壁、石壁、土壁、树干等处行攀缘生活。如两趾向前、两趾朝后的啄木鸟、杜鹃，四趾朝前的雨燕，三、四趾基部并连的戴胜 (*Upupa epops*)、翠鸟等均属于攀禽。

北京地区的 32 种攀禽因种而异，分布在不同环境区域。常年留居的啄木鸟从城市平原至海拔 1400 米的山地都有分布，它们的栖息地均为林地。夏候鸟普通雨燕 (*Apus apus*) 分布在城郊各区的古建筑及立交桥周围，利用建筑物的孔洞造巢繁殖，夏候鸟大杜鹃 (*Cuculus canorus*) 集中在有东方大苇莺 (*Acrocephalus orientalis*) 繁殖的湿地苇丛地带，四声杜鹃 (*Cuculus micropterus*) 则寻找城市公园及中低山有鸦科鸟类繁殖的地方，它们分别利用苇莺、灰喜鹊 (*Cyanopica cyanus*) 等鸟类完成巢寄生。戴胜一年四季都能见到，它们选择在城市公园、河流湖泊边、低山等有树洞或柴垛缝隙的地方造巢繁殖。普通翠鸟 (*Alcedo atthis*) 在城市及京郊各区有水面的地方都能见到，它们捕食水中的小鱼和大型昆虫，在土壁上掘洞营巢繁殖。

4.Scansorial birds

The feet and toes of scansorial birds are diverse, and are adapted to use cliffs, rock walls, soil walls, tree trunks, etc. Scansorial birds include woodpeckers and cuckoos with two toes forward and two toes backwards; swifts with four toes forward; hoopoes and kingfishers with third and fourth toes connected at the base.

Scansorial birds 32 species in Beijing occur in different environmental areas.

Resident woodpeckers are seen in woodlands from urban plains to mountains at an altitude of 1,400m. Summer migratory Common Swift (*Apus apus*) is found around ancient buildings and overpasses in suburbs. It uses the holes of the building to build nests and breed. Common Cuckoo (*Cuculus canorus)* occurs in reeds and wetlands where Oriental Reed Warbler (*Acrocephalus orientalis*) breeds, while Indian Cuckoo (*Cuculus micropterus*) occurs in urban parks and in the middle and low mountains where Corvidae breed. Both use the nests of warblers and Azure-winged Magpie (*Cyanopica cyanus*) for nest parasitism. Common Hoopoe (*Upupa epops*) can be seen all year round, and builds nests and breeds at tree holes or firewood gaps in urban parks, rivers and lakes, low mountains, etc. Common Kingfisher (*Alcedo atthis*) can be seen at open water in urban or suburbs. They prey on small fish and large insects in the water and dig holes in the soil wall to build nests and breed.

5. 猛 禽

猛禽的喙、爪锐利带钩，视觉器官发达，飞翔能力强，多具有捕杀动物为食的习性。羽色较暗淡，常以灰色、褐色、黑色、棕色为主要体色。代表种类有日行性的金雕 (*Aquila chrysaetos*)、雀鹰 (*Accipiter nisus*)、红隼 (*Falco tinnunculus*)、猎隼 (*Falco cherrug*) 和夜行性的雕鸮 (*Bubo bubo*) 等。

北京地区的猛禽，在迁徙季节有鹰形目、隼形目共 49 种飞越北京上空，最集中的迁徙通道当属小西山（太行山余脉至燕山山脉）上空，最多时，日通过量超过千只。各区山地高海拔处有金雕、秃鹫 (*Aegypius monachus*)、雕鸮等常年留居并繁殖，也有苍鹰 (*Accipiter gentilis*)、赤腹鹰 (*Accipiter soloensis*)、日本松雀鹰 (*Accipiter gularis*)、红隼、燕隼 (*Falco subbuteo*)、红脚隼 (*Falco amurensis*)、北领角鸮 (*Otus semitorques*)、红角鸮 (*Otus sunia*) 在海拔 1000 米以下中低山至平原繁殖。城市中尚有红隼、纵纹腹小鸮 (*Athene noctua*) 繁殖。冬候鸟大鵟 (*Buteo hemilasius*)、毛脚鵟 (*Buteo lagopus*) 喜在城外各区的旷野上空盘旋，长耳鸮 (*Asio otus*) 分布于城市公园、平原、低山的林地，而短耳鸮 (*Asio flammeus*) 栖息于城区以外各区的荒地草丛之中。

5.Raptors

Raptors have sharp beaks and claws with hooks, well-developed visual organs, strong flying ability, and most of them hunt animals for food. Their plumage is generally dull, and often in grey, brown, and black. Representative species are diurnal Golden Eagle (*Aquila chrysaetos*), Eurasian Sparrowhawk (*Accipiter nisus*), Common Kestrel (*Falco tinnunculus*), Saker Falcon (*Falco cherrug*), and nocturnal Eurasian Eagle Owl (*Bubo bubo*).

There are forty-nine species of raptors migrating across Beijing, with up to more than a thousand individuals per day at Xiaoxi Mountain (branch of Taihang Mountains to Yanshan Mountains). There are Golden Eagle, Cinereous Vulture (*Aegypius monachus*), Eurasian Eagle Owl breeding at high elevation of mountains. Northern Goshawk (*Accipiter gentilis*), Chinese Sparrowhawk (*Accipiter soloensis*), Japanese Sparrowhawk (*Accipiter gularis*), Common Kestrel, Eurasian Hobby (*Falco subbuteo*), Amur Falcon (*Falco amurensis*), Japanese Scops Owl (*Otus semitorques*), and Oriental Scops Owl (*Otus sunia*) breed in the middle and low mountains below 1,000 m to the plains. A few raptors like Common Kestrel and Little Owl (*Athene noctua*) even breed in the urban area. Also, Beijing has many wintering raptors such as Upland Buzzard (*Buteo hemilasius*) and Rough-legged Buzzrard (*Buteo lagopus*) over open countries, Short-eared Owl (*Asio flammeus*) in rural grasslands and Long-eared Owl (*Asio otus*) in woodlands of urban parks.

6. 鸣 禽

鸣禽种类繁多，鸣叫器官（鸣肌和鸣管）发达。它们善于鸣叫，巧于营巢，繁殖时有复杂多变的行为，身体为中、小型，雏鸟均为晚成，在巢中得到亲鸟的哺育才能正常发育。代表种类有喜鹊 (*Pica pica*)、乌鸦、山雀、百灵、家燕、椋鸟、麻雀 (*Passer montanus*) 等。

北京地区的 240 种鸣禽种类最为丰富，在全市各区都有许多种类分布。在北京分布的中国特有种共 8 种，除去褐马鸡外，其他 7 种均为鸣禽：作为留鸟的山噪鹛（*Garrulax davidi*）、银喉长尾山雀（*Aegithalos glaucogularis*）主要分布于各区的中低海拔山地，中华朱雀（*Carpodacus davidianus*）主要分布于北京西部

和北部山区海拔 1100 米以上的亚高山林地，乌鸫 (*Turdus mandarinus*) 近 20 年来进入北京，主要分布于城市公园、城区周边的平原地带。夏候鸟宝兴歌鸫 (*Turdus mupinensis*) 在海拔 1200 米以上亚高山林地繁殖。黄腹山雀 (*Pardaliparus venustulus*) 繁殖在各区中低海拔山地阔叶林及针阔混交林，其他季节扩散到平原及城市公园，有些个体到长江以南越冬。冬候鸟贺兰山红尾鸲 (*Phoenicurus alaschanicus*)，在门头沟（东灵山）亚高山林缘灌丛地带越冬。还有喜鹊、灰喜鹊 (*Cyanopica cyanus*)、秃鼻乌鸦 (*Corvus frugilegus*)、麻雀等伴人鸟种常年留居于各区的居民区或在民居附近造巢繁殖。

6.Passerines

Passerines are diverse set of species, with well-developed song organs (muscles and tubes). They are good at singing and nesting, with complicated and various behaviours during mating. Their bodies are medium to small. The chicks are altrices and fed by parents in the nest. Representative species include magpies, crows, tits, larks, swallows, starlings and sparrows, etc.

Passerines have the highest species richness in Beijing 240 species compared with other functional groups. This group of birds are distributed in all districts. Among the endemic bird species of China, there are eight occuring in Beijing. Except for the Brown Eared Pheasant, the other seven are all passerines: resident Plain Laughingthrush (*Pterorhinus davidi*), Silver-throated Bushtit (*Aegithalos glaucoguls*) in the low and middle mountains; Chinese Beautiful Rosefinch (*Carpodacus davidianus*) in the subalpine woodland above 1,100 m in the north and east of Beijing; Chinese Blackbird (*Turdus mandarinus*) in parks and the plains surrounding the city; summer breeder Chinese Thrush (*Turdus mupinensis*) breeds in subalpine woodlands above 1,200 m. In addition, Yellow-bellied Tit (*Pardaliparus venustulus*) breeds in broad-leaved forests and coniferous and broad-leaved mixed forests in low and middle mountains, and spreads to plains and urban parks, with some individuals wintering south of the Yangtze River. Winter visitor Alashan Redstart (*Phoenicurus alaschanicus*) winters in the shrublands in the edge of subalpine forests in Mentougou District (Dongling Mountain). There are also Common Magpie (*Pica pica*), Azure-winged Magpie (*Cyanopica cyanus*), Rook (*Corvus frugilegus*), Eurasian Tree Sparrow (*Passer montanus*), etc. inhabiting and breeding in residential areas.

北京市地形、水系图

Topographic and drainage system of Beijing

注：海陀山海拔数据，来自北京市延庆区人民政府网站2021年2月20日发布的《延庆概况》（http://www.bjyq.gov.cn/yanqing/mlyq/yqgk60/2860134/index.shtml）。雾灵山海拔数据，来自河北省生态环境厅网站2015年6月9日发布的《河北雾灵山国家级自然保护区》（http://hbepb.hebei.gov.cn/hbhjt/zwgk/fdzdgknr/zdlyxxgk/ziranshengtaibaohu/101593685477674.html）。

其余海拔数据来自北京师范学院出版社1989年9月出版的《北京自然地理》（霍亚贞主编，ISBN：7-81014-159-7）

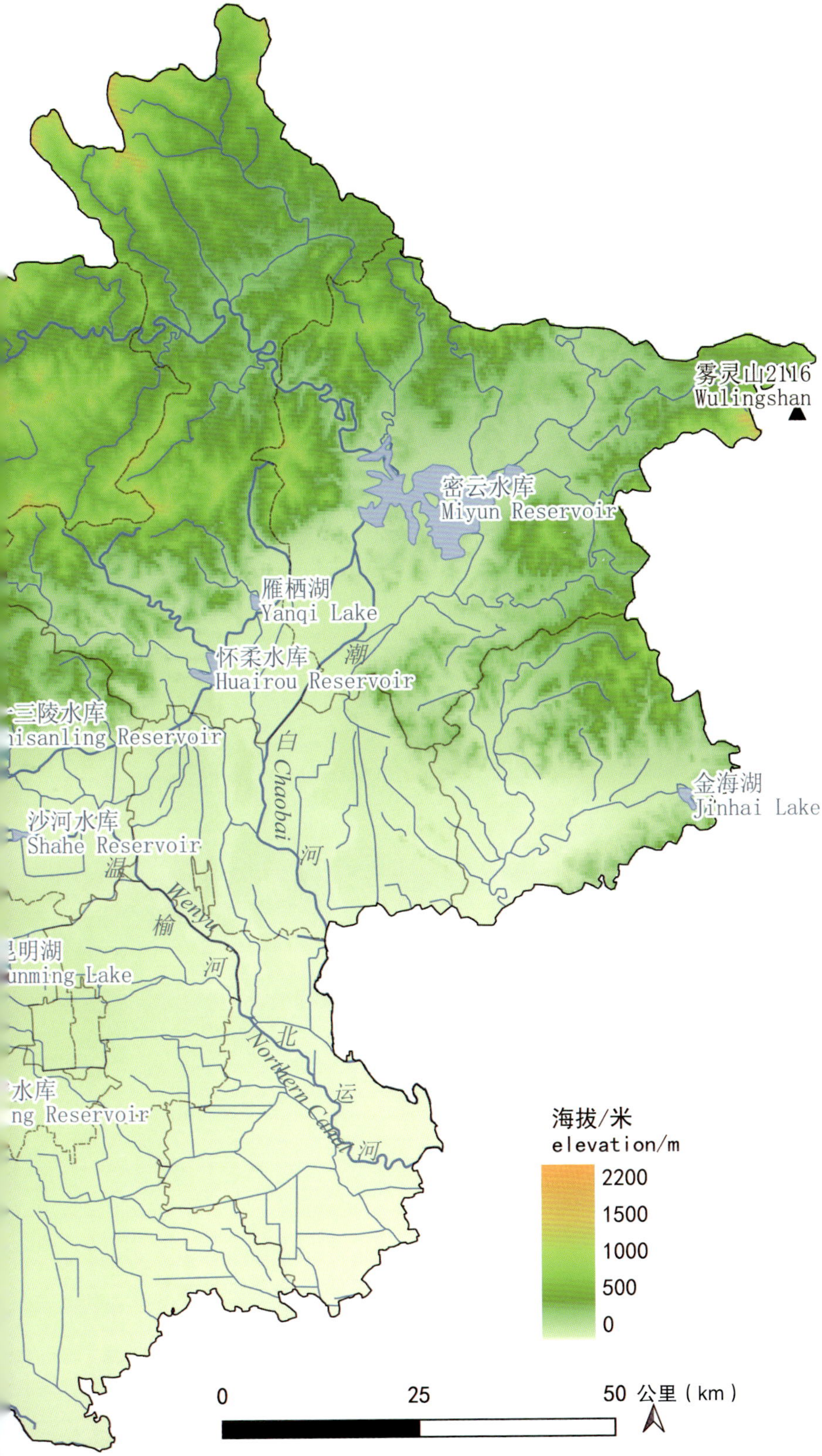
雾灵山2116
Wulingshan
密云水库
Miyun Reservoir
雁栖湖
Yanqi Lake
怀柔水库
Huairou Reservoir
潮
白
Chaobai
河
三陵水库
isanling Reservoir
沙河水库
Shahe Reservoir
温
Wenyu
榆
河
明湖
unming Lake
北
Northern Canal
运
河
水库
ng Reservoir
金海湖
Jinhai Lake
海拔/米
elevation/m
2200
1500
1000
500
0
0
25
50 公里（km）

第四章

Chapter 4

BIRDS OF BEIJING

北京地区鸟类

I

北京鸟类图谱

陆禽
LANDFOWLS

BIRDS OF BEIJING

一、鸡形目 GALLIFORMES

花尾榛鸡	*Tetrastes bonasia*	Hazel Grouse	雉科 Phasianidae

形态特征 中型陆禽，体长 33~40 厘米。具短而略上翘的冠羽，前额白色，边缘黑色。雄鸟头部褐色，缀有淡褐色横斑，眼上方红色，颏、喉黑色，周围有一白色带自颊和颈侧而下；上体余部褐色，具灰白色横斑；尾羽灰褐色，具灰色端斑和黑色次端斑；下体白色，具黑色斑点。雌鸟似雄鸟，但颏、喉为皮黄色。

分布与生境 中国见于东北、西北地区。北京甚罕见于怀柔、密云山区林地，具体居留型尚不明确。

习性与生态 喜集小群，甚至育雏期间亦集群活动。冬、春季常上树取食芽苞。除繁殖期外活动范围甚不固定，有季节性垂直迁移现象。

Morphological characteristics A medium-sized landfowl (33-40 cm) with short erectile crest and white forehead with a black edge. Male has finely barred buff on brown head and a red spot above eye; chin and throat black; a white streak down the side of head and neck; upperparts brown with grey bars; tail greyish-brown with grey tip and black subterminal band; white below, spotted black. Female is like male but with a buff chin and throat.

Distribution and habitat Northern China. In Beijing, very rare in the montane forests of Huairou and Miyun District. Residential status: unclear.

Behaviour and ecology Tends to gather in small flocks, sometimes even during the breeding season. Population in mountains may move to lower land in winter. Often seen feeding together on tree buds in winter and early spring.

	国家二级保护野生动物	IUCN 红色名录 无危 (LC) 2016.10.1

雄鸟 Male

雌鸟 Female

石鸡	*Alectoris chukar*	Chukar Partridge	雉科 Phasianidae

形态特征 中型陆禽，体长 27~37 厘米。雌雄同色。喙红色，甚鲜艳。额基与贯眼纹黑色，经头侧、颈侧向下延伸，两侧黑色带至前颈汇合，颊、颏、喉皮黄色，头顶至枕部及上体余部灰褐色。胸部灰色，腹部和尾下覆羽棕色，两胁为较淡的皮黄色，具黑色与栗色相间的显著横斑。跗跖和趾皆为淡红色。

分布与生境 广泛分布于除东北外的中国北部地区。北京区域性常见于低山丘陵地带的岩石、砂石坡地，偶见于山脚的农田区域，为留鸟。

习性与生态 一般在山坡处集群活动，受惊时多迅速奔跑，较少飞行。早晚时雄鸟喜在岩石上高声鸣叫。营巢于有植被或较大岩石遮挡的地面；窝卵数 7~17 枚。

Morphological characteristics A medium-sized landfowl (27-37 cm). Sexes alike. Adult has distinctive red bill and a black stripe from forehead through the eye and down the neck-sides to the breast; two stripes on each side join above the upper breast; cheek, chin and throat white or creamy white; upperparts greyish-brown, breast grey; belly and undertail-coverts buff; flanks creamy white with black and chestnut barring; tarsi and toes are reddish pink.

Distribution and habitat Northern China (except for the northeast). In Beijing, locally common in dry hill country, particularly on rocky outcrops. Occasionally seen in farmlands. Residential status: resident.

Behaviour and ecology Small to large flocks (up to more than 100 individuals) gather on slopes. Tends to run at speed rather than fly when disturbed. Males often seen calling from prominent rocks in early mornings. The nest is on the ground and sheltered by vegetation or rocks. Clutch size: 7-17.

北京市二级保护野生动物		IUCN 红色名录 无危 (LC) 2018.8.9

斑翅山鹑	*Perdix dauurica*	Daurian Partridge	雉科 Phasianidae

雄鸟 Male

幼鸟 Juvenile

北京市二级保护野生动物

IUCN 红色名录 无危 (LC) 2016.10.1

形态特征 小型陆禽，体长 25~31 厘米。雄鸟头顶褐色，额至头侧棕黄色，喉侧羽略延长且尖，似须，故又名“须山鹑”；上体及翼上覆羽大致为灰褐色，具显著的栗色斑点；前颈至胸为淡褐色或灰褐色，下体余部淡棕色，腹部中央具一显著的马蹄形黑斑。雌鸟似雄鸟，但体羽不如雄鸟鲜艳，腹部黑斑较小或几不可见。

分布与生境 分布于中国北部。北京广泛分布于西部、北部的丘陵及山地，非繁殖季亦见于平原。多集群于草地、石滩等有零散植被的开阔生境活动，为不常见留鸟。

习性与生态 受惊时大多通过快速奔跑保持距离观望，较少飞行；飞行时振翼快速、贴近地面。于地面筑巢；窝卵数 14~17 枚。

Morphological characteristics A small landfowl (25-31 cm). Male has grey cap and rufous face; long feathers on the chin and throat form a "beard", hence an alternative name "Bearded Partridge"; upperparts and upper wing-coverts greyish-brown with dark brown vermiculation; underparts generally slate-grey with an orange central stripe and a black U-shaped patch on the belly; female is duller than the male and has a smaller belly patch.

Distribution and habitat Northern China. In Beijing, founds on grassland or rocky outcrops with sparse shrubs on low hills in the west and north of Beijing. Tends to move to lowland plains in winter. Residential status: uncommon resident.

Behaviour and ecology Tends to run or keep hidden rather than fly when disturbed. Flies with fast wing-beats close to the ground. Ground-nester, clutch size: 14-17.

鹌鹑	*Coturnix japonica*	Japanese Quail	雉科 Phasianidae

形态特征 小型陆禽，体长 14~20 厘米，体型紧凑。雄鸟繁殖羽头顶至后颈褐色，具白色顶冠纹和白色眉纹，头侧至喉皆为棕红色，上体余部褐色，具黑色横斑和淡黄色细纵纹；胸、腹部大致为白色。雌鸟繁殖羽头侧为褐色，颏、喉为皮黄色，其余似雄鸟繁殖羽。雌、雄鸟非繁殖羽皆似雌鸟繁殖羽，但上体栗色较淡而偏黄褐色，喉白色。

分布与生境 广布于中国东部地区。北京多见于较低海拔有低矮植被的农田和草地等处，为区域性常见旅鸟和冬候鸟。

习性与生态 多成对或集 5 只以内的小群活动，受惊时多贴地潜伏，常在人走到极近时突然飞出。繁殖期雄鸟好斗。多在草地上营巢；窝卵数 9~10 枚。

Morphological characteristics A small and plump landfowl (14-20 cm) with a brown body and white median-crown stripe and supercilium. Breeding male has distinctive reddish-brown face, chin and throat; upperparts barred black and finely striped creamy white; breast and belly generally white or creamy white. Breeding female has brown face and buff throat. Non-breeding adult is like breeding female but with paler upperparts and white throat.

Distribution and habitat Eastern and southern China. In Beijing, locally common in open lowland areas such as farmland and grassland. Residential status: migrant and winter visitor.

Behaviour and ecology Usually encountered in pairs or small flocks (usually less than five individuals). Tends to keep hidden or creep away when disturbed, only flushes at very short distance. Male is aggressive in breeding season. Usually nests on grasslands and farmlands. Clutch size: 9-10.

北京市二级保护野生动物 | IUCN 红色名录 近危 (NT) 2016.10.1

雌鸟 Female

雄鸟 Male

雌鸟 Female

勺鸡	*Pucrasia macrolopha*	Koklass Pheasant	雉科 Phasianidae

雄鸟 Male

雌鸟 Female

雄鸟 Male

	国家二级保护野生动物	IUCN 红色名录 无危 (LC) 2016.10.1

形态特征 中型陆禽，体长 53~63 厘米。雄鸟头部为具金属光泽的暗绿色，具显著的黑褐色细长冠羽；颈侧具一块白斑，白斑下方至胸为棕黄色；上体余部、胸侧至两胁羽毛大致呈披针状，银灰色，具黑色细纵纹；尾上覆羽和一对中央尾羽灰色，具黑、栗相间的纵纹；其余尾羽灰白色，具宽阔的黑色次端斑；下体中央栗色。雌鸟头部黄褐色，冠羽较短，亦为黄褐色，全身大致呈褐色。

分布与生境 分布于自华北至珠江以北的区域。北京不常见于海拔 1000~1800 米的林地，繁殖期亦出没于亚高山草甸；冬季有时迁移至海拔较低处，为留鸟。

习性与生态 多单只或成对活动。一般筑巢于灌丛下的地面或高山草甸；窝卵数 5~7 枚。

Morphological characteristics A medium-sized landfowl (53-63 cm). Male has dark green head and conspicuous long crest; neck-sides white; breast buff or brown; silver body is heavily streaked black or dark brown; grey uppertail-coverts and central tail feathers with black and chestnut streaks; outer tail feathers grey with black band near the tip; chestnut stripe on the breast and belly. Female has brown head and body with a shorter crest.

Distribution and habitat Southern and eastern China to North China. In Beijing, uncommon in forested hills (from 1,000-1,800 m); also occurs on sub-alpine grassland in breeding season. Tends to move to lower altitudes in winter. Residential status: resident.

Behaviour and ecology Usually found singly or in pairs. Nesting in forests, shrubs or on grasslands. Clutch size: 5-7.

褐马鸡	*Crossoptilon mantchuricum*	Brown Eared Pheasant	雉科 Phasianidae

形态特征 大型陆禽，体长 82~110 厘米。雌雄同色。头顶羽毛呈天鹅绒状，黑色，头侧具红色裸皮，且具特别延长的白色耳羽簇。颊、颏亦为白色，颈部黑褐色，上背、两翼和下体为暗棕褐色，下背、腰、尾上覆羽和尾羽皆为银白色，但尾羽端部为黑色。尾羽较长，外翈羽支发散为丝状下垂，似马尾。下体大致为深褐色。跗跖红色。

分布与生境 中国特有鸟类，分布于华北山地。北京仅见于西部山区（东灵山、小龙门林场、百花山等），多见于 1200~2000 米海拔范围的林地或林缘灌丛，为不常见留鸟。

习性与生态 繁殖期成对，非繁殖季集小群活动。繁殖期间雄鸟甚好斗。繁殖期常于林间发出甚响亮的鸣叫，似“ge－a”。营巢于林间地面；窝卵数 6~9 枚。

Morphological characteristics A large landfowl (82-110 cm). Sexes alike. Adult has black crown and red bare facial skin; white moustachial stripe extends to form stiff ear tufts; body generally dark brown; back, rump and uppertail-coverts silver; prominent drooping tail feathers white with black tips. Underparts dark brown; tarsi red.

Distribution and habitat North China (endemic). Uncommon in montane forests and shrubs (usually 1,200-2,000 m) in western Beijing. Residential status: resident.

Behaviour and ecology Usually in pairs in breeding season and in flocks in non-breeding season. Male often gives loud calls "ge – a" during mornings in breeding season. Male is aggressive in breeding season. Nests on the ground. Clutch size: 6-9.

国家一级保护野生动物 | IUCN 红色名录 易危 (VU) 2016.10.1

环颈雉 | *Phasianus colchicus* | Common Pheasant | 雉科 Phasianidae

雄鸟（左上）和雌鸟 Male (upperleft) and females

北京市二级保护野生动物 | | IUCN 红色名录 无危 (LC) 2016.10.1

形态特征 大型陆禽，体长 58~90 厘米。雄鸟头顶灰色，具白色眉纹，头侧具鲜艳的红色裸皮，头、颈余部金属墨绿色，北京地区的亚种颈部皆具白环；上背棕色为主，具白色点斑，下背和腰蓝灰色；尾上覆羽棕黄色，尾羽甚长，为黄褐色，具深色横斑；两翼内侧、翼上覆羽与上背羽色大致相同，外侧覆羽蓝灰色，飞羽褐色，具白色横斑，下体大致为栗色。雌鸟全身皆为黄褐色，上体具深色斑，尾羽短于雄鸟。

分布与生境 广布于中国各地。北京亦常见于林地、灌丛、农田等各种生境，为留鸟。

习性与生态 非繁殖期喜集群活动（冬季尤甚）。在占域时发出响亮的“ge－ge”声。营巢于周围有植被或岩石遮蔽的地面凹坑处；窝卵数 6~14 枚。

Morphological characteristics A large landfowl (58-90 cm) with prominent long tail. Male has grey crown, white supercilium and red bare facial skin; remainder of head green. All subspecies in Beijing have white collar on neck; mantle brown with white spots and bars; lower back and rump slate-grey; uppertail-coverts brown; tail is also brown but barred dark; inner upperwing coverts brown; outer coverts grey; brown remiges barred dark; underparts generally dark chestnut. Female is shorter-tailed; body is generally brown with black scales on mantle.

Distribution and habitat Widespread in China. In Beijing, common in grassland, farmland, shrubs and woodland edge up. Residential status: resident.

Behaviour and ecology Found singly or in small flocks, particularly in winter. Calls loudly "ge – ge" when displaying. Nests in a shallow depression in the ground surrounded by vegetation or rocks. Clutch size: 6-14.

二、鸽形目 COLUMBIFORMES

原鸽	*Columba livia*	Rock Pigeon	鸠鸽科 Columbidae

IUCN 红色名录 无危 (LC) 2016.10.1

形态特征 中型陆禽。体长 31~36 厘米。雌雄相似。头、颈、胸、上背呈深灰色，颈部、胸部的羽毛具有悦目的红绿色金属光泽，翼上具两条黑斑，下体浅灰色，尾部有较宽的黑色端斑。

分布与生境 野生种群分布于中国西北部和喜马拉雅山脉。栖息于山地至平原的岩石峭壁之上。北京所见为半野化或家养，为留鸟。

习性与生态 喜集群活动。飞行时振翼较快。在北京城区的野化种群数量和生存状况暂不明确。

Morphological characteristics A medium-sized landfowl (31-36 cm). Sexes alike. Head, neck, breast and upper back dark grey, glossed with metallic rufous-green on neck and breast; two black bars on wings; underparts pale grey and a black terminal band on the tail.

Distribution and habitat Wild population found in northwest China and the Himalayas. Found near rocks and cliffs at higher elevations. In Beijing, only feral individuals of domestic origin.

Behaviour and ecology In flocks. Wingbeats fast. Population and reproduction of the local feral Rock Pigeon in urban Beijing are unclear.

岩鸽	*Columba rupestris*	Hill Pigeon	鸠鸽科 Columbidae

形态特征 中型陆禽，体长 30~35 厘米。羽色似家鸽但尾羽差异显著。雌雄相似。喙黑；虹膜呈暗褐色；头顶、颏及喉部呈蓝灰色，颈与上胸有红绿色金属光泽。两翼折合时有两道黑色带斑。腰部白色，尾次端有一宽阔白色横带。跗跖暗红。

分布与生境 广泛分布于中国北方地区。北京区域性常见于西北部山区的崖壁处，为留鸟。

习性与生态 喜集群，常飞至山谷、平原觅食，以植物种子、浆果、谷物为食。营巢于山岩缝隙间；巢材由枯枝、树叶、杂草、羽毛构成；窝卵数 2 枚。

Morphological characteristics A medium-sized landfowl (30-35 cm). Very similar to domestic pigeon but tail pattern diagnostic. Sexes alike. Bluish-grey on crown, chin and throat; metallic purple green on neck and upper chest; two black bars on folded wings; rump white; tail has a black terminal band with a prominent white subterminal band. Black bill, dark brown iris, dark red tarsi.

Distribution and habitat Northern China. In Beijing, locally common in mountainous areas in the north and west. Residential status: resident.

Behaviour and ecology Tends to be encountered in small groups or flocks. Forages for seeds, berries and grains. Nests in rock crevice with twigs, leaves, grass, and feathers. Clutch size: 2.

北京市二级保护野生动物 | | IUCN 红色名录 无危 (LC) 2016.10.1

LANDFOWLS

山斑鸠	*Streptopelia orientalis*	Oriental Turtle Dove	鸠鸽科 Columbidae

IUCN 红色名录 无危 (LC) 2016.10.1

形态特征 中型陆禽，体长 31~35 厘米。雌雄相似。体型略显壮实。颈侧部具黑色与蓝灰色条纹的块状斑。翼上覆羽羽缘棕红色，呈鳞片状。尾羽黑褐色，具灰白色端斑。

分布与生境 中国广布。北京常见于浅山区和远郊，城市中少见。栖息于山区和山坡丘陵多树木地带或平原旷野，为留鸟。

习性与生态 常集小群活动，多在开阔农耕区、村庄及房前屋后取食于地面，边走边找食物。主要以各种作物种子、草籽和野生浆果为食，亦兼食昆虫。营巢于乔木上，用稀疏枯枝构成平盘状巢；窝卵数 2 枚。

Morphological characteristics A medium-sized, stocky landfowl (31-35 cm). Sexes alike. Scaly patterned rufous fringes to wing coverts and a black-and-white streaked patch on the sides of the neck. Tail black with greyish-white tips.

Distribution and habitat Widespread in China. In Beijing, common in lower mountain area and outer suburbs, rare in the city centre. Found in wooded mountain slopes or hills, also in lowland plains and fields. Residential status: resident.

Behaviour and ecology Tends to feed in small flocks. Often forages on the ground for seeds, berries and sometimes insects. Builds plate-like nest on trees with twigs. Clutch size: 2.

灰斑鸠	*Streptopelia decaocto*	Eurasian Collared Dove	鸠鸽科 Columbidae

形态特征 小型陆禽，体长 28~30 厘米。雌雄相似。虹膜暗红色。周身灰色调为主，喙黑色，颈基部有一道黑领环，上体羽自背肩至尾上覆羽均呈灰褐色，飞羽黑褐，尾下覆羽偏蓝灰。跗跖呈暗粉红色。

分布与生境 除台湾岛、海南岛及青藏高原外全国均有分布，多见于北方。北京常见于山林、郊区村落房顶，城市中很少出现，为留鸟。

习性与生态 常到田间与其他斑鸠混群取食，以谷物、野果、草籽及昆虫为食。营巢于乔木分枝上，巢用枯枝筑成皿形；窝卵数 2 枚。

Morphological characteristics A small landfowl (28-30 cm). Sexes alike. Overall grey plumage; black bill, narrow black collar, greyish-brown back from mantle to tail coverts; black remiges and bluish-grey undertail -coverts. Dark pinkish tarsi.

Distribution and habitat Widespread except for Taiwan, Hainan and the Tibetan Plateau. In Beijing, locally common in open woodland and in suburban villages but rare in the city centre. Often seen on rooftops or wires. Residential status: resident.

Behaviour and ecology Sometimes mixed with other dove flocks in fields. Feeds on the ground where it forages for grains, seeds, berries, and insects. Builds dish-shaped nest on trees with dry twigs. Clutch size: 2.

IUCN 红色名录 无危 (LC) 2019.8.14

火斑鸠	*Streptopelia tranquebarica*	Red Turtle Dove	鸠鸽科 Columbidae

雄鸟 Male
雌鸟 Female

IUCN 红色名录 无危 (LC) 2018.8.7

形态特征 小型陆禽，体长 22~24 厘米。喙黑色。雄鸟头顶、后颈、腰及尾上覆羽为蓝灰色，其余体羽呈红褐色，颈基部有一道黑色半领环，尾下覆羽为白色；跗跖黑色。雌鸟似灰斑鸠而小。

分布与生境 分布于中国东部、中部和南部广大地区。北京见于郊区水域附近林地，为罕见旅鸟和夏候鸟。

习性与生态 在地面奔走觅食，以植物果实、种籽、谷物为主，偶食昆虫。在乔木分枝上营巢，以枯枝、干草筑成皿状；窝卵数 2 枚。

Morphological characteristics A small landfowl (22-24 cm). Male is bluish-grey from crown to neck, reddish-brown body with black collar at the base of neck, rump and uppertail-coverts bluish-grey, undertail-coverts whitish. Female similar to Eurasian Collared Dove but smaller.

Distribution and habitat Widespread in eastern, central and southern China. In Beijing, rare in wooded areas near water. Residential status: passage migrant and summer breeder.

Behaviour and ecology Often seen on the ground, where it forages for seeds, grain and, occasionally, insects. Builds dish-like nest with twigs and dry grass on the fork of branches in trees. Clutch size: 2.

珠颈斑鸠	*Streptopelia chinensis*	Spotted Dove	鸠鸽科 Columbidae

形态特征 小型陆禽，体长 27~30 厘米。雌雄相似。喙淡褐，整体为灰粉色调；颈侧部有明显白色珍珠状点斑，颏喉和腹部呈浅白色；跗跖紫红色。未成年个体无此珍珠状点斑。

分布与生境 遍布中国中部和南部。北京甚常见于城市绿地、公园、小区中，为留鸟。

习性与生态 常以小群在地面觅食，以谷物、野果和杂草种子为食，亦食昆虫。鸣声通常为三声一度的轻柔“咕”声。营巢于树上，用稀疏枯枝构成盘状巢，亦可见于楼房护栏、阳台、空调室外机旁繁殖；窝卵数 2 枚。

Morphological characteristics A small landfowl (27-30 cm). Sexes alike. Overall greyish-pink plumage; distinctive white pearly spotted patch on the sides of neck; chin, throat and underparts off white. Pale brown bill, purplish-red tarsi. Juvenile lacks pearly spots on neck.

Distribution and habitat Widespread in central and southern of China. In Beijing, abundant in city parks, gardens and residential areas. Residential status: resident.

Behaviour and ecology Often feeds on the ground, singly or in small groups, where it forages for grains, berries and seeds and insects. Coos three purring notes. Nest made of twigs, usually in a tree, but sometimes on building guardrails or balconies. Clutch size: 2.

IUCN 红色名录 无危 (LC) 2016.10.1

三、沙鸡目 PTEROCLIFORMES

毛腿沙鸡	*Syrrhaptes paradoxus*	Pallas's Sandgrouse	沙鸡科 Pteroclidae

形态特征 中型陆禽，体长 35~40 厘米。雄鸟全身羽毛呈沙黄色，背部杂以黑色横斑，胸部灰黄色，有细横斑，形成黑胸带，腹部中央具黑色块斑，初级飞羽特别延长，中央一对尾羽长而尖；喙蓝灰色；跗跖被以淡土黄色短羽。雌鸟似雄鸟而下胸无横斑，胸侧杂以黑色点斑。

分布与生境 广布于中国北方。在北京，某些年份冬季集大群见于干涸的河滩地及荒地，为不常见冬候鸟。

习性与生态 栖于开阔沙地或草原，喜集群飞翔，擅奔走觅食。飞行快速。主食植物种子和嫩芽、昆虫。遇险时紧贴地面凭毛色隐蔽。营巢于沙地浅窝；巢内铺草叶；窝卵数 2~4 枚。

Morphological characteristics A medium-sized landfowl (35-40 cm). Male is overall buff with black barring on back, finely barred band on greyish-yellow chest; large black mark on central belly; long primaries, central tail-feathers are long and pointed; bill bluish-grey and pale buff short feathers on tarsi. Female is like male but lacks barring on chest and grey upper breast.

Distribution and habitat Northern China. In Beijing, an uncommon, irruptive winter visitor, sometimes in large numbers, to open, dry grassland areas or heath.

Behaviour and ecology Feeds in open sandy land or prairies on seeds, buds and insects. Often flies in large flocks. Nests in sand with grass and leaves. Clutch size: 2-4.

北京市一级保护野生动物 | IUCN 红色名录 无危 (LC) 2016.10.1

雄鸟 Male

雄鸟 Male

雌鸟 Female

II

北京鸟类图谱

游禽
NATATORES

BIRDS OF BEIJING

四、雁形目 ANSERIFORMES

栗树鸭	*Dendrocygna javanica*	Lesser Whistling Duck	鸭科 Anatidae

形态特征 小型游禽，体长 37~42 厘米。雌雄同色。体型略似鹅。喙深灰色；头、颈褐色，头顶深褐色；背部深褐色，具栗褐色鳞状斑，中、小覆羽栗色，大覆羽、初级覆羽和飞羽黑褐色，尾上覆羽栗色，尾羽黑色；颏、喉白色或皮黄色，下体栗棕色，尾下覆羽乳白色。

分布与生境 分布于中国长江以南，喜红树林、沼泽等植被茂盛的湿地。北京为甚罕见迷鸟，仅于 2017 年 6 月在通州区北运河有一次记录。

习性与生态 多集群活动。白天常停歇于近水的树上或草地里，并于夜间觅食。飞行时常发出两声或三声一度的轻啸声。

Morphological characteristics A small natatore (37-42 cm). Sexes alike. Body shape resembles goose. Bill is blackish-grey; head and neck brown but darker on crown; back is dark brown with rufous fringes; lesser and median coverts are rufous-brown, while the greater and primary covers and remiges are blackish-brown; uppertail-coverts rufous; tail black; chin and throat white or buff; underparts largely rufous-brown; undertail-coverts white.

Distribution and habitat At mangrove forests, marshes and grasslands in southern China. Very rare vagrant in Beijing. Only one record at the Beijing-Hangzhou Grand Canal in Tongzhou District (June 2017).

Behaviour and ecology Usually in flocks. Rests in trees or grass near water in daytime; feeds at night. Often give two or three notes whistling call in flight.

国家二级保护野生动物 | IUCN 红色名录 无危 (LC) 2016.10.1

鸿雁	*Anser cygnoid*	Swan Goose	鸭科 Anatidae

北京市一级保护野生动物　国家二级保护野生动物　IUCN 红色名录　易危 (VU) 2016.10.1

形态特征　大型游禽，体长 81~94 厘米。雌雄同色。喙较长而粗壮，喙形、头部和颈部形态略似大天鹅。头顶至后颈为深褐色，头侧皮黄色，前颈及颈侧白色，与后颈形成鲜明对比。上体羽褐色，具白色羽缘，停歇时形成较细的白色横斑。胸部皮黄色，至腹部变淡至白色，尾下覆羽亦为白色。跗跖橙色。

分布与生境　中国广泛分布。北京见于水库、湖泊和较大的河流等开阔水面及其附近的草地、荒地、农田等处，为不常见旅鸟。

习性与生态　喜集群活动。迁徙过境时集成十余只至数百只的大群并与其他雁类及天鹅混群。筑巢于茂密的苇丛中；窝卵数 5~6 枚。

Morphological characteristics　A large natatore (81-94 cm). Sexes alike. Body shape resembles Whooper Swan. Bill long and heavy; dark brown from crown to hindneck; buff on cheek; white on foreneck; upperparts brown with white fringes; breast is buff and the belly and undertail coverts are white; tarsi orange.

Distribution and habitat　Widespread in China. In Beijing, uncommon at reservoirs, lakes, large rivers, and farmland and grassland near water. Residential status: passage migrant.

Behaviour and ecology　Often in flocks or joins mixed flocks with other geese and swans. Nests in extensive reedbeds. Clutch size: 5-6.

豆雁	*Anser fabalis*	Taiga Bean Goose	鸭科 Anatidae

形态特征 大型游禽，体长 80~90 厘米。雌雄同色。喙黑色，近端处橙色，全身大致为褐色，上体具较细的皮黄色横斑，尾上覆羽白色，胸、腹部为灰色，至下腹和尾下覆羽转为白色。跗蹠橙色。

分布与生境 中国广泛分布。北京见于水库、湖泊和宽阔的河面，亦见于湿地附近的农田等地，为区域性常见旅鸟及不常见冬候鸟。

习性与生态 典型的雁类习性，喜集群，迁徙过境时的大群可至数千只。筑巢于沼泽或湿地附近的地面，窝卵数 3~4 枚。

Morphological characteristics A large natatore (80-90 cm). Sexes alike. Bill black with an orange subterminal band. Body largely brown with some buff bars on upperparts; uppertail-coverts white; breast and belly grey; vent and undertail-coverts white; tarsi orange.

Distribution and habitat Widespread in China. In Beijing, locally common at reservoirs, lakes and large rivers, as well as farmlands and grasses near wetlands. Residential status: passage migrant and rare winter visitor.

Behaviour and ecology Typical geese behaviour. Often forms flocks with hundreds or thousands of individuals. Nests on the ground near wetlands. Clutch size: 3-4.

北京市二级保护野生动物 | IUCN 红色名录 无危 (LC) 2018.8.9

短嘴豆雁	*Anser serrirostris*	Tundra Bean Goose	鸭科 Anatidae

北京市二级保护野生动物 *		IUCN 红色名录　无危 (LC) 2018.8.9

形态特征　大型游禽，体长 70~88 厘米。雌雄相似。喙黑色，近端处有一小段橙色区域。上体褐色、具浅色斑纹，尾上覆羽白色；胸、腹部灰色或灰褐色，尾下覆羽白色。跗跖橙色。甚似豆雁，但体型略小，且喙、颈皆较短。

分布与生境　中国广泛分布。北京见于有开阔水面的湿地区域以及附近的农田。为区域性常见旅鸟及罕见冬候鸟。

习性与生态　同豆雁，迁徙时亦常混群于豆雁中。

＊注：《北京市二级保护野生动物名录》及 IUCN 红色名录中未收录（更新 / 接受）短嘴豆雁和豆雁的物种划分，故此处沿用豆雁的数据。

Morphological characteristics　A large natatore (70-88 cm). Sexes alike. Bill black with orange subterminal band; upperparts dark brown with buff fringes on mantle; uppertail-coverts white; breast and belly grey or greyish-brown; undertail-coverts white; tarsi orange. Similar to Taiga Bean Goose but has smaller body size and shorter bill and neck.

Distribution and habitat　Widespread in China. Locally common in large wetlands in Beijing. Residential status: passage migrant and rare winter visitor.

Behaviour and ecology　Similar to Taiga Bean Goose, with which it is often seen in mixed flocks.

灰雁	*Anser anser*	Greylag Goose	鸭科 Anatidae

形态特征 大型游禽，体长 76~89 厘米。雌雄同色。喙粉色或橙黄色，全身大致为灰褐色，头、颈羽色略深，颈部羽毛通常形成显著的纵向“沟槽”，上体具淡色横斑，胸、腹部灰白色，两胁深色，具白色横斑，尾下覆羽白色。跗跖粉色。

分布与生境 广布于中国。北京多见于各水库、湖泊等地，为不常见旅鸟；偶然出现于城区公园中的更可能为逃逸个体。

习性与生态 迁徙季节喜集群活动或与其他雁类混群。营巢于苇丛中；窝卵数 4~5 枚。

Morphological characteristics A large natatore (76-89 cm). Sexes alike. Bill pink or yellow-orange; body largely greyish-brown, darker on head and neck; neck feathers form heavy "grooves"; upperparts have white fringes; breast and belly light grey; flanks dark with distinctive white bars; undertail-coverts white; tarsi pink.

Distribution and habitat Widespread in China. Uncommon passage migrant at reservoirs and lakes in rural Beijing. A few individuals at urban lakes may be escapees from zoos or other collections.

Behaviour and ecology Forms flocks while migrating. Nests in reedbeds. Clutch size: 4-5.

北京市二级保护野生动物

IUCN 红色名录 无危 (LC) 2018.8.9

白额雁	*Anser albifrons*	Greater White-fronted Goose	鸭科 Anatidae

	国家二级保护野生动物	IUCN 红色名录 无危 (LC) 2016.10.1

形态特征 大型游禽，体长 65~86 厘米。雌雄同色。喙橙黄色或粉色，喙基和前额基部白色。头、颈褐色，上体余部灰褐色，具皮黄色横纹，尾上覆羽白色，尾羽黑褐色，具白色端斑，胸、腹部灰褐色，腹部有形状不规则的黑色斑块，下腹及尾下覆羽白色。跗跖橙色。

分布与生境 中国广泛分布。北京多见于水库、湖泊和宽阔的河流，为不常见旅鸟。

习性与生态 非繁殖季节喜集群或混群于其他雁类中。于灌丛和草地较丰富的湿地营巢；窝卵数 4~5 枚。

Morphological characteristics A large natatore (65-86 cm). Sexes alike. Bill is yellow-orange or pink; distinctive white patch on forehead; head and neck brown; upperparts greyish-brown with buff fringes; uppertail-coverts white; tail blackish-brown with white terminal band; breast and belly greyish-brown with variable black patches; vent and undertail-coverts white; tarsi orange.

Distribution and habitat Widespread in China. Uncommon passage migrant at reservoirs, lakes and rivers in Beijing.

Behaviour and ecology Usually in flocks or mixed flocks with other geese in non-breeding season. Nests on the ground in wetlands with rich grass or shrubs. Clutch size: 4-5.

游禽

II

小白额雁	*Anser erythropus*	Lesser White-fronted Goose	鸭科 Anatidae

形态特征 中型游禽，体长 56~60 厘米。雌雄同色。喙较短，粉色；喙基和前额白色。眼圈黄色。头、颈深褐色，上体余部褐色，具淡褐色横纹。尾上覆羽白色，尾羽黑褐色，端斑白色，胸、腹部褐色，腹部具黑色斑块，尾下覆羽白色。跗跖橙黄色。似白额雁，但本种体型更小、颈短，且前额白色面积更大。

分布与生境 中国广泛分布。北京见于开阔水面，为罕见旅鸟。

习性与生态 迁徙季节多集群活动，但于北京记录较少，多为单只或小群。繁殖习性似白额雁。

Morphological characteristics A medium-sized natatore (56-60 cm). Sexes alike. Bill short, pink; prominent white forehead and yellow eye-ring; head and neck dark brown; upperparts largely brown with off-white fringes; uppertail-coverts white; tail blackish-brown with white terminal band; underparts brown with black patches on belly; undertail-coverts white; tarsi yellow-orange. Similar to Greater White-fronted Goose but significantly smaller with shorter neck and larger white patch on forehead.

Distribution and habitat Widespread in China. Rare passage migrant at large wetlands in Beijing.

Behaviour and ecology Behaviours generally similar to Greater White-fronted Goose. Records in Beijing are usually of single individuals or small flocks.

北京市二级保护野生动物 | 国家二级保护野生动物 | IUCN 红色名录 易危 (VU) 2018.8.6

斑头雁	*Anser indicus*	Bar-headed Goose	鸭科 Anatidae

IUCN 红色名录 无危 (LC) 2018.8.9

形态特征 大型游禽，体长 71~76 厘米，雌雄同色。喙橙黄色，喙端黑色。头白色，具两道显著的新月形黑斑。前颈、后颈黑色，颈侧白色；上体、下体大致为浅灰色；两胁深色；尾下覆羽白色。跗跖橙黄色。

分布与生境 中国广泛分布，但于西部地区更常见。北京见于各大湿地的开阔水面和水边草地，为罕见旅鸟。城区公园内记录的斑头雁多为逃逸个体。

习性与生态 于中国西部多集群活动。北京的记录则多为单只或 10 只以内的小群。一般繁殖于高原草甸；窝卵数 4~6 枚。

Morphological characteristics A large natatore (71-76 cm). Sexes alike. Bill yellow-orange with dark tip; head white with two distinctive black crescents; foreneck and hindneck greyish-black; neck-sides white; body largely pale grey; flanks dark grey; undertail-coverts white; tarsi yellow-orange.

Distribution and habitat Widespread in China but more common in the west. In Beijing, rare passage migrant along rivers, lakes or grassland near water. A few individuals at urban lakes may be escapees from zoos or other collections.

Behaviour and ecology Forms large flocks in western China. Records in Beijing are usually of single individuals or small flocks (less than ten individuals). Nests in wet grasslands. Clutch size: 4-6.

雪雁	*Anser caerulescens*	Snow Goose	鸭科 Anatidae

形态特征 大型游禽，体长 66~80 厘米。雌雄同色。喙和跗跖粉红色。白色型成鸟通体白色，仅初级飞羽黑色；深色型（亦称蓝色型）成鸟头部及颈上部白色，颈下部至上体余部蓝黑色或黑褐色；飞羽黑色，覆羽灰色，尾上覆羽白色，尾羽黑褐色，端部白色，下体大致为蓝黑色，尾下覆羽白色。

分布与生境 中国东部和中部偶有越冬个体记录。北京甚罕见于郊区湿地或近水农田。居留型不明。

习性与生态 甚喜集群活动，甚至于繁殖期亦集大群（北美）。但中国并非雪雁的主要越冬地，因此国内的记录多为 1~2 只个体，常与其他雁类混群。

Morphological characteristics A large natatore (66-80 cm). Sexes alike. Bill and tarsi pink. White morph is largely white with black primaries. Dark morph (or blue morph) is white on head and upper neck; lower neck and upperparts dark blue or blackish-brown; primaries black; upperwing-coverts grey; uppertail-coverts white; tail dark brown with white tip; underparts blackish-blue; undertail-coverts white.

Distribution and habitat Scarce winter records in central and eastern China. Very rare at large wetlands and farmlands in Beijing. Residential status: unclear.

Behaviour and ecology Tends to form large flocks in both breeding and non-breeding seasons (North America). However, records in China are usually of single individuals in mixed flocks with other geese.

IUCN 红色名录 无危 (LC) 2018.8.9

黑雁	*Branta bernicla*	Brant Goose	鸭科 Anatidae

IUCN 红色名录 无危 (LC) 2016.10.1

形态特征 中型游禽，体长 56~66 厘米。雌雄同色。喙黑色。头、颈、胸黑色，颈侧具一白色斑，上体大致呈灰褐色，尾上覆羽白色，尾羽黑色，上腹灰褐色，下腹至尾下覆羽白色。跗跖黑色。

分布与生境 多见于中国东部沿海。在北京为甚罕见旅鸟，仅于 2014 年 4~5 月在怀柔水库有一次记录。

习性与生态 性活跃，常集群活动，飞行时振翼迅速有力。

Morphological characteristics A medium-sized natatore (56-66 cm). Sexes alike. Bill, head, neck and breast black; white stripe on breast-sides; upperparts largely greyish-brown; uppertail-coverts white; tail black; belly greyish-brown; vent and undertail-coverts white; tarsi black.

Distribution and habitat Coastal areas in eastern China. Very rare passage migrant in Beijing: only one record from Huairou Reservoir in April and May 2014.

Behaviour and ecology Active. Tends to form flocks. In flight, wing beats fast and powerful.

疣鼻天鹅	*Cygnus olor*	Mute Swan	鸭科 Anatidae

形态特征 大型游禽，体长 130~155 厘米。雌雄同色。成鸟喙红色，喙基和喙端黑色；前额具黑色疣状突起，雌鸟的疣状突较雄鸟小；眼先黑色，通体白色，颈长，头、颈沾淡黄色；跗跖和爪皆为黑色。幼鸟全身为淡棕色，无疣状突。

分布与生境 分布于中国北方至东南沿海。北京多记录于水草丰盛的开阔湖泊和水库，为不常见旅鸟。

习性与生态 游水时常将头部低垂、颈部弯曲成"S"形、两翼拱起，姿态优雅而独特。体型较笨重，因此起飞前需在水面长距离助跑。繁殖期雌、雄鸟攻击性皆极强。窝卵数 5~6 枚；常将雏鸟背负于背上。

Morphological characteristics A large natatore (130-155 cm). Sexes alike. Adult has red bill, with black tip and base and prominent black knob at base of bill, larger on the male; lores black; body largely white; head and neck sometimes pale rusty-yellow; tarsi black. Juvenile greyish-brown, lacking knob on bill.

Distribution and habitat From northern China to southeast China. Seen in large areas of open water, such as reservoirs and lakes. Uncommon passage migrant in Beijing.

Behaviour and ecology On water, neck usually held in a graceful S-shape; wings raised slightly from back. Very heavy body, so it must run across water to take-off. Both male and female are aggressive in breeding season. Clutch size: 5-6. Cygnets often ride on adults' back.

国家二级保护野生动物　IUCN 红色名录　无危 (LC) 2016.10.1

小天鹅	*Cygnus columbianus*	Tundra Swan	鸭科 Anatidae

形态特征 大型游禽，体长 110~135 厘米。雌雄同色。成鸟喙黑色，喙基两侧具黄斑（黄斑不越过鼻孔）；通体白色；跗跖和爪黑色。亚成鸟喙粉色，喙端和喙基黑色，全身灰褐色。

分布与生境 广布于中国北部，越冬区域南至长江流域。北京见于水库、湖泊等开阔水域，为区域性常见旅鸟。

习性与生态 迁徙和越冬时多集家族群或数十只的小群过境。起飞前需助跑。繁殖于草丛或苇丛中；窝卵数 2~5 枚。

Morphological characteristics A large natatore (110-135 cm). Sexes alike. Adult has black bill, with prominent yellow patch on side of bill base, not extending beyond nostril; body all white; tarsi black. Juvenile has pink bill with black tip and base; body greyish-brown.

Distribution and habitat Widespread from northern China to the Yangtze River. Locally common passage migrant at large reservoirs and lakes in Beijing.

Behaviour and ecology Forms small to large flocks in non-breeding season. Must run to take-off. Nests in wet grass or reeds. Clutch size: 2-5.

	国家二级保护野生动物	IUCN 红色名录 无危 (LC) 2016.10.1

成鸟 Adult

亚成鸟 Immature

大天鹅	*Cygnus cygnus*	Whooper Swan	鸭科 Anatidae

成鸟 Adult

成鸟 Adult

亚成鸟 Immature

国家二级保护野生动物　IUCN 红色名录　无危 (LC)　2016.10.1

形态特征　大型游禽，体长 120~165 厘米。雌雄同色。成鸟喙黑色，喙基黄色。全身皆为白色。跗跖和爪黑色。亚成鸟喙淡黄色，端部黑色。与小天鹅相似，但本种体型较大，颈较细长，且喙基的黄色部分面积较大，向喙端延伸过鼻孔。

分布与生境　广布于中国北部、中部和东部。北京多活动于有开阔水面的湿地，亦见于圆明园、颐和园等城市公园水域，为区域性常见的旅鸟。

习性与生态　游水时颈向上伸直，垂直于水面，极少弯曲成“S”形。喜集群活动，迁徙过境时常集家族群或数十只至上百只的群体。起飞前需在水面助跑。繁殖于苇丛中；窝卵数 4~7 枚。

Morphological characteristics A large natatore (120-165 cm). Sexes alike. Adult has black bill with yellow base; body plumage all white; tarsi black. Juvenile has pale yellow bill with black tip. Resembles Tundra Swan but larger with a larger yellow patch on bill, extending beyond nostril.

Distribution and habitat Widespread in northern, central and eastern China. In Beijing, locally common passage migrant at wetlands with open water. Sometimes also seen on lakes of urban parks.

Behaviour and ecology Neck is long and held vertical when swimming. Tends to form flocks or family groups. Runs to take-off. Nests in reeds at large wetlands. Clutch size: 4-7.

翘鼻麻鸭	*Tadorna tadorna*	Common Shelduck	鸭科 Anatidae

形态特征 中型游禽，体长 55~63 厘米。雄鸟喙红色，上喙基部有一显著的红色疣状突；头、颈暗绿色，具金属光泽；上、下体白色为主，有一宽阔的栗棕色环自背部延伸至胸部，翼上覆羽白色，初级飞羽黑色，次级飞羽金属绿色，三级飞羽栗色；尾羽白色，具黑色端斑，腹部中央至尾下覆羽棕色；跗跖粉红色。雌鸟似雄鸟，但喙基无疣状突，栗色环带亦较窄。

分布与生境 广布于中国。北京主要见于郊区各湿地，为区域性常见旅鸟。

习性与生态 喜集小群过境。繁殖于近湿地的洞穴中；窝卵数 8~10 枚；雌雄鸟共同育雏。

Morphological characteristics A medium-sized natatore (55-63 cm). Male has red bill, with a red knob at base; head and neck glossed dark green; largely white on upperparts and underparts; distinctive chestnut-brown band extends from back to breast; upperwing-coverts white; primaries black; secondaries glossed-green; tertials chestnut; tail white with black terminal band; brown from central belly to undertail coverts; tarsi pinkish-red. Female similar to male but lacks knob on bill and has narrower band on breast.

Distribution and habitat Widespread in China. Locally common passage migrant at rural wetlands in Beijing.

Behaviour and ecology Forms small flocks in non-breeding season. Nests in holes near water. Clutch size: 8-10. Male and female share the parenting.

北京市二级保护野生动物 | IUCN 红色名录 无危 (LC) 2019.8.14

雄鸟 Male

雌鸟 Female

雄鸟 Male

赤麻鸭	*Tadorna ferruginea*	Ruddy Shelduck	鸭科 Anatidae

雄鸟 Male

雌鸟 Female

北京市二级保护野生动物 | IUCN 红色名录 无危 (LC) 2016.10.1

形态特征 中型游禽，体长 58~68 厘米。雄鸟喙黑色，通体大致为赤褐色；头部羽色较淡，为棕白色，具一较细而清晰的黑色颈环；翼上覆羽白色，飞羽黑色，次级飞羽具绿色金属光泽；尾羽黑色；跗跖黑色。雌鸟似雄鸟，但羽色略淡，且无黑色颈环。

分布与生境 广布于中国。于北京各湿地为常见旅鸟和冬候鸟。亦常离开水域，进入近水草地或农田觅食、休息。偶见于城市公园水域。

习性与生态 喜集大群活动，于北京郊区主要湿地中常见数百只的迁徙过境和越冬集群。巢多置于树洞、崖壁岩洞或地洞中；窝卵数 8~10 枚。

Morphological characteristics A medium-sized natatore (58-68 cm). Male has black bill; body plumage largely orange-brown but paler on head and neck, with a narrow but distinctive black collar on neck; upperwing-coverts white; remiges black, with green gloss on secondaries; tail and tarsi black. Female has paler plumage and lacks black collar.

Distribution and habitat Widespread in China. In Beijing, it is a common passage migrant and winter visitor. Usually seen at reservoirs, lakes, rivers, sometime seen feeding on farmland or grassland near water; occasionally seen in urban wetlands.

Behaviour and ecology Tends to form large flocks in non-breeding season. Nests in cliff cavities, tree holes or underground cavities. Clutch size: 8-10.

鸳鸯	*Aix galericulata*	Mandarin Duck	鸭科 Anatidae

形态特征 小型游禽，体长 41~51 厘米。雄鸟繁殖羽喙红色；前额绿色，眉纹白色，甚宽阔，头顶后部橙色，颈部羽毛橙棕色且呈须状，上体大致为褐色；飞羽黑色，次级飞羽具绿色金属光泽，端部白色，最内侧三级飞羽特化为帆状，直立于背上；胸部深紫色，下胸具两道白色带；两胁红褐色，下体余部白色；跗跖黄色。雌鸟喙褐色，全身大致为灰褐色，具白色眼圈和眼纹，跗跖灰色。雄鸟非繁殖羽似雌鸟但喙仍为红色。

分布与生境 分布于中国东部和中部。在北京为区域性常见的旅鸟和夏候鸟及留鸟。春、秋季多见于城区、郊区湿地的开阔水面，繁殖季则多在部分城区公园和山地溪流、池塘活动。

习性与生态 在近水的树洞中筑巢；窝卵数 7~12 枚。

Morphological characteristics A small natatore (41-51 cm). Breeding male has bright red bill; crown and hindcrown orange; distinctive white supercilium and orange whiskers; upperparts largely brown; remiges black; secondaries glossed green; enlarged tertials orange, sticking up like sails; breast purple, with two white bands on lower breast; belly white; flanks reddish-brown; tarsi yellow. Female has brown bill; body largely greyish-brown, with white eye-ring and eye-stripe; tarsi grey. Non-breeding male similar to female, but with red bill.

Distribution and habitat Widespread in eastern and central China. Locally common summer breeder, passage migrant and some resident in Beijing. Seen in lakes and rivers in both urban and rural Beijing. Breeds in urban park wetlands and mountainous streams and ponds in forests.

Behaviour and ecology Nests in tree holes near water. Clutch size: 7-12.

	国家二级保护野生动物	IUCN 红色名录 无危 (LC) 2018.8.9

雄鸟 Male

雌鸟 Female

棉凫	*Nettapus coromandelianus*	Cotton Pygmy Goose	鸭科 Anatidae

北京市二级保护野生动物 | 国家二级保护野生动物 | IUCN 红色名录 无危 (LC) 2016.10.1

形态特征 小型游禽，体长 31~38 厘米。雄鸟喙灰色；虹膜红色；头顶黑色，头侧至颈部白色，颈基部具宽阔的黑色颈环，略具绿色金属光泽；上体黑褐色，翼上覆羽翠绿色，初级飞羽黑色，各羽中段白色，次级飞羽翠绿色，端部白色，尾上覆羽白色；下体大致为白色。雌鸟虹膜为深褐色，头顶褐色，具黑色贯眼纹；胸部具黑色斑点，两胁淡皮黄色。雄鸟蚀羽与雌鸟相似。

分布与生境 分布于中国南部。北京地区甚罕见，主要活动于水生植物丰富的湖泊、河流、池塘和沼泽等地，推断为迷鸟。

习性与生态 在分布区内多成对或集小群活动，在北京则为单只记录。在近水树洞中繁殖；窝卵数 8~14 枚。

Morphological characteristics A small natatore (31-38 cm). Male has grey bill and red eyes; crown black; white from head-sides to neck; broad black collar on lower neck; upperparts blackish-brown; upperwing-coverts green; primaries black with white patch; secondaries green with white tip; uppertail-coverts and underparts white. Female has dark brown eyes and black eye-stripe; brown on crown; underparts largely white, with black spots on breast and pale buff on flanks. Eclipse male resembles female.

Distribution and habitat Southern China. Found at lakes, ponds, rivers and marshes with abundant vegetation. Very rare in Beijing. Residential status: rare vagrant.

Behaviour and ecology Forms small flocks or in pairs in southern China. Records in Beijing are mostly of singles. Nests in tree holes near water. Clutch size: 8-14.

赤膀鸭	*Mareca strepera*	Gadwall	鸭科 Anatidae

形态特征 中型游禽，体长 46~55 厘米。雄鸟繁殖羽喙黑灰色；头顶棕色，头侧、颈部灰色，背部及小覆羽灰褐色，中覆羽和外侧大覆羽栗色，内侧大覆羽黑色；初级飞羽褐色，外侧次级飞羽黑色，内侧次级飞羽白色。腰、尾上和尾下覆羽黑色，尾羽褐色；胸部褐色，具白色鳞状斑；腹部白色，两胁褐色。雌鸟喙橙色，喙峰黑色，全身大致为褐色，具黑色鳞状斑。

分布与生境 中国广泛分布。北京见于水库、湖泊和河流的开阔水面，为常见旅鸟。

习性与生态 典型的河鸭习性，迁徙过境时喜集群活动，常与绿头鸭等其他过境鸭类混群。常以头下脚上的姿态倒栽在水中取食水生植物。于接近湿地的草丛或灌丛中筑巢；窝卵数 8~12 枚。

Morphological characteristics A medium-sized natatore (46-55 cm). Breeding male has blackish-brey bill; crown brown; head-sides and neck grey; mantle greyish-brown; rump, uppertail- and undertail-coverts black; tail is brown; primaries brown; outer secondaries black and inner secondaries white; outer greater coverts and median coverts chestnut-brown, inner greater coverts black; breast brown, barred white; belly white; flanks brown. Female has orange bill with black culmen; body largely brown, scaled black.

Distribution and habitat Widespread in China. In Beijing, common at reservoirs, rivers and lakes. Residential status: passage migrant.

Behaviour and ecology Tends to be found in flocks in non-breeding season. Often forms mixed flocks with other freshwater ducks such as mallard. Up-ends when feeding in water. Nests in grass or shrubby habitat near wetlands. Clutch size: 8-12.

北京市二级保护野生动物 | IUCN 红色名录 无危 (LC) 2016.10.1

雄鸟 Male

雌鸟 Female

罗纹鸭	*Mareca falcata*	Falcated Duck	鸭科 Anatidae

雄鸟 Male

雌鸟 Female

北京市二级保护野生动物 | | IUCN 红色名录 近危 (NT) 2016.10.1

形态特征 中型游禽，体长 46~54 厘米。雄鸟繁殖羽喙黑色；头顶栗褐色，前额基部具一白色斑点，头两侧翠绿色，颏、喉至颈部白色，具一绿色颈环；上体余部及下体大致为白色，密布黑色鳞状斑纹；翼上覆羽和初级飞羽灰褐色，次级飞羽暗绿色，端部白色，三级飞羽延长且向下弯曲，尾下覆羽黑色，两侧具三角形淡黄色斑块。雌鸟头部褐色，颏、喉和颈部白色，具褐色短纵纹，上体和下体余部褐色，具黑色“V”形斑。雄鸟非繁殖羽似雌鸟。

分布与生境 分布于中国东部至中部。北京地区为区域性常见旅鸟，迁徙季节见于湿地开阔水域。

习性与生态 集群迁徙和越冬。于植被丰富的水边草地、灌丛下营巢；窝卵数 6~10 枚。

Morphological characteristics A medium-sized natatore (46-54 cm). Breeding male has black bill and chestnut-brown crown with a white spot on forehead; head-sides glossy green; chin, throat and neck white; green collar on lower neck; upperparts and underparts largely white, finely scaled black; upperwing-coverts and primaries greyish-brown; secondaries dark green with white tips; tertials long and drooping; undertail-coverts black, with creamy-white sides. Female largely brown; chin, throat and neck greyish-white, streaked brown. Eclipse male similar to female.

Distribution and habitat Eastern to central China. Locally common passage migrant at large reservoirs, lakes and rivers in Beijing.

Behaviour and ecology Forms flocks in non-breeding season. Nests in grassland or under shrubs near water. Clutch size: 6-10.

赤颈鸭	*Mareca penelope*	Eurasian Wigeon	鸭科 Anatidae

形态特征 小型游禽，体长 42~50 厘米。雄鸟繁殖羽喙灰色，端部黑色；前额至头顶棕白色，头侧至颈部棕红色；上体灰白色，具细密的暗褐色鳞状斑；初级飞羽灰褐色；中覆羽、大覆羽白色；次级飞羽翠绿色，端部黑色；三级飞羽特别延长，黑色，具白色外缘；尾上、尾下覆羽皆为黑色；胸部棕色，腹部白色，两胁灰色。雌鸟上体大致为褐色，具深色斑点，尾下覆羽白色，具褐色斑点。雄鸟非繁殖羽似雌鸟。

分布与生境 广布于中国各地（除青藏高原外）。北京见于有开阔水面的湿地，为常见旅鸟。

习性与生态 典型的河鸭习性，迁徙过境时喜集群或与其他鸭类混群。营巢于湿地附近的草丛或灌丛中；窝卵数 7~11 枚。

Morphological characteristics A small natatore (42-50 cm). Breeding male has grey bill with black tip; head and neck chestnut; forehead and crown creamy-buff; upperparts greyish-white, with dark brown scales; primaries brown; median and greater coverts white, forming large white patch on the wing; secondaries glossed green, with black bases; tertials long and black, with white fringes; uppertail- and undertail-coverts black; breast brown; belly white, flanks grey. Female brown with dark spots and bars; undertail-coverts white, with brown spots. Eclipse male similar to female.

Distribution and habitat Widespread in China, except for the Tibetan Plateau. Common passage migrant in Beijing. Usually seen at wetlands with open water.

Behaviour and ecology In the non-breeding season forms flocks or joins mixed flocks with other duck species. Nests in grassland or shrubby habitat in wetlands. Clutch size: 7-11.

北京市二级保护野生动物		IUCN 红色名录 无危 (LC) 2016.10.1

前雄后雌 Male (front), Female (rear)

雄鸟 Male

绿眉鸭	*Mareca americana*	American Wigeon	鸭科 Anatidae

前雄后雌 Male (front), Female (rear)

IUCN 红色名录 无危 (LC) 2016.10.1

形态特征 中型游禽，体长 45~56 厘米。雄鸟繁殖羽喙灰色，喙端黑色；头顶棕白色，眼后方有一宽阔的绿色带延伸至颈侧，眼先、颊至颈部为浅灰色，具细密的黑色斑点；上体余部、胸部和两胁大致为红褐色，尾上覆羽和尾下覆羽黑色，腹部白色。雌鸟全身大致为红褐色，具黑色斑点。非繁殖羽雄鸟似雌鸟。

分布与生境 主要分布于北美，偶有迷鸟至亚洲和欧洲沿海，中国在东部沿海有零星记录。甚罕见于北京地区，仅于 2020 年春在昌平十三陵水库有过一次迷鸟记录。

习性与生态 非繁殖期喜集群，亚洲的迷鸟个体常与赤颈鸭混群，习性似赤颈鸭。

Morphological characteristics A medium-sized natatore (45-56 cm). Breeding male has pale grey bill with a black tip; prominent buff blaze on forehead and crown; broad green band extends from eye to neck-sides; lores, cheeks and neck grey with black spots; upperparts brown; breast and flanks rufous; belly white; uppertail- and undertail-coverts black. Female has pale grey head and neck, and brown flanks. Eclipse male resembles female.

Distribution and habitat Mainly North America. Scarce in east China. Very rare in Beijing, only one vagrant record at the Ming Tombs Reservoir, Changping District (Spring 2020).

Behaviour and ecology Forms flocks in non-breeding season. Vagrants in China often join mixed flocks of other duck, especially Eurasian Wigeon.

绿头鸭	*Anas platyrhynchos*	Mallard	鸭科 Anatidae

形态特征 中型游禽，体长 50~62 厘米。雄鸟繁殖羽喙黄色；头、颈翠绿色，具金属光泽，有一白色颈环，背部褐色；翼上覆羽和初级飞羽褐色，次级飞羽蓝紫色，具白色端斑和黑色次端斑；腰、尾上覆羽、尾下覆羽皆为黑色；中央两枚尾上覆羽向上卷曲；胸部栗棕色，腹部灰白色。雌鸟喙橙黄色，喙端和喙中部黑色，全身大致为黄褐色，具黑色斑纹。雄鸟非繁殖羽似雌鸟，但喙为黄色。

分布与生境 广布于中国各地。北京地区见于各种类型的湿地，包括城区公园、高校校园的水域，为甚常见的夏候鸟、冬候鸟和旅鸟。

习性与生态 典型的河鸭习性。非繁殖期集群于开阔水面活动。繁殖于水生植被丰富的湿地；窝卵数 7~11 枚。

Morphological characteristics A medium-sized natatore (50-62 cm). Breeding male has yellow bill; head and neck glossy green, with a distinctive white collar; mantle, upperwing-coverts and primaries brown; secondaries glossy blue with white terminal band; uppertail- and undertail-coverts black, with the central uppertail-coverts upcurled; breast purplish-chestnut and belly pale grey. Female has orange bill with black culmen and centre; body brown with black streaks. Eclipse male similar to female but with yellow bill.

Distribution and habitat Widespread in China. Abundant summer breeder, winter visitor and passage migrant in Beijing. Seen in all wetland habitats.

Behaviour and ecology Forms large flocks in non-breeding season. Nests at wetlands with abundant vegetation. Clutch size: 7-11.

北京市二级保护野生动物		IUCN 红色名录 无危 (LC) 2016.10.1

雄鸟 Male

雌鸟 Female

斑嘴鸭	*Anas zonorhyncha*	Eastern Spot-billed Duck	鸭科 Anatidae

北京市二级保护野生动物 | IUCN 红色名录 无危 (LC) 2018.8.6

形态特征 中型游禽，体长 53~64 厘米。雄鸟喙黑色，近端处黄色；头顶深褐色，头侧乳白色，具深黑色贯眼纹，颈部淡褐色；上体褐色，翼上覆羽和初级飞羽深褐色，次级飞羽金属蓝绿色，具白色端斑和黑色次端斑；胸白色，具深色点斑；腹部及尾下覆羽深褐色。雌鸟与雄鸟相似，但下体羽色较淡。

分布与生境 中国广泛分布于除西部之外的其他地区。北京见于城区和郊区植被丰富的湿地，为常见夏候鸟、旅鸟和罕见冬候鸟。

习性与生态 营巢于湿地草丛、灌丛和苇丛中；窝卵数 8~14 枚。飞行时稍显笨重，振翼较慢。

Morphological characteristics A medium-sized natatore (53-64 cm). Male has black bill, tipped yellow; crown is dark brown; creamy-white supercilium contrasts strongly with dark eye-stripe; neck pale brown; upperparts brown; upperwing-coverts and primaries dark brown; secondaries glossy-blue with distinctive white terminal band and black subterminal band; breast white, spotted dark; remainder of underparts dark brown. Female similar to male but paler below.

Distribution and habitat Widespread in China, except for western China. Seen at both urban and rural wetlands with rich vegetation. Common summer breeder, passage migrant and rare winter visitor in Beijing.

Behaviour and ecology Nests in wet grass, reeds and shrubs. Clutch size: 8-14. In flight, wing beats slow and heavy.

针尾鸭	*Anas acuta*	Northern Pintail	鸭科 Anatidae

形态特征 中型游禽，体长 61~72 厘米。雄鸟繁殖羽喙蓝灰色，喙峰和喙端黑色；头部暗褐色，前颈和颈侧白色，后颈暗褐色；背部及尾上覆羽灰白色，密布深色横斑；肩羽较长，黑色，羽缘银灰色；次级飞羽铜绿色，端部白色；中央尾羽黑色，显著延长，外侧尾羽灰白色；胸部、腹部白色，两肋具细密的灰色波状斑纹，尾下覆羽黑色。雌鸟头部棕褐色，上体暗褐色，具皮黄色“V”形斑，两肋与上体同色，下体白色。雄鸟非繁殖羽似雌鸟但喙色不变。

分布与生境 中国除青藏高原外广泛分布。北京主要见于郊区湿地开阔水面，为区域性常见旅鸟。

习性与生态 典型的河鸭习性。于北京过境时喜与其他鸭类混群。营巢于近水的草丛中；窝卵数 6~11 枚。

Morphological characteristics A medium-sized natatore (61-72 cm). Breeding male has greyish-blue bill with black culmen and tip; head and neck largely dark brown but foreneck and neck-sides white, forming a white stripe on neck; upperparts pale grey, with fine dark scales; scapulars long and black with silver fringes; speculum glossy green, with white and brown borders; central rectrices long and pointed; tail-sides pale greyish-white; underparts largely white; flanks vermiculated dark grey and undertail coverts black. Females largely brown, with buff scales on upperparts and white below. Eclipse male resembles female but retains bill pattern.

Distribution and habitat Widespread in China except for the Tibetan Plateau. In Beijing, locally common at rural lakes, reservoirs and rivers. Residential status: passage migrant.

Behaviour and ecology Typical dabbling duck behaviour. Often joins mixed flocks with other ducks. Nests in grass near water. Clutch size: 6-11.

北京市二级保护野生动物 | IUCN 红色名录 无危 (LC) 2019.8.13

雄鸟 Male

雌鸟 Female

左雄右雌 Male (left), female (right)

绿翅鸭 *Anas crecca* Eurasian Teal

鸭科 Anatidae

雄鸟 Male

雄鸟 Male

雌鸟 Female

雌鸟 Female

北京市二级保护野生动物

IUCN 红色名录 无危 (LC) 2020.8.13

形态特征 小型游禽，体长 34~38 厘米。雄鸟繁殖羽喙黑色；前额至头顶栗棕色，头侧眼上方绿色，眼下方至颈部栗棕色；上体和两胁密布黑白相间的横纹，大覆羽端部白色，外侧次级飞羽黑色，内侧次级飞羽翠绿色，皆具白色端斑，形成显著的翼镜；尾上覆羽和尾羽黑褐色；胸部淡黄色，具黑色点斑；腹部白色，臀至尾下覆羽黑色，尾下两侧具乳白色块斑。雌鸟头及上体大致为褐色。雄鸟非繁殖羽似雌鸟。

分布与生境 中国广泛分布。北京地区见于水库、湖泊和河流的开阔水面，为常见旅鸟。

习性与生态 迁徙过境时集群或与其他鸭类混群，白天在开阔水面休息，黄昏和夜间在浅水处取食。营巢于湿地草丛或灌丛中；窝卵数 8~11 枚。

Morphological characteristics A small natatore (34-38 cm). Breeding male has black bill; chestnut-brown from forehead to crown; green above eyes; chestnut-brown from cheek to neck, with buff borders on head-sides; upperparts and flanks vermiculated black and white; greater coverts have white tips; outer secondaries black, inner secondaries glossed green, both with white terminal band, forming prominent speculum; uppertail-coverts and tail blackish-brown; breast pale yellow, with black spots; belly white; vent and undertail-coverts black, with cream tail-sides. Female largely brown. Eclipse male resembles female.

Distribution and habitat Widespread in China. In Beijing, common passage migrant at reservoirs, lakes and rivers.

Behaviour and ecology Often joins mixed flocks with other Anas duck. Usually feeds at night in non-breeding season. Nests in shrubs and grass in wetlands. Clutch size: 8-11.

琵嘴鸭 *Spatula clypeata* Northern Shoveler

鸭科 Anatidae

形态特征 小型游禽，体长 44~52 厘米。雌、雄鸟皆具特别扩大呈铲状的喙。雄鸟繁殖羽喙黑色；虹膜黄色；头、颈绿色；上体暗褐色，翼上小覆羽和中覆羽蓝灰色，大覆羽褐色，端部白色，初级覆羽和初级飞羽深褐色，次级飞羽翠绿色，具黑色端斑，三级飞羽黑色，尾上覆羽暗绿色，尾羽白色；胸部白色，腹部栗色，尾下覆羽黑色。雌鸟喙黄褐色，虹膜褐色，头、颈褐色，上体和两胁深褐色，具皮黄色鳞状斑纹。雄鸟非繁殖羽似雌鸟。

分布与生境 中国广泛分布。北京主要见于郊区湿地开阔水面，为区域性常见旅鸟。

习性与生态 喜以其铲状的喙在水面滤食水生动物和水藻等。迁徙时多单只或集小群过境，常与其他鸭类混群。筑巢于近水草丛中；窝卵数 7~13 枚。

Morphological characteristics A small natatore (44-52 cm). Both sexes have very long and broad bill. Breeding male has black bill and yellow eyes; head and neck green; upperparts dark brown; lesser and median coverts greyish-blue; greater coverts brown with white tip; primaries and primary coverts dark brown; secondaries glossed green, with black terminal band; tertials black; uppertail coverts dark green; tail white; breast white; belly chestnut; undertail coverts black. Female has orange-brown bill and brown eyes; body largely brown with buff scales. Eclipse male resembles female.

Distribution and habitat Widespread in China. Locally common passage migrant in reservoirs and lakes in rural Beijing.

Behaviour and ecology Filters for food with its large bill in shallow water when feeding. Forms small flocks or joins mixed flocks in non-breeding season. Nests on the ground in grassland near water. Clutch size: 7-13.

北京市二级保护野生动物 | IUCN 红色名录 无危 (LC) 2019.8.14

雄鸟 Male

雌鸟 Female

白眉鸭	*Spatula querquedula*	Garganey	鸭科 Anatidae

雄鸟 Male

雌鸟 Female

雄鸟 Male

北京市二级保护野生动物 | IUCN 红色名录 无危 (LC) 2016.10.1

形态特征 小型游禽，体长 34~41 厘米。雄鸟繁殖羽前额至头顶黑褐色，具清晰且宽阔的白色眉纹，头侧至颈部棕褐色，具白色短纵纹；上体黑褐色，内侧肩羽较长，呈绿色，翼上覆羽蓝灰色，大覆羽具宽阔的白色端斑，初级飞羽深褐色，次级飞羽金属绿色、端部白色；胸黄褐色，具黑色波状斑，腹部棕白色，两胁灰白色，具黑色波状细纹，尾下覆羽棕白色具黑斑。雌鸟头部黄褐色，眉纹白色。雄鸟非繁殖羽似雌鸟。

分布与生境 中国除青藏高原外广泛分布。北京见于植被丰富的湿地，为不常见旅鸟。

习性与生态 常集小群迁徙过境。于湿地草丛、灌丛下的地面筑巢；窝卵数 8~12 枚。

Morphological characteristics A small natatore (34-41 cm). Breeding male has a distinctive white supercilium; dark brown from forehead to crown; head-sides and neck brown, with some fine white streaks; upperparts blackish-brown; scapulars long and dark green; upperwing-coverts pale greyish-blue; primaries blackish-brown; secondaries glossed green, with white terminal band; breast brown, vermiculated black; belly off-white; flanks pale grey, with fine dark vermiculations. Female has pale brown head and white supercilium. Eclipse male resembles female.

Distribution and habitat Widespread in China except for the Tibetan Plateau. Found in wetlands with plentiful vegetation. Uncommon passage migrant in Beijing.

Behaviour and ecology Usually forms small flocks in non-breeding season. Nests on the ground in grassland in wetlands. Clutch size: 8-12.

花脸鸭	*Sibirionetta formosa*	Baikal Teal	鸭科 Anatidae

形态特征 小型游禽，体长 37~44 厘米。雄鸟繁殖羽喙黑色；头顶深褐色，头两侧具乳黄色斑块、绿色带斑和白色细纹形成的艳丽花纹；上体大致为褐色，肩羽细长，外翈黑色，内翈白色，羽缘栗色；翼上覆羽褐色，大覆羽端部栗色，初级飞羽褐色，外侧次级飞羽黑色，内侧次级飞羽翠绿色，皆具白色端斑；胸浅棕色，具褐色斑点，胸两侧各具一白色带。雌鸟全身大致为褐色，似白眉鸭雌鸟但无白色眉纹。雄鸟非繁殖羽似雌鸟。

分布与生境 分布于中国东部和南部。北京多见于附近有农田的湿地，为不常见旅鸟。

习性与生态 常集小群或与其他雁鸭混群过境。多于草丛中筑巢；窝卵数 6~9 枚。

Morphological characteristics A small natatore (37-44 cm). Breeding male has black bill; crown dark brown; distinctive head pattern includes a cream patch, green band and white streaks on head-side; upperparts largely brown; scapulars long, with black outer vane, white inner vane and chestnut fringes; upperwing-coverts brown; greater coverts tipped chestnut; primaries brown; outer secondaries black; inner secondaries glossy green; all secondaries have white terminal band; breast pale brown with dark spots with a white stripe on breast-sides. Female largely brown, resembles female Garganey but has pale brown supercilium. Eclipse male resembles female.

Distribution and habitat Eastern and southern China. In Beijing, uncommon passage migrant at reservoirs, lakes, ponds and rivers, especially those with farmlands nearby.

Behaviour and ecology In non-breeding season in flocks or joins mixed flocks with other dabbling duck. Usually nests in grassland near wetlands. Clutch size: 6-9.

北京市一级保护野生动物 | 国家二级保护野生动物 | IUCN 红色名录 无危 (LC) 2016.10.1

雄鸟 Male

雌鸟 Female

赤嘴潜鸭	*Netta rufina*	Red-crested Pochard	鸭科 Anatidae

雄鸟 Male

雌鸟 Female

左雄右雌 Male (left), female (right)

北京市二级保护野生动物 | IUCN 红色名录 无危 (LC) 2016.10.1

形态特征 中型游禽，体长 51~57 厘米。雄鸟繁殖羽喙红色；虹膜红色；头部栗棕色，较同体型鸭类显得膨大；背部褐色，翼上覆羽与背部同色，飞羽白色，端部褐色，腰及尾上覆羽黑色，尾羽褐色，具白色羽缘；下体大致为黑色，两胁白色。雌鸟喙深褐色，端部黄色，虹膜褐色，头顶褐色，颊至前颈白色，上、下体余部大致为褐色。雄鸟非繁殖羽似雌鸟，但喙仍为红色。

分布与生境 主要见于中国西部至中部。北京见于郊区各大水库、湖泊、河流，为不常见旅鸟。时有逃逸个体在紫竹院公园等市中心水域活动。

习性与生态 多成对或集群活动，主食水生植物，既可潜水，亦可头下尾上似河鸭状觅食。于苇丛中筑巢；窝卵数 7~9 枚。

Morphological characteristics A medium-sized natatore (51-57 cm). Breeding male has red bill and eyes; large head bright orange-brown; upperparts brown; primaries white, with brown tips; rump and uppertail-coverts black; tail brown, with white fringes; underparts largely black; flanks white. Female has orange-yellow tipped bill, blackish-brown at base; brown eyes; crown is chocolate-brown, contrasting with white cheek and upper neck; body largely brown. Eclipse male resembles female but retains red bill.

Distribution and habitat Widespread from west to central China. Uncommon passage migrant at large reservoirs, lakes and rivers in rural Beijing. Escapees sometimes seen in the central part of the city such as Zizhuyuan Park.

Behaviour and ecology Usually in pairs or flocks. Feeds mainly on aquatic plants. When feeding, can dive under water and dabble. Nests in reeds. Clutch size: 7-9.

红头潜鸭	*Aythya ferina*	Common Pochard	鸭科 Anatidae

形态特征 小型游禽，体长 42~49 厘米。雄鸟繁殖羽喙基和喙端黑色，喙中段灰色；虹膜红色；头、颈红色；背部和两翼灰白色，腰、尾上和尾下覆羽黑色，尾羽灰褐色；胸黑色，腹部灰白色。雌鸟虹膜深褐色，头、颈、胸褐色，上体和下体灰褐色。

分布与生境 广泛分布于中国各地。在北京活动于水生植物丰富的湿地或开阔水面，为城区和郊区水域的常见旅鸟。

习性与生态 常集小群或与其他鸭类混群。擅潜水。筑巢于苇丛中；窝卵数 6~9 枚。

Morphological characteristics A small natatore (42-49 cm). Breeding male has black bill with grey on centre; eyes bright red; head and neck chestnut-red; upperparts greyish-white; black from rump to uppertail- and undertail-coverts; tail is greyish-brown; breast black; belly pale greyish-white. Female has dark eyes and greyish-brown body with darker head, neck and breast.

Distribution and habitat Widespread in China. Common passage migrant at well-vegetated wetlands and large reservoirs and lakes in both urban and rural Beijing.

Behaviour and ecology Usually seen in small flocks or mixed flocks with diving duck. Nests on the ground in reeds near water. Clutch size: 6-9.

北京市二级保护野生动物 | IUCN 红色名录 易危 (VU) 2016.10.1

雄鸟 Male

雌鸟 Female

青头潜鸭	*Aythya baeri*	Baer's Pochard	鸭科 Anatidae

雄鸟 Male

雄鸟 Male

雄鸟 Male

北京市二级保护野生动物 | 国家一级保护野生动物 | IUCN 红色名录 极危 (CR) 2019.8.15

形态特征 小型游禽，体长 42~47 厘米。雄鸟繁殖羽喙灰色，端部黑色；虹膜白色；头、颈黑色，具绿色金属光泽；背部和翼上覆羽褐色，飞羽白色，端部黑褐色，腰及尾上覆羽黑色；胸栗褐色，两胁白色与褐色交杂，腹部及尾下覆羽白色。雌鸟眼先栗褐色，头、颈黑褐色。

分布与生境 中国见于东部至南部。为北京地区有稳定记录的罕见旅鸟，多见于较大的水库、河流、湖泊等地。在北京水生植被丰富的湿地中可能有繁殖，需关注其 5~7 月的在京记录。

习性与生态 迁徙时常与其他潜鸭混群。于湿地苇丛和草丛中营巢；窝卵数 6~15 枚。近年来种群数量锐减。

Morphological characteristics A small natatore (42-47 cm). Breeding male has grey bill with black tip; eyes are white; head and neck black, with green gloss; mantle and upperwing-coverts brown; remiges white, tipped blackish-brown; rump and uppertail-coverts black; breast brown; flanks brown and white; belly and undertail-coverts white. Female has blackish-brown head and neck and chestnut-brown lores.

Distribution and habitat Eastern and southern China. Scarce but regular passage migrant at large reservoirs, lakes and rivers in Beijing. Possibly breeds in wetlands with abundant vegetation in rural Beijing.

Behaviour and ecology In non-breeding season, often joins mixed flocks of other Aythya duck. Nests on the ground in reeds and grass at wetlands. Clutch size: 6-15. Population has declined significantly in recent decades.

白眼潜鸭	*Aythya nyroca*	Ferruginous Duck	鸭科 Anatidae

形态特征　小型游禽，体长 33~42 厘米。雄鸟繁殖羽喙灰色，尖端黑色；虹膜白色；头、颈、胸及两胁栗色，上体深褐色，飞羽白色，端部深褐色，腰和尾上覆羽黑色，腹部至尾下覆羽白色。雌鸟虹膜褐色，全身大致为褐色，羽色不及雄鸟鲜艳。

分布与生境　广布于中国（除东北地区外）。北京地区见于各类开阔水域及水生植物丰富的湿地，主要为不常见旅鸟；近年来记录增多，于延庆野鸭湖有繁殖记录。

习性与生态　典型的潜鸭习性，擅潜水觅食，迁徙时单只或集小群过境，亦有时与其他潜鸭混群。营巢于湿地苇丛或草丛中；窝卵数 7~11 枚。

Morphological characteristics　A small natatore (33-42 cm). Breeding male has grey bill with black tip; eyes white; chestnut from head to breast and flanks; upperparts blackish-brown; remiges white with dark tips; rump and uppertail-coverts black; belly and undertail-coverts white. Female has dark eyes; body largely brown, duller than male.

Distribution and habitat　Widespread except for the northeast China. Uncommon but increasing migrant at reservoirs, lakes, rivers and other well-vegetated waters with breeding records at Yeyahu, Yanqing District.

Behaviour and ecology　Dives for food. In non-breeding season, usually in small flocks, sometimes with other Aythya duck. Nests at wetlands with abundant reeds and grass. Clutch size: 7-11.

北京市二级保护野生动物　　IUCN 红色名录　近危 (NT)　2019.6.12

雄鸟 Male
雌鸟 Female

凤头潜鸭 | *Aythya fuligula* | Tufted Duck | 鸭科 Anatidae

雄鸟 Male

雌鸟 Female

北京市二级保护野生动物 | IUCN 红色名录 无危 (LC) 2016.10.1

形态特征 小型游禽，体长 40~47 厘米。雄鸟繁殖羽喙灰色，喙端黑色；虹膜黄色；头部具黑色冠羽披于头后；全身大致黑白相间，头、颈、上体、胸和尾下覆羽黑色，尾羽褐色，腹部和两胁白色。雌鸟冠羽显著较雄鸟短，通体大致为暗褐色，喙基近白色。雄鸟非繁殖羽似雌鸟，但两胁为灰色。

分布与生境 中国广泛分布。北京见于城区和郊区的水库、河流、湖泊、池塘等水域，为常见旅鸟。

习性与生态 喜集群过境，极擅潜水，主食水生无脊椎动物和小鱼、蝌蚪等。营巢于近水的草丛或灌丛中；窝卵数 8~10 枚。

Morphological characteristics A small natatore (40-47 cm). Breeding male has grey bill, tipped black; instinctive yellow eyes and long, black crest; head, neck, upperparts, breast and undertail-coverts all-black, contrasting strongly with white belly and flanks. Female has much shorter crest; body largely dark brown; flanks brown; bill base sometimes off-white. Eclipse male has grey or greyish-brown flanks.

Distribution and habitat Widespread in China. In Beijing: common passage migrant at reservoirs, lakes, rivers and ponds. Seen at both urban and wetlands in Beijing.

Behaviour and ecology Forms flocks in non-breeding season. Dives for fish, amphibians, aquatic invertebrates and aquatic plants. Nests on the ground at grassland or shrubby habitat near water. Clutch size: 8-10.

斑背潜鸭	*Aythya marila*	Greater Scaup	鸭科 Anatidae

形态特征 小型游禽，体长 42~51 厘米。雄鸟繁殖羽喙灰色，喙端黑色；虹膜黄色；头、颈黑色，具绿色金属光泽；上体白色，具黑色波状斑纹，腰和尾上覆羽黑色，尾羽深褐色；胸、尾下覆羽黑色，腹部及两胁白色。雌鸟喙基白色，上体深褐色，具不明显的灰白色波状细纹，下体大致为褐色。

分布与生境 分布于中国东部及南部，在华北地区多选择沿海湿地过境。在北京为罕见旅鸟，见于水库、湖泊等各种开阔水域。

习性与生态 擅潜水捕食。迁徙时常与凤头潜鸭等其他潜鸭混群。

Morphological characteristics A small natatore (42-51 cm). Breeding male has long, grey bill with black nail; eyes yellow; head and neck black with dark green gloss; upperparts white with fine black vermiculations; rump and uppertail-coverts black; tail dark brown; breast and undertail-coverts black; belly and flanks white. Female has distinctive white patch at bill base; body largely brown; mantle darker, vermiculated grey.

Distribution and habitat Eastern and southern China. Mainly seen in coastal areas in non-breeding season. Rare passage migrant at large reservoirs and lakes in Beijing.

Behaviour and ecology Dives for food. Usually joins mixed flocks with other diving ducks.

北京市二级保护野生动物

IUCN 红色名录 无危 (LC) 2018.8.7

雄鸟 Male

雌鸟 Female

丑鸭	*Histrionicus histrionicus*	Harlequin Duck	鸭科 Anatidae

IUCN 红色名录 无危 (LC) 2018.8.7

形态特征 小型游禽，体长 38~45 厘米。雄鸟繁殖羽喙铅灰色；前额至头顶黑色，头两侧至颈部大致为黑色，但眼先和耳部各具一圆形白色块斑，颈侧有一较细的白色带，颈基部有一较宽白色带；上体蓝黑色，肩羽白色，尾上、尾下覆羽黑色，尾羽深灰色；下体深蓝色，胸侧具白色条纹，两胁栗色。雌鸟通体大致为褐色，眼先上下各有一块白斑，耳部亦有一圆形白斑。

分布与生境 中国于北方地区和东部有零星记录。多见于沿海地区以及水流湍急的河流中。北京地区甚罕见，仅于 2017 年 2 月在元大都护城河朝阳区北土城段有一次记录，居留型暂不明确。

习性与生态 单独或集小群活动。喜在急流中频繁潜水，以水生无脊椎动物为食物。

Morphological characteristics A small natatore (38-45 cm). Breeding male has grey bill; black from forehead to crown; head-sides and neck mainly dark blue, with white patches on face, ear-coverts and lower neck; upperparts blackish-blue; scapulars white; uppertail- and undertail-coverts black; tail dark grey; underparts largely dark blue, with white stripe, bordered black, on breast-sides; flanks chestnut. Female is mainly brown, with two white patches above and below eyes, and white spot on ear-coverts.

Distribution and habitat Scarce in northern and eastern China. Usually seen at coastal areas and rushing rivers or streams. Very rare in Beijing: only one record at the moat of Khanbaliq (Chaoyang District) in February 2017. Residential status: unclear.

Behaviour and ecology Solitary or in small flocks. Prefers to dive for aquatic invertebrates in fast-flowing streams.

斑脸海番鸭	*Melanitta fusca*	Velvet Scoter	鸭科 Anatidae

形态特征 中型游禽，体长 51~58 厘米。雄鸟繁殖羽喙红色，喙基有一黑色瘤状突；全身大致为黑色，眼后有一新月形白斑，次级飞羽白色，飞行时甚显著。雌鸟喙黑色，通体褐色，仅眼先、耳羽和次级飞羽白色。

分布与生境 在中国主要分布于东部和南部沿海，于内陆地区亦偶有记录。北京见于较大水库、河流、湖泊的开阔水面，为罕见旅鸟。

习性与生态 擅潜水取食水生无脊椎动物和鱼类。多为单只、成对或小群。营巢于近水的苔原、林缘或疏林地面；窝卵数 7~9 枚。

Morphological characteristics A medium-sized natatore (51-58 cm). Breeding male has red bill with black knob at base; body mainly glossy black, with a white crescent below eye; secondaries white, forming distinctive speculum. Female has black bill; body largely brown, with white on lores, ear-coverts and speculum.

Distribution and habitat Mainly coastal areas in eastern and southern China, scarce in inland. Rare passage migrant at large reservoirs, lakes and rivers in Beijing.

Behaviour and ecology Dives for aquatic invertebrates and fish. Usually solitary, in pairs or in small flocks. Nests on the ground in woodland or tundra near water. Clutch size: 7-9.

北京市二级保护野生动物 | | IUCN 红色名录 无危 (LC) 2018.8.7

雄鸟 Male

雌鸟 Female

长尾鸭	*Clangula hyemalis*	Long-tailed Duck	鸭科 Anatidae

雄鸟 Male

雌鸟 Female

北京市二级保护野生动物 | IUCN 红色名录 易危 (VU) 2018.8.7

形态特征 小到中型游禽，体长37~47（雌）或51~60（雄）厘米。雄鸟喙黑色，近端处粉色。雄鸟冬羽头顶、枕部及颈部白色，头两侧灰色，颈侧具一黑色块斑；背部白色，腰及尾上覆羽黑色，中央尾羽黑色，特别延长；胸黑色，腹部及尾下覆羽白色。雄鸟夏羽头、颈、胸黑色，头侧有一显著的白色块斑，上体大致为深褐色，下体白色，中央尾羽仍延长。雌鸟喙灰色，无延长的中央尾羽，上体大致为深褐色，但头侧、颈部和下体白色。

分布与生境 分布于中国东部沿海，偶见于内陆。北京地区见于水库、湖泊、河流和池塘的开阔水面，为罕见旅鸟。

习性与生态 非繁殖期多集群于海面活动，但于内陆则多为单只记载，潜水取食小型鱼类和无脊椎动物。

Morphological characteristics A small to medium-sized natatore (male: 51-60 cm; female: 37-47 cm). Male has black bill with pink near tip. Male in winter and spring plumage is white from crown to nape, as well as neck; head-sides grey with black patch on neck-sides; mantle white; rump and uppertail-coverts black; black central rectrices long and pointed; breast black; remainder of underparts white. Male in summer retains long tail; mainly black from head to breast with white facial patch; dark brown above and white below. Female has grey bill and lacks long tail; head, neck, breast and upperparts largely brown, with white on face, neck and underparts.

Distribution and habitat Coastal areas in China. Scarce inland. Rare passage migrant at reservoirs, lakes and rivers in Beijing.

Behaviour and ecology Forms flocks at sea in non-breeding season but usually solitary inland. Dives for small fish and aquatic invertebrates.

鹊鸭	*Bucephala clangula*	Common Goldeneye	鸭科 Anatidae

形态特征 小型游禽，体长 40~48 厘米，与其他鸭类相比显得头大、颈短、体型紧凑。雄鸟繁殖羽喙黑色，虹膜黄色；头及上颈黑色，具绿色金属光泽，喙基有一圆形白斑；上体黑色为主，翼上小覆羽和初级飞羽黑色，中覆羽、大覆羽和次级飞羽白色；下体白色，仅尾下覆羽黑褐色。雌鸟喙深褐色，喙端橙黄色，头部至上颈棕褐色，上体、胸及两胁灰色，腹部白色。

分布与生境 中国广泛分布。北京见于城区、郊区各种湿地的开阔水面，为区域性常见旅鸟及冬候鸟。

习性与生态 非繁殖期多集成数十只的群体，擅潜水捕食。筑巢于树洞中；窝卵数 8~12 枚。

Morphological characteristics A small natatore (40-48 cm). A compact diving duck with relatively large head and short neck. Breeding male has black bill and yellow eyes; head and upper neck green-glossed black with a distinctive white facial patch; upperparts largely black; median coverts, greater coverts and secondaries white, forming a prominent white speculum; underparts mostly white with black undertail-coverts. Female has dark brown bill with orange-yellow tip; head and upper neck brown; upperparts, breast and flanks largely grey; belly white.

Distribution and habitat Widespread in China. In Beijing: locally common passage migrant and winter visitor at urban and rural wetlands with open water.

Behaviour and ecology Forms small flocks in non-breeding season. Dives for food. Nests in tree holes. Clutch size: 8-12.

北京市二级保护野生动物		IUCN 红色名录 无危 (LC) 2018.8.7

左雌右雄 Female (left), male (right)

雄鸟 Male

雌鸟 Female

斑头秋沙鸭	*Mergellus albellus*	Smew	鸭科 Anatidae

雄鸟 Male

雌鸟 Female

北京市二级保护野生动物　　国家二级保护野生动物　　IUCN 红色名录　无危 (LC) 2016.10.1

形态特征　小型游禽，体长 38~44 厘米。雄鸟繁殖羽喙深灰色，头、颈白色为主，具短冠羽，眼先及眼周有一黑色圆形斑，枕部黑色；上体及两翼黑色，翼上覆羽具白斑，胸两侧具黑色细纹；下体余部白色，两胁淡灰色。雌鸟头顶、头侧、枕部至后颈栗棕色，眼先黑色，颊、颏、喉及前颈白色，上体及两胁灰色，下体白色。雄鸟非繁殖羽似雌鸟。

分布与生境　除青藏高原外广布于中国各地。北京见于湖泊、水库、河流、池塘等水域，为区域性常见旅鸟和冬候鸟。

习性与生态　性羞怯，易被惊飞。擅潜水捕食鱼类和水生无脊椎动物，非繁殖期多单只或集小群活动。繁殖期营巢于树洞中；窝卵数 6~10 枚。

Morphological characteristics A small natatore (38-44 cm). Breeding male has dark grey bill and short crest. Head and neck largely white with a black facial patch from lores to eyes; nape and upperparts black; breast is white with fine black lines on sides; remainder of underparts mainly white; flanks pale grey. Female is chestnut-brown on crown, nape and hindneck, darker on lores, contrasting with white chin, throat, cheek and neck; upperparts and flanks grey; underparts white. Eclipse male resembles female with white crest and breast.

Distribution and habitat Widespread in China except for the Tibetan Plateau. Locally common passage migrant and winter visitor at reservoirs, lakes, rivers and ponds in Beijing.

Behaviour and ecology Shy and easily disturbed. Dives for small fish and aquatic invertebrates. In non-breeding season, usually seen in small flocks. Nests in tree holes. Clutch size: 6-10.

普通秋沙鸭 | *Mergus merganser* | Common Merganser | 鸭科 Anatidae

形态特征 中型游禽，体长 58~68 厘米，为秋沙鸭中体型最大、分布最广的一种。喙暗红色，喙尖而长，端部下弯呈钩状。雄鸟繁殖羽具短冠羽，头及上颈黑色，具绿色金属光泽；上体灰褐色，翼上小覆羽灰色，大、中覆羽和次级飞羽白色，初级飞羽黑色，下颈、胸及其余下体白色。雌鸟头及上颈棕褐色，颏、喉白色，上体及两胁灰色，下体白色。雄鸟非繁殖羽似雌鸟。

分布与生境 中国广泛分布。北京地区见于城区和郊区的河流、湖泊和水库中，为常见旅鸟和冬候鸟。

习性与生态 非繁殖期常集数十只至数百只的群体活动。擅潜水捕鱼。起飞前需在水面助跑。

Morphological characteristics A medium-sized natatore (58-68 cm). The largest and most widely distributed sawbill. Both sexes have long, hook-tipped red bill. Breeding male has short crest; head and upper neck green-glossed black; upperparts greyish-brown; lesser coverts grey; greater coverts, median coverts and secondaries white; primaries black; lower neck, breast and remainder of underparts all-white. Female has brown head and upper neck; chin and throat white; upperparts and flanks grey; underparts white. Eclipse male resembles female.

Distribution and habitat Widespread in China. Common passage migrant and winter visitor at reservoirs, lakes and rivers in urban and rural Beijing.

Behaviour and ecology Forms flocks in non-breeding season. Feeds on fish or other small animals. Runs across water for take-off.

北京市二级保护野生动物 | | IUCN 红色名录 无危 (LC) 2018.8.7

雄鸟 Male

雌鸟 Female

红胸秋沙鸭	*Mergus serrator*	Red-breasted Merganser	鸭科 Anatidae

左雌右雄 Female (left), male (right)

北京市二级保护野生动物　　IUCN 红色名录 无危 (LC) 2018.8.7

形态特征 中型游禽，体长 52~60 厘米。喙红色，较细长，喙端无钩。雄鸟繁殖羽头绿色，具上翘的冠羽；背部黑色，小覆羽和初级飞羽灰褐色，大覆羽、中覆羽和次级飞羽白色，形成翼上的大块白斑；腰、尾上覆羽及尾羽灰褐色；上颈白色，下颈至胸锈红色，胸侧黑色，具白色斑点，下体余部白色，两胁密布深灰色与白色相间的波状细纹。雌鸟头部棕褐色，眼先、颏、喉白色，颈部、上体及两胁灰色，胸淡褐色，下体白色。

分布与生境 中国除西南地区外广泛分布。非繁殖期于沿海地区更多见。在北京地区一般活动于面积大且水深的水库、湖泊及河流，为不常见旅鸟。

习性与生态 典型的秋沙鸭习性，擅潜水捕捉鱼类和水生无脊椎动物，非繁殖期多成对或集小群活动。

Morphological characteristics A medium-sized natatore (52-60 cm). Both sexes have long, slender, red bill lacking hook on tip. Breeding male has green head, with a prominent crest; mantle black; lesser coverts and primaries greyish-brown; greater coverts, median coverts and secondaries white, forming a distinctive white patch on the wing; rump, uppertail-coverts and tail greyish-brown; upper neck white; lower neck and breast rusty-brown; breast-sides black with white spots; belly white; flanks vermiculated dark grey and white. Female has brown head but white lores, chin and throat; neck, upperparts and flanks mostly grey; breast pale brown; underparts white.

Distribution and habitat Widespread in China except for the southwest China. Mostly seen in coastal areas in non-breeding season. Uncommon passage migrant at large reservoirs, lakes and rivers in Beijing.

Behaviour and ecology Dives for fish and aquatic invertebrates. In pairs or small flocks in non-breeding season.

中华秋沙鸭 | *Mergus squamatus* | Scaly-sided Merganser | 鸭科 Anatidae

形态特征 中型游禽，体长 52~62 厘米。喙红色，长而直。雌、雄鸟具较长的双冠状冠羽。雄鸟繁殖羽头及上颈黑色，具暗绿色光泽，上背黑色，两翼灰色，翼上具两块白斑，下背、腰、尾上覆羽和两胁白色，具黑色鳞状斑，胸、腹及尾下覆羽白色。雌鸟头部及上颈黄褐色，上体灰色，下体白色，两胁具灰色鳞状斑。

分布与生境 除西部地区外在中国广泛分布。北京地区见于郊区较大的河流、水库和湖泊中，为罕见旅鸟。

习性与生态 非繁殖期多单只、成对或集小群活动。在林间溪流附近的树洞中筑巢；窝卵数 8~12 枚。近年来种群数量下降至较低程度（全球数量 2400~4500 只，IUCN，2016）。

Morphological characteristics A medium-sized natatore (52-62 cm). Both sexes have long and straight red bill and long, double hindcrest. Breeding male has black head and upper neck with dark green gloss; mantle black; wings grey with two white patches; rump, uppertail-coverts and flanks white with black scales; breast, belly and undertail coverts white. Female has brown head and upper neck; grey above and white below; flanks scaled grey.

Distribution and habitat Widespread in China except for the west. Rare passage migrant at large reservoirs, lakes and rivers in Beijing.

Behaviour and ecology In non-breeding season usually seen solitarily, in pairs or in small flocks. Nests in tree holes near forest streams. Clutch size: 8-12. The global population has declined to an estimated 2,400 to 4,500 individuals (IUCN, 2016).

国家一级保护野生动物 | IUCN 红色名录 濒危 (EN) 2016.10.1

雄鸟 Male

雌鸟 Female

五、鸊鷉目 PODICIPEDIFORMES

小鸊鷉	*Tachybaptus ruficollis*	Little Grebe	鸊鷉科 Podicipedidae

繁殖羽 Breeding

非繁殖羽 Non-breeding

北京市二级保护野生动物 | IUCN 红色名录 无危 (LC) 2016.10.1

形态特征 小型游禽，体长 25~32 厘米。雌雄同色。成鸟繁殖羽喙黑色，喙端白色，喙基具黄色椭圆形斑点；虹膜淡黄色；头顶、后颈及上体黑褐色，颊、前颈及颈侧栗色，胸及两胁灰褐色，腹部白色。成鸟非繁殖羽全身大致为褐色，颏、喉白色，颊及颈侧淡黄色。幼鸟头、颈部具黑白相间的纵纹。

分布与生境 中国广泛分布。北京见于城、郊区的湖泊、池塘、水库、河流等各类湿地水域，为甚常见夏候鸟和旅鸟，于冬季不封冻的水域亦有少量个体越冬。

习性与生态 擅潜水捕食。非繁殖期集小群活动。鸣声为一连串响亮而尖锐的颤音。于水生植物丰富的湿地造浮巢繁殖；窝卵数 4~7 枚；有将低龄雏鸟背负于背上的习性。

Morphological characteristics A small natatore (25-32 cm). Sexes alike. Breeding adult has black bill with white tip and a yellow spot at base; eyes pale yellow; crown, nape and upperparts largely blackish-brown; cheek, foreneck and neck-sides rufous; breast and flanks greyish-brown; belly white. Non-breeding adult brown above and buff below; cheek, throat and neck white or buff instead of rufous. Juvenile streaked black on head and neck.

Distribution and habitat Widespread in China. In Beijing: abundant at reservoirs, lakes, rivers and ponds, including wetlands in urban parks. Residential status: mainly summer breeder and passage migrant; also a few wintering individuals in unfrozen water.

Behaviour and ecology Dives into water and pursues fish and invertebrates. Forms small flocks in non-breeding season. Often gives a series of high-pitched trills. Builds floating nest at well-vegetated wetlands. Clutch size: 4-7. Young birds often ride on the backs of parents.

赤颈䴙䴘 | *Podiceps grisegena* | Red-necked Grebe | 䴙䴘科 Podicipedidae

繁殖羽 Breeding

亚成鸟 Immature

过渡羽 Transitional plumage

国家二级保护野生动物 | IUCN 红色名录 无危 (LC) 2018.8.7

形态特征 小型游禽，体长 40~53 厘米。雌雄同色。成鸟繁殖羽喙暗灰色，喙基黄色；头顶及枕部黑色，具甚短的黑色冠羽，颊、颏、喉白色，后颈及上体深褐色，两翼深褐色，次级飞羽白色；前颈至胸栗红色，下体白色。成鸟非繁殖羽前颈和胸白色。幼鸟头、颈具黑白纵纹，体灰色。

分布与生境 中国见于东部和西北地区。在北京活动于较大的水库、湖泊和河流，为罕见旅鸟。

习性与生态 常单只或成对活动。于湿地苇丛中营浮巢繁殖；窝卵数 4~5 枚；有将低龄雏鸟背负于背上的习性。

Morphological characteristics A small natatore (40-53 cm). Sexes alike. Breeding adult has dark grey bill with yellow tip and a very short black crest; crown to nape black, contrasting with white cheek, chin and throat; upperparts and wings mainly dark brown; secondaries white; rufous from neck to breast; underparts white. Non-breeding adult has white foreneck and breast. Juvenile has black and white stripes on head and neck; body mainly grey.

Distribution and habitat Eastern and northwest China. Rare passage migrant at large reservoirs, lakes and rivers in Beijing.

Behaviour and ecology Usually solitary or in pairs. Nests in reeds or in other floating vegetation. Clutch size: 4-5. Young birds often ride on the backs of parents.

凤头䴙䴘	*Podiceps cristatus*	Great Crested Grebe	䴙䴘科 Podicipedidae

形态特征 中型游禽，体长 45~55 厘米。喙长而尖锐，头顶黑色，头顶两侧的羽毛延长，形成两束较显著的黑色冠羽，头后侧至喉具略延长的棕色饰羽，颈长，上体黑褐色，肩羽至次级飞羽有显著白斑，仅飞行时可见。前颈、胸、腹皆为白色。冬羽头侧棕色部分大多被白色取代。似赤颈䴙䴘但前颈为白色。

分布与生境 广布于中国各地。常见于北京郊区水库、河流、湖泊等开阔而水流平缓的湿地，亦见于颐和园、圆明园等有较大面积水域的城市公园。在北京主要为夏候鸟和旅鸟。

习性与生态 擅潜水捕鱼和其他水生动物。以芦苇和水草营造浮巢；每巢 4~6 枚卵；雌雄鸟共同参与孵卵和育雏，亲鸟常将刚孵化不久的雏鸟放于其背上活动。

Morphological characteristics A medium sized natatore (45-55 cm) with long bill and slender neck. Sexes alike. Breeding adult has black crest and prominent brown ruff; entire upper plumage is blackish-brown; breast and belly white. In flight, two distinctive white patches on the wings. Non-breeding adult has shorter crest and no ruff; similar to Red-necked Grebe but white foreneck.

Distribution and habitat Widespread in China. Common summer breeder and passage migrant in wetlands of Beijing. Seen regularly on lakes of urban parks.

Behaviour and ecology Dives for fish and aquatic invertebrates. Builds floating nest with waterweeds. Clutch size: 4-6. Both male and female involved in incubation and feeding chicks. Young chicks are often carried on the backs of adults.

北京市一级保护野生动物 | IUCN 红色名录 无危 (LC) 2019.8.11

繁殖羽 Breeding

幼鸟（右）Juvenile (right)

非繁殖羽 Non-breeding

NATATORES

角䴙䴘 | *Podiceps auritus* | Horned Grebe | 䴙䴘科 Podicipedidae

繁殖羽 Breeding

非繁殖羽 Non-breeding

国家二级保护野生动物　IUCN 红色名录　易危 (VU) 2018.8.7

形态特征　小型游禽，体长 31~39 厘米。雌雄同色。成鸟繁殖羽喙黑色，喙端白色，喙较短而粗壮；虹膜红色；头、后颈及上体黑色，自眼先至头后方具显著的金黄色饰羽。前颈、颈侧、胸及两胁栗红色；腹部白色。非繁殖羽头顶、后颈及上体黑色，头两侧、前颈、胸及下体余部白色。

分布与生境　分布于中国东部及西北地区。北京见于较大的湖泊、水库和河流，为罕见旅鸟。

习性与生态　一般单只或成对活动于距离岸边较远之处。具䴙䴘典型的繁殖习性：在水生植物丰富之处筑浮巢；窝卵数 4~5 枚。

Morphological characteristics　A small natatore (31-39 cm). Sexes alike. Breeding adult has short black bill with white tip; eyes red; largely black from head, hindneck and upperparts; prominent yellow band from lores to rear of head; foreneck, neck-sides, breast and flanks chestnut; belly white. Non-breeding adult has black on crown, hindneck and upperparts; head-sides, foreneck, breast and underparts white.

Distribution and habitat　Eastern and northwest China. Rare passage migrant at large reservoirs, lakes and rivers in Beijing.

Behaviour and ecology　Usually solitary or in pairs. Typical grebe breeding behaviours: builds floating nest in well-vegetated wetlands. Clutch size: 4-5.

黑颈䴙䴘	*Podiceps nigricollis*	Black-necked Grebe	䴙䴘科 Podicipedidae

形态特征 小型游禽，体长 25~34 厘米。雌雄同色。成鸟繁殖羽喙黑色，略上翘；虹膜红色；头、颈及上体皆为黑色，眼后具一扇形金黄色饰羽；两翼大致为褐色，但次级飞羽和内侧初级飞羽白色，形成一块飞行时可见的白斑；胸、腹部白色，胸侧及两胁栗红色。非繁殖羽上体大致为黑色，颊、颏、喉、胸及下体余部大致为白色，前颈灰色。

分布与生境 中国除青藏高原外广泛分布。北京见于水库、湖泊、河流、池塘等湿地的开阔水面，为不常见旅鸟。

习性与生态 多单独、成对或集小群活动，擅潜水觅食。巢浮于水面，但固定于芦苇等挺水植物中或其他浮水植物之上；窝卵数 4~6 枚。

Morphological characteristics A small natatore (25-34 cm). Sexes alike. Breeding adult has upcurved black bill and red eyes; head, neck and upperparts mostly black with bright yellow fan-shaped ear plumes; breast and belly white; breast-sides and flanks chestnut; wings largely brown, but secondaries and inner primaries white, forming a distinctive white patch which can be seen in flight. Non-breeding adult is black above and white below; foreneck grey instead of black.

Distribution and habitat Widespread in China except for the Tibetan Plateau. In Beijing: uncommon passage migrant at wetlands with open water, such as reservoirs, lakes, rivers and ponds.

Behaviour and ecology Usually solitary, in pairs or in small flocks. Dives for small fish and aquatic invertebrates. Constructs floating nest anchored to reeds or other aquatic vegetation. Clutch size: 4~6.

北京市一级保护野生动物 | 国家二级保护野生动物 | IUCN 红色名录 无危 (LC) 2018.8.7

繁殖羽 Breeding

非繁殖羽 Non-breeding

NATATORES

六、潜鸟目 GAVIIFORMES

红喉潜鸟	*Gavia stellata*	Red-throated Diver	潜鸟科 Gaviidae

形态特征 中型游禽，体长 53~69 厘米。雌雄同色。成鸟繁殖羽喙黑色，略上翘；虹膜红色；头、颏至颈侧灰色，后颈白色，具黑色纵纹；上体和两翼黑褐色；喉、前颈至上胸栗红色，下体余部白色，两胁具黑色纵纹。非繁殖羽头顶、后颈及上体深褐色，具白色斑点。颏、喉、颊、前颈、胸及下体余部皆为白色。

分布与生境 分布于中国东、南部沿海地区。在北京较大的水库、河流中有零星记录，为甚罕见旅鸟。

习性与生态 非繁殖期多单独在水面活动，远离岸边，擅潜水捕鱼和其他水生动物。

Morphological characteristics A medium-sized natatore (53-69 cm). Sexes alike. Breeding adult has slightly upcurved black bill and red eyes; plain grey from head and chin to neck-sides; hindneck white, steaked black; upperparts and wings largely blackish-brown; throat, foreneck and upper breast red; underparts largely white; flanks streaked black. Non-breeding adult is dark brown with white spots on crown, hindneck and upperparts; chin, throat, cheek, foreneck, breast and remainder of underparts white.

Distribution and habitat Coastal areas in east and south China. Very rare passage migrant at large reservoirs and rivers in Beijing.

Behaviour and ecology Usually solitary in non-breeding season; away from edge on reservoirs or lakes. Dives deep to pursue fish and other small aquatic animals.

IUCN 红色名录 无危 (LC) 2018.8.7

黑喉潜鸟	*Gavia arctica*	Black-throated Diver	潜鸟科 Gaviidae

IUCN 红色名录 无危 (LC) 2018.8.7

形态特征 中型游禽，体长 56~75 厘米。雌雄同色。成鸟繁殖羽喙黑灰色，不上翘；虹膜暗红色；头顶至后颈灰色；上体黑色，具白色斑块；颏、喉、颊及前颈黑色，具不显著的绿色金属光泽；颈侧和胸侧具黑白相间的纵纹；胸及下体白色。非繁殖羽喙灰色，喙峰及喙端黑色；头、后颈、上体深灰色；颊、颏、喉、前颈、胸及下体白色。

分布与生境 见于中国东部沿海。在北京地区活动于水库、河流、湖泊的开阔水面，为甚罕见旅鸟。

习性与生态 擅潜水，可长时间在水下追捕鱼群。非繁殖期多单只记录。

Morphological characteristics A medium-sized natatore (56-75 cm). Sexes alike. Breeding adult has straight, dark blackish-grey bill and eyes deep red; grey from crown to hindneck; upperparts black, squared white; black from cheek, chin and throat to foreneck; neck-sides and breast-sides streaked black and white; breast and underparts white. Non-breeding adult has grey bill with black culmen and tip; plain grey above and white below.

Distribution and habitat Coastal areas in eastern China. Very rare passage migrant at reservoirs, lakes and rivers in Beijing.

Behaviour and ecology Dives deep into water for fish. Usually solitary in non-breeding season.

II

游禽

太平洋潜鸟	*Gavia pacifica*	Pacific Diver	潜鸟科 Gaviidae

形态特征 中型游禽，体长 63~66 厘米。雌雄同色。成鸟繁殖羽喙黑色，不上翘。虹膜暗红色。头顶至后颈灰白色。上体黑色，具长方形白色斑块。颏、喉、前颈黑色，略具紫色光泽，颈侧和胸侧白色，具黑色纵纹，胸、腹部白色。非繁殖羽头、后颈及上体黑褐色。颏、喉、胸及下体白色，喉部具一黑色横纹。

分布与生境 零星分布于中国东部和南部沿海。于北京为甚罕见旅鸟，仅在 2008 年 5 月记录于怀柔水库。

习性与生态 典型的潜鸟习性，擅潜水捕鱼。非繁殖期常单独在开阔水面活动，起飞前需助跑，在陆地行走缓慢而困难。

Morphological characteristics A medium-sized natatore (63-66 cm). Sexes alike. Breeding adult has straight, black bill and deep read eyes; pale plain grey from crown to hindneck; upperparts black with white squares; chin, throat and foreneck black with purple gloss; black and white streaks from neck-sides to breast-sides; breast and belly white. Non-breeding adult is blackish-brown from head, nape and hindneck to upperparts; chin, throat, breast and remainder of underparts white, with a black line on throat.

Distribution and habitat Scarce along the east and south coast of China. Very rare passage migrant in Beijing, with only one record at Huairou Reservoir, Huairou District in May 2008.

Behaviour and ecology Dives deep for fish, like other divers. Solitary in non-breeding season. Runs heavily above water to take-off. Ungainly on the ground.

IUCN 红色名录 无危 (LC) 2018.8.7

七、鲣鸟目 SULIFORMES

白斑军舰鸟	*Fregata ariel*	Lesser Frigatebird	军舰鸟科 Fregatidae

	国家二级保护野生动物	IUCN 红色名录　无危 (LC) 2018.8.10

形态特征　大型游禽，体长 71~80 厘米。雄鸟几乎全身黑色，略具蓝绿色金属光泽；喉囊红色，腹侧部具白色斑块。雌鸟上体黑褐色，缺少光泽；胸、上腹及两肋白色。尾具很深的开叉。亚成鸟头颈部棕黄色。

分布与生境　中国见于东南沿海地区，偶至内陆。北京为甚罕见迷鸟，见于开阔水域（野鸭湖、密云水库、沙河、南海子等）空中。

习性与生态　主要栖息于热带海洋岛屿，内陆多见其高空飞过，极少停落。营巢于海岛树上或灌丛上；巢由植物枝叶搭建而成；窝卵数 1 枚。

Morphological characteristics　A large natatore (71-80 cm). Male is black overall, glossed bluish-green, with red gular pouch and white flank patches. Female has blackish-brown upperparts, with white at the centre of the underparts. Juvenile resembles female with tawny head and neck.

Distribution and habitat　Southeast coast of China, rare inland. In Beijing, very rare vagrant to areas of open water (e.g.: Yeyahu, Miyun and Shahe Reservoirs and Nanhaizi Park). Usually flies high over open water.

Behaviour and ecology　Found mainly on tropical islands. Nest made of twigs and leaves, usually in trees or shrubs. Clutch size: 1.

普通鸬鹚 *Phalacrocorax carbo* Great Cormorant

鸬鹚科 Phalacrocoracidae

形态特征 大型游禽，体长 79~92 厘米。雌雄相似。喙长而前端钩状，喙端深灰色，喙基较淡。成鸟繁殖羽虹膜翠绿色，眼周和喉侧裸皮黄色；颊白色；周身黑色具铜绿色金属光泽，头顶至颈侧白色，两肋具显著的白斑。成鸟非繁殖羽似繁殖羽，但头、颈无白色区域，两肋无白斑。

分布与生境 广泛见于中国各地水域。北京常见于开阔水域或岸边树上，主要为旅鸟，夏季在沙河水库、野鸭湖等地有度夏个体（可能为亚成鸟）但未见繁殖；于冬季不冻结水域有少数越冬。

习性与生态 擅潜水捕鱼，南方多有驯养，谓之“鱼鹰”。因为尾脂腺不甚发达，常见其在岸边或树上展翅晾晒羽毛。游水时颈向上直伸，身体大部沉入水下。常集大群，飞行时排成“V”字队形。集群营巢于水边树上；窝卵数 3~5 枚。

Morphological characteristics A large natatore (79-92 cm). Sexes alike. Long bill hooked at tip, mostly dark grey but paler at base. Breeding adult has green eyes with yellow facial and gular skin; cheek white; body mostly black, with dark green gloss; variable white patch from crown to neck-sides; distinctive white flank patch. Non-breeding adult resembles breeding adult but black on head, neck and flanks.

Distribution and habitat Widely spread in China. In Beijing, common in open water areas, such as reservoirs or large lakes and rivers, as well as on trees near water. Residential status: mostly a passage migrant, a few oversummer (possibly immature) but no evidence of breeding and some spend the winter if there is open water.

Behaviour and ecology Dives for fish. Often seen perched on shore or tree with wings outspread. Swims with body mostly submerged. Often flies in v-formation when in large flocks. Nests in trees on shore. Clutch size: 3-5.

北京市二级保护野生动物 | IUCN 红色名录 无危 (LC) 2018.8.9

繁殖羽 Breeding

非繁殖羽 Non-breeding

亚成鸟 Immature

绿背鸬鹚	*Phalacrocorax capillatus*	Japanese Cormorant	鸬鹚科 Phalacrocoracidae

亚成鸟 Immature

繁殖羽 Breeding

IUCN 红色名录 无危 (LC) 2018.8.8

形态特征 大型游禽，体长 80~84 厘米。雌雄相似。喙较细长，尖端呈钩状，大致呈灰色。喙基附近具黄色裸皮。颊白色。全身大致为黑色，背部及翼上覆羽具暗绿色金属光泽。成鸟繁殖羽头、颈部具白色丝状羽，形成白色纵纹或更大面积的白色区域；两胁具白斑。甚似普通鸬鹚但颊白色区域面积较大，且上体为暗绿色而非铜绿色。

分布与生境 中国见于辽东半岛、东南沿海和邻近岛屿及海岸上。多在礁石区域活动。北京地区为甚罕见旅鸟。

习性与生态 常成群生活。通常几对在一起成小群繁殖，亦有单对独立繁殖。多营巢于海岸和海岛岸边；巢由枯草和海草构成；窝卵数 4~5 枚。

Morphological characteristics A large natatore (80-84 cm). Sexes alike. Grey bill long and slender, hooked at tip; facial skin yellow with white cheek patch; Body largely black; glossed deep green from mantle to upperwing-coverts. Breeding adult has white plumes from head to neck and white patch on flanks. Closely resembles Great Cormorant but has deep green upperparts and in breeding plumage larger white check patch.

Distribution and habitat East coast and islands of China. Occurs in rocky areas. Very rare passage migrant in Beijing.

Behaviour and ecology Feeds in groups. Breed in small colonies; build nest with dry grass and sea grass on rocky coast or islands. Clutch size: 4-5.

八、鹈形目（一）PELECANIFORMES

白鹈鹕	*Pelecanus onocrotalus*	Great White Pelican	鹈鹕科 Pelecanidae

形态特征　大型游禽，体长 140~175 厘米。雌雄相似。通体白色，黑色的眼位于粉黄色的裸皮上极为醒目。喙长而粗直，铅灰至粉红色，喙下有一橙黄色皮囊。头后部有一簇白色冠羽。初级飞羽和次级飞羽黑褐色，头、颈和长的冠羽缀有粉黄色，有时扩展至上背和肩。跗跖肉色，为全蹼足。

分布与生境　中国主要见于新疆、青海等地，在东部仅有零星记录。北京为甚罕见迷鸟，见于郊区较大且开阔的湿地。

习性与生态　飞行缓慢，擅游泳。常在开阔静水面游水，缓慢而优雅。营巢于湖中小岛和湖边芦苇浅滩、河流岸边及沼泽地上、芦苇丛中浅水处，也在树上营巢；巢结构较为庞大，由树枝、枯草和水生植物构成；窝卵数 2~3 枚。

Morphological characteristics　A large natatore (140-175 cm). Sexes alike. Overall white plumage with black primaries and secondaries, white tufted nuchal crest; conspicuous black eye on pinkish facial skin; long straight stocky bill greyish-pink with an orange-yellow pouch; pinkish tarsi and webbed feet.

Distribution and habitat　Mainly in Xinjiang and Qinghai Province. Scarce in eastern China. Very rare vagrant in Beijing. Seen at large wetlands in rural areas.

Behaviour and ecology　Often seen swimming on open lakes. Slow in flight. Nest made of twigs, grass and aquatic plants, usually built on islands, marsh or river shore. Clutch size: 2-3.

国家一级保护野生动物　IUCN 红色名录　无危 (LC) 2018.8.9

卷羽鹈鹕	*Pelecanus crispus*	Dalmatian Pelican	鹈鹕科 Pelecanidae

	国家一级保护野生动物	IUCN 红色名录　近危 (NT) 2017.10.1

形态特征　大型游禽，体长 160~180 厘米。雌雄相似。头上的冠羽呈卷曲状。颊部和眼周裸露的皮肤均为乳黄色或肉色。喙长而粗，上喙灰褐色，前端具一爪状弯钩，下喙黄色且具可收缩的大型黄色喉囊。体羽主要为白色，飞羽羽尖黑色。跗跖近灰色，为全蹼足。

分布与生境　在中国繁殖于北方地区，冬季主要在东南一带集群越冬，该越冬种群数量仅 50~100 只。迁徙期在北京见于开阔水面（沙河、颐和园等水域），为罕见旅鸟。

习性与生态　喜群居和游泳，常在开阔静水面游水，缓慢优雅。主要以鱼类、甲壳类、软体动物、两栖动物等为食。营巢于近水的树上；窝卵数 3~4 枚。

Morphological characteristics　A large natatore (160-180 cm). Sexes alike. White overall with fluffy curly tufts and black tipped primaries; bill long and thick, dark grey upper mandible with a hooked tip, yellow lower mandible with a collapsible large pouch; grey tarsi and webbed feet.

Distribution and habitat　Breeds in Northern China, winter in the southeast, with about 50-100 individuals in this wintering population. In Beijing, a rare migrant to areas of open water, for example large lakes and reservoirs such as Shahe reservoir and the Summer Palace.

Behaviour and ecology　Often swims in groups in open water. Feeds on fish, crustaceans, molluscs and amphibians. Nests on trees near water. Clutch size: 3-4.

三趾鸥	*Rissa tridactyla*	Black-legged Kittiwake	鸥科 Laridae

繁殖羽 Breeding

非繁殖羽 Non-breeding

非繁殖羽 Non-breeding

IUCN 红色名录 易危 (VU) 2018.8.8

形态特征 小型游禽，体长 37~43 厘米。雌雄同色。成鸟繁殖羽喙黄绿色，头、颈、胸至下体白色，背、腰及两翼灰色，外侧初级飞羽端部黑色，次级飞羽端部白色，尾上覆羽及尾羽白色，跗跖黑褐色。成鸟非繁殖羽似繁殖羽，但头后部至后颈灰色。幼鸟喙黑色，眼后及后颈有黑色斑块，外侧初级飞羽黑色，部分翼上覆羽亦为黑色，两翼展开时形成“M”形黑色带斑，尾羽白色，端部黑色。

分布与生境 分布于中国东部及南部沿海，零星见于内陆地区。北京罕见于较大的水库、湖泊、河流等湿地，为迷鸟。

习性与生态 集群营巢于海岸和海岛的岩石上；非繁殖期在京记录多为单只。

Morphological characteristics A small natatore (37-43 cm). Sexes alike. Breeding adult has short, greenish-yellow bill; mantle and wings mostly grey; outer primaries tipped black; secondaries tipped white; remainder of plumage white; tarsi dark brown. Non-breeding adult similar to breeding adult but grey on nape and hindneck. Juvenile has black bill and black patches on ear-coverts and hindneck; in flight, the black outer primaries and upperwing-coverts form a distinctive black "M" on wings; tail white with black terminal band.

Distribution and habitat Coastal areas in east and south China. Scarce inland. In Beijing: rare vagrant at large reservoirs, lakes and rivers.

Behaviour and ecology Nests in large numbers on sea cliffs and other coastal rocky areas. Records in Beijing are usually of singles.

细嘴鸥 | *Chroicocephalus genei* | Slender-billed Gull | 鸥科 Laridae

形态特征 小型游禽，体长 42~47 厘米。雌雄相似。成鸟繁殖羽喙暗红色；虹膜淡黄色；全身大致为白色，背部及两翼灰色，外侧初级飞羽具黑色端斑和羽缘，胸、腹沾淡粉色；跗跖红色。非繁殖羽喙红色；耳部具淡灰色斑块。幼鸟似非繁殖羽成鸟但翼上覆羽具褐色斑，尾羽具较窄的深色次端斑。

分布与生境 于中国东部和西南地区有零星记录。多在沿海活动。北京地区为甚罕见迷鸟，仅于 2014 年 9 月和 2015 年 9 月在密云水库有过两次单只记载。

习性与生态 于主要分布区内（欧亚大陆西部及中部）一般集群活动，在中国则多为单只记录且与红嘴鸥等其他鸥类混群。

Morphological characteristics A small natatore (42-47 cm). Sexes alike. Breeding adult has deep red bill and pale yellow eyes; head and body largely white; mantle and wings grey; outer primaries have black tips and fringes; breast and belly sometimes faint pale pink; tarsi red. Non-breeding adult has bright red bill and a pale grey spot on ear-coverts. Juvenile similar to non-breeding adult but with paler bill; brown on upperwing-coverts and blackish-brown subterminal band on tail.

Distribution and habitat Scarce in eastern and southwest China. Prefers coastal wetlands in non-breeding season. Very rare vagrant in Beijing, with only two records at Miyun Reservoir (September 2014, September 2015).

Behaviour and ecology Within its main range (i.e. western and central Eurasia), often in flocks. In China, usually solitary and joins mixed gull flocks, often with Black-headed Gulls.

IUCN 红色名录 无危 (LC) 2019.8.14

棕头鸥	*Chroicocephalus brunnicephalus*	Brown-headed Gull	鸥科 Laridae

繁殖羽 Breeding

繁殖羽 Breeding

幼鸟 Juvenile

IUCN 红色名录 无危 (LC) 2018.8.8

形态特征 小型游禽，体长 41~45 厘米。雌雄同色。成鸟繁殖羽喙暗红色；虹膜淡黄色；头部棕褐色，具有缺口的白色眼圈；颈、胸、腰、尾及下体皆为白色，背部及两翼大致为灰色，内侧初级飞羽白色，端部黑色，最外侧两枚初级飞羽黑色，具白色次端斑。成鸟非繁殖羽喙淡红色或橙红色，端部黑色；头白色，耳部具黑色斑，其余似繁殖羽个体。幼鸟似非繁殖羽成鸟，但翼上覆羽具棕色斑点，最外侧初级飞羽全为黑色，无白色次端斑，尾羽端部黑色。

分布与生境 分布于中国西部至中部。见于内陆湖泊、河流。北京地区为罕见迷鸟。

习性与生态 在主要分布区内喜集群繁殖和越冬，非繁殖期常与红嘴鸥混群。集大群营巢于近水的裸地或草地上；窝卵数 3 枚。

Morphological characteristics A small natatore (41-45 cm). Sexes alike. Breeding adult has deep red bill and pale yellow eyes; head brown with broken white eye-ring; body largely white; mantle and wings grey with black tips to the primaries; outer primaries have white mirrors. Non-breeding adult has pale red or orange-red bill with black tip; blackish-brown spot on ear-coverts. Juvenile resembles non-breeding adult but has all-black outer primaries, brown upperwing-coverts and black terminal tail-band.

Distribution and habitat Western to central China. Prefers inland lakes and rivers. Rare vagrant in Beijing.

Behaviour and ecology In main range, encountered in large flocks; in Beijing, vagrants usually join flocks of Black-headed gulls. Colonial breeder; nests on the ground near water. Clutch size: 3.

红嘴鸥 | *Chroicocephalus ridibundus* | Black-headed Gull | 鸥科 Laridae

形态特征 小型游禽，体长 37~43 厘米。雌雄同色。成鸟繁殖羽喙暗红色；虹膜暗褐色；头深棕色，具断开的白色眼圈；全身大部分为白色，背部及两翼灰色，外侧初级飞羽白色，端部黑色，其余飞羽尖端白色，跗跖红色。成鸟非繁殖羽喙红色，喙端黑色；头白色，耳羽黑色。幼鸟似成鸟非繁殖羽，但喙和跗跖颜色较淡，翼上覆羽具褐色斑点，尾羽具黑色端斑。似棕头鸥但外侧初级飞羽为白色而非黑色。

分布与生境 广布于中国的沿海及内陆各种水域。北京地区为城区、郊区湿地的常见旅鸟。

习性与生态 性喧闹。集群繁殖、迁徙和越冬。窝卵数 2~4 枚。

Morphological characteristics A small natatore (37-43 cm). Sexes alike. Breeding adult has deep red bill and dark brown eyes; head dark brown with prominent broken white eye-ring; mantle and wings largely grey but outer primaries white with black tips, forming prominent leading edge; inner primaries tipped white; remainder of plumage white; tarsi red. Non-breeding adult has red bill, tipped black; head is white with a black spot behind the eye. Juvenile similar to non-breeding adult but has paler bill, brown upper-wing coverts and black terminal band on tail. Resembles Brown-headed gull but has white leading edge on primaries.

Distribution and habitat Widespread in both coastal and inland areas in China. Common passage migrant at urban and rural wetlands in Beijing.

Behaviour and ecology Noisy. Forms flocks in both breeding season and non-breeding season. Clutch size: 2-4.

IUCN 红色名录　无危 (LC) 2018.8.17

繁殖羽 Breeding

非繁殖羽 Non-breeding

幼鸟 Juvenile

黑嘴鸥	*Saundersilarus saundersi*	Saunders's Gull	鸥科 Laridae

非繁殖羽 Non-breeding

繁殖羽 Breeding

国家一级保护野生动物　　IUCN 红色名录　易危 (VU) 2018.8.8

形态特征　小型游禽，体长 29~34 厘米。雌雄同色。成鸟繁殖羽喙黑色；虹膜暗褐色；头黑色，具断开的显著白色眼圈；颈、胸、腰、尾及下体白色，背部及翼上覆羽浅灰色，初级飞羽灰白色，端部近黑色，跗跖红色。成鸟非繁殖羽头部白色，头顶具灰黑色横斑，耳羽具一黑色斑点。幼鸟似非繁殖羽之成鸟，但所有飞羽端部皆为黑色，翼上覆羽褐色，尾羽端部黑色。

分布与生境　分布于中国东部及南部，多见于沿海湿地。北京地区为甚罕见旅鸟，仅于 2020 年 3 月在密云水库有一次记载。

习性与生态　喜在开阔的沿海滩涂和潮间带捕食水生无脊椎动物，喜集小群活动，于内陆的零星记载则多为单独个体，或与红嘴鸥等混群。集小群营巢于开阔的滩涂；窝卵数 2~4 枚。

Morphological characteristics　A small natatore (29-34 cm). Sexes alike. Breeding adult has black bill and dark eyes; head black with broken white eye-ring; mantle and wings largely pale grey; outer primaries white with black tips, forming prominent leading edge; remainder of plumage white; tarsi red. Non-breeding adult has white head, with faint grey bars on crown and a dark spot behind the eyes. Juvenile similar to non-breeding adult but with black-tipped remiges, brown upperwing-coverts and a dark terminal tail-band.

Distribution and habitat　Eastern and southern China. Prefers coastal wetlands. Very rare passage migrant in Beijing: only one record – an adult at Miyun Reservoir (March 2020).

Behaviour and ecology　Feeds on invertebrates such as crabs and worms at estuaries and mudflats. Usually forms flocks. Rare inland, where it joins flocks of other gulls. Nests colonially on coastal flats. Clutch size: 2-4.

小鸥	*Hydrocoloeus minutes*	Little Gull	鸥科 Laridae

形态特征 小型游禽，体长 25~29 厘米。雌雄同色。体型较中国分布的其他鸥类小而紧凑。成鸟繁殖羽、喙及整个头部皆为黑色，无白色眼圈；背部、翼覆羽及飞羽浅灰色，各级飞羽端部皆为白色，翼下深灰色。上、下体余部白色。成鸟非繁殖羽头白色，仅头顶后部和耳羽黑色。幼鸟似成鸟非繁殖羽，但翼上覆羽沾褐色，且尾羽端部黑色。

分布与生境 中国西北和东北有繁殖种群，非繁殖季一般向西迁徙，但于内陆和东部有零星记录。多活动于植被丰富的淡水湿地。在北京为甚罕见旅鸟，于延庆野鸭湖、密云水库和沙河水库偶有记录。

习性与生态 喜集小群在水面上空轻快地飞行。于湿地苇丛和草丛中的地面上营巢；窝卵数 2~3 枚。

Morphological characteristics A small natatore (25-29 cm). Sexes alike. Rather small and compact compared with other gull species. Breeding adult has black bill and head; mantle and wings pale grey; remiges have white tips; underwing largely dark grey; remainder of plumage white. Non-breeding adult has white head; rear crown grey and a black spot behind the eyes. Juvenile resembles non-breeding adult but with brown upperwing-coverts and a black terminal tail-band.

Distribution and habitat Breeds in northwest and northeast China. Migrates via western China. Scarce in central and eastern China. Usually seen in well-vegetated freshwater wetlands. In Beijing: very rare passage migrant at large wetlands such as Yeyahu, Miyun Reservoir and Shahe Reservoir.

Behaviour and ecology Flight light and agile; often in small flocks. Nests on the ground in reeds or grass at wetlands. Clutch size: 2-3.

国家二级保护野生动物 | IUCN 红色名录 无危 (LC) 2018.8.8

非繁殖羽 Non-breeding

繁殖羽 Breeding

遗鸥	*Ichthyaetus relictus*	Relict Gull	鸥科 Laridae

繁殖羽 Breeding

幼鸟 Juvenile

国家一级保护野生动物

IUCN 红色名录 易危 (VU) 2017.10.1

形态特征 小型游禽，体长 39~46 厘米。雌雄同色。成鸟繁殖羽喙暗红色；头黑色，具前后断开的白色眼圈；颈、胸、腰、尾及整个下体皆为白色，背部及两翼大致为灰白色，外侧初级飞羽黑色，具白色端斑。成鸟非繁殖羽喙黑色，头白色，枕部至后颈具黑色细纹。幼鸟似非繁殖羽成鸟，但翼上覆羽具灰褐色斑点，外侧初级飞羽黑色，其余飞羽具黑色端斑，尾羽端部黑色。

分布与生境 分布于中国北部及东南部沿海。北京一般见于郊区较大湿地，为不常见旅鸟。

习性与生态 非繁殖期喜集群活动。集群营巢于荒漠、半荒漠湿地的湖心小岛上；窝卵数 2~3 枚。

Morphological characteristics A small natatore (39-46 cm). Sexes alike. Breeding adult has red bill; head black with broken white eye-ring; mantle and wings largely pale grey; outer primaries black with white tips; remainder of plumage white. Non-breeding adult has black bill and white head; nape and neck finely streaked. Juvenile resembles non-breeding adult but upperwing-coverts greyish-brown; outer primaries lack white tips; tail has black terminal band.

Distribution and habitat From North China to coastal areas in east and south China. Uncommon passage migrant at large reservoirs, lakes and rivers in Beijing.

Behaviour and ecology Forms flocks in non-breeding season. Nests colonially on the ground near saline or semi-saline lakes. Clutch size: 2-3.

渔鸥	*Ichthyaetus ichthyaetus*	Pallas's Gull	鸥科 Laridae

形态特征 中型游禽，体长 63~72 厘米。雌雄同色。喙长而粗厚，基部黄色，端部具黑色与红色环带。繁殖羽头黑色，具断开的白色眼圈；颈、胸、腰、尾及下体白色，背部及翼上覆羽灰色，外侧初级飞羽白色，具黑色次端斑，内侧初级飞羽及次级飞羽灰色，端部白色。成鸟非繁殖羽头白色，眼后及枕部具褐色纵纹。未成年个体似成鸟非繁殖羽，但喙色较淡，枕部至后颈有显著深色纵纹，尾羽具宽阔的黑色端斑。

分布与生境 中国除东北外广泛分布。北京地区见于郊区有开阔水域的水库和河流，为罕见旅鸟。

习性与生态 飞行缓慢而笨重。在北京的记载大多为单只或小群。喜集群在近水地面上营巢；窝卵数 2~4 枚。

Morphological characteristics A medium-sized natatore (63-72 cm). Sexes alike. Breeding adult has long, strong, bill with yellow base and red and black bands at tip; black head and broken white eye-ring; mantle and upper-wing coverts grey; outer primaries white, with black band near tips; inner primaries and secondaries grey, tipped white; remainder of plumage white. Non-breeding adult has white head, streaked dark behind eye and on nape. Juvenile similar to non-breeding adult but has paler bill, heavy streaks on neck and broad black terminal band on tail.

Distribution and habitat Widespread in China except for the northeast, more common in western China. In Beijing: rare passage migrant at large reservoirs and rivers in rural areas.

Behaviour and ecology Flight slow and heavy. Usually solitary or in small groups in Beijing. Nests colonially on the ground near wetlands. Clutch size: 2-4.

IUCN 红色名录 无危 (LC) 2018.8.8

繁殖羽 Breeding

非繁殖羽 Non-breeding

黑尾鸥	*Larus crassirostris*	Black-tailed Gull	鸥科 Laridae

繁殖羽 Breeding
亚成鸟 Immature

IUCN 红色名录　无危 (LC) 2018.8.8

形态特征　小型游禽，体长 43~51 厘米。雌雄同色。喙黄色，喙端红色，近端处黑色。繁殖羽头、颈、腰、尾及下体白色，背部及两翼深灰色，外侧初级飞羽黑色，内侧初级飞羽及次级飞羽具白色端斑；跗跖黄色。成鸟非繁殖羽头后部至后颈灰褐色。幼鸟喙粉色，喙端黑色；全身大致为褐色；跗跖粉色或淡黄色。

分布与生境　分布于中国东部至南部沿海。北京地区见于水库、湖泊、河流的开阔水面，为罕见旅鸟。

习性与生态　常集群活动，或与银鸥、红嘴鸥等其他鸥类混群。集群于海岸岩石上营巢；窝卵数一般为 2 枚。

Morphological characteristics　A small natatore (43-51 cm). Sexes alike. Breeding adult largely white; bill long, yellow at base, red at tip and black near tip; mantle and wings dark grey; outer primaries black; inner primaries and secondaries tipped white; tarsi yellow. Non-breeding adult resembles breeding adult but rear crown, nape and hindneck streaked greyish-brown. Juvenile has pink bill with black tip; body largely brown; tarsi pink or pale yellow.

Distribution and habitat　Coastal areas from east to south China. In Beijing: rare passage migrant at reservoirs, lakes and rivers.

Behaviour and ecology　Usually in flocks or joins mixed flocks with Black-headed Gull, Lesser Black-backed Gull or Siberian Gull. Nests colonially on coastal rocky areas or cliffs. Clutch size: 2.

普通海鸥	*Larus canus*	Mew Gull	鸥科 Laridae

形态特征 小型游禽，体长 40~51 厘米。雌雄同色。成鸟繁殖羽喙黄色或黄绿色；除背部和两翼外全身体羽皆为白色，背部及两翼翼上覆羽灰色，外侧初级飞羽黑色，具白色次端斑，内侧初级飞羽和次级飞羽灰色，端部白色；跗跖黄色。成鸟非繁殖羽头、颈密布褐色纵纹，其余似繁殖羽个体。幼鸟喙淡黄色或粉色，近端部黑色，上体具灰褐色斑纹，尾羽具宽阔的黑色端斑。

分布与生境 中国除西部外广泛分布，但繁殖期多在沿海活动。北京为不常见旅鸟，见于各种湿地的开阔水面。

习性与生态 常集小群活动，飞行灵活而轻快。集群于近水的地面上营巢；窝卵数 2~3 枚。

Morphological characteristics A small natatore (40-51 cm). Sexes alike. Breeding adult has yellow or greenish-yellow bill; mantle and upperwing-coverts grey; outer primaries black, with white mirror near tip; inner primaries and secondaries grey, tipped white; remainder of plumage white; tarsi yellow. Non-breeding adult has heavy brown streaks on head and neck. Juvenile has pale yellow or pink bill, with a black band near tip; upperparts brown with pale fringes; tail has broad black terminal band.

Distribution and habitat Widespread in China except for the west. Prefers coastal wetlands in breeding season. In Beijing, an uncommon passage migrant at wetlands with open water.

Behaviour and ecology Often in small flocks. Flight light and agile. Nests colonially on the ground near water. Clutch size: 2-3.

IUCN 红色名录 无危 (LC) 2018.8.8

繁殖羽 Breeding

过渡羽 Transitional plumage

幼鸟 Juvenile

北极鸥	*Larus hyperboreus*	Glaucous Gull	鸥科 Laridae

成鸟 Adult
亚成鸟 Immature

IUCN 红色名录　无危 (LC) 2018.8.7

形态特征　大型游禽，体长 64~80 厘米。雌雄同色。成鸟喙黄色，下喙端部红色，甚为粗壮。繁殖羽头、颈、腰、尾和下体皆为白色，背部和两翼灰白色，所有飞羽端部白色；跗跖粉色。非繁殖羽头、颈及上胸密布灰褐色纵纹。幼鸟喙黑色，通体大致呈浅咖啡色。

分布与生境　在非繁殖期于中国东部沿海有稳定记录。在北京为甚罕见迷鸟，仅于 2011 年 11 月在延庆野鸭湖有过一次幼鸟记载。

习性与生态　多单只、成对或集小群活动，有时亦与其他鸥类（如银鸥）混群。飞行时振翼缓慢而略显笨重。

Morphological characteristics　A large natatore (64-80 cm). Sexes alike. Breeding adult largely white with strong bill, yellow at base with red gonys; mantle and wings pale grey; flight-feathers tipped white; tarsi pink. Non-breeding adult has extensive greyish-brown streaks from head and neck to upper breast. Juvenile has black bill; body mostly coffee-coloured.

Distribution and habitat　Regularly seen along coasts in north and east China, rare inland. In Beijing: very rare vagrant; recorded only once (juvenile) at Yeyahu, Yanqing District in November 2011.

Behaviour and ecology　Usually solitary, in pairs or in small flocks. Sometimes joins mixed-species flocks of gulls. Flight slow and heavy.

小黑背银鸥	*Larus fuscus*	Lesser Black-backed Gull	鸥科 Laridae

形态特征 中型游禽，体长 51~65 厘米。雌雄同色。成鸟喙黄色，下喙端部红色。繁殖羽头、颈、胸、腰、尾及下体皆为白色；背部及两翼翼上覆羽深灰色，外侧初级飞羽黑色，具白色端斑，内侧初级飞羽和次级飞羽深灰色，端部白色；跗跖黄色。非繁殖羽头部和后颈密布深褐色纵纹。幼鸟喙黑色；上体具褐色斑纹，外侧飞羽全为黑色，尾羽具宽阔的黑色端斑，下体大致为灰白色；跗跖粉色。

分布与生境 广布于中国东部和南部，主要见于沿海和长江流域。北京地区主要见于水库、湖泊、河流等开阔湿地，为不常见旅鸟。

习性与生态 典型的银鸥习性，常单独、成对或集小群活动。飞行时振翼较慢，常利用气流在空中盘旋。

Morphological characteristics A medium-sized natatore (51-65 cm). Sexes alike. Breeding adult has yellow bill with red gonys spot; dark grey on mantle and wings; outer primaries black, with white spots at tips; inner primaries and secondaries have white tips; remainder of plumage white; tarsi yellow. Non-breeding adult has heavy dark streaks on head and hindneck. Juvenile has black bill; upperparts coffee-coloured; outer primaries all-black; tail has prominent black terminal band; underparts mostly pale greyish-white; tarsi pink.

Distribution and habitat Widespread in eastern and southern China. Mainly along coast and along the Yangtze River. Uncommon passage migrant at reservoirs, lakes and rivers in Beijing.

Behaviour and ecology Often solitary, in pairs or in small flocks. Wingbeats slow and steady. Commonly hover.

IUCN 红色名录 无危 (LC) 2018.8.8

西伯利亚银鸥 *Larus smithsonianus* Siberian Gull

鸥科
Laridae

IUCN 红色名录　无危 (LC) 2018.8.8

形态特征　中型游禽，体长 55~68 厘米。雌雄同色。成鸟喙黄色，下喙具红色端斑。繁殖羽背部和两翼灰色至浅灰色，翼尖黑色，最外侧初级飞羽具白色端斑，全身余部皆为白色。非繁殖羽后颈具淡褐色短纵纹。幼鸟喙黑色，上体咖啡色，尾羽端部黑色。似小黑背银鸥，但上体灰色较淡。

分布与生境　分布于中国北方至东南沿海。北京多见于较大湿地的开阔水域，*mongolicus* 亚种为区域性常见旅鸟；*vegae* 亚种于内陆罕见。

习性与生态　行为特征与小黑背银鸥相似。非繁殖期常集群活动。繁殖期营巢于近水和沼泽地面；窝卵数 2~3 枚。

Morphological characteristics　A medium-sized natatore (55-68 cm). Sexes alike. Breeding adult has yellow bill with red gonys; grey or pale grey on mantle and wings; outer primaries black with white tips; remainder of plumage all-white. Non-breeding adult has fine brown streaks on hindneck. Juvenile has black bill and coffee-coloured upperparts; tail has black terminal band. Closely resembles Lesser Black-backed Gull but upperparts paler.

Distribution and habitat　Widespread from northern China to southeast China. Prefers coastal wetlands. Subspecies *mongolicus* locally common passage migrant at large reservoirs, lakes and rivers in Beijing; *vegae* rare inland.

Behaviour and ecology　Behaviours similar to Lesser Black-backed Gull. Usually forms small to large flocks in non-breeding season. Nests colonially on the ground near lakes and rivers, or in marshes. Clutch size: 2-3.

鸥嘴噪鸥	*Gelochelidon nilotica*	Gull-billed Tern	鸥科 Laridae

形态特征 小型游禽，体长 31~43 厘米。雌雄同色。喙黑色，较其他燕鸥短而粗厚。繁殖羽头顶、头顶两侧（眼部以上）及枕部黑色；背、腰及两翼翼上覆羽大致为浅灰色，外侧初级飞羽内翈深灰色；尾羽白色，呈浅叉状；头部眼线以下、颏、喉、颈及整个下体皆为白色；跗跖黑色。非繁殖羽头顶白色，耳部具一黑色斑。幼鸟似非繁殖羽成鸟但翼上覆羽沾棕色。

分布与生境 分布于中国东部、南部沿海和西北地区。见于各种内陆湿地及沿海湿地和河口。在北京为不常见旅鸟。

习性与生态 喜集松散的小群活动。在水面、近水草地或沼泽上空敏捷地飞行，以捕捉小型鱼类和无脊椎动物等。营巢于近水的裸露地面；窝卵数 2~4 枚。

Morphological characteristics A small natatore (31-43 cm). Sexes alike. Bill short and heavy, all-black. Breeding adult is black from cap to nape; mantle, rump and wings largely grey; outer primaries have dark grey inner vanes; tail slightly forked; remainder of plumage white; tarsi black. Non-breeding adult lacks black cap but has black patch on ear-coverts. Juvenile resembles non-breeding adult but upperwing-coverts pale brown.

Distribution and habitat Along coasts of east and south China. Also in northwest and northeast China. Seen at both inland and coastal wetlands and estuaries. Uncommon passage migrant in Beijing.

Behaviour and ecology Tends to form loose flocks. Agile flight above water, grassland and marshes from where it takes invertebrates and small fish. Nests colonially in scrapes on ground near water. Clutch size: 2-4.

IUCN 红色名录 无危 (LC) 2019.8.14

繁殖羽 Breeding

非繁殖羽 Non-breeding

红嘴巨燕鸥 *Hydroprogne caspia* Caspian Tern

鸥科 Laridae

非繁殖羽 Non-breeding

繁殖羽 Breeding

IUCN 红色名录 无危 (LC) 2018.8.7

形态特征 中型游禽，体长 48~56 厘米。雌雄同色。与其他燕鸥相比，本种具粗壮的喙和较大的头部。成鸟繁殖羽喙红色；头顶至枕部黑色，具较短的冠羽；全身大部分为白色，仅背部和两翼为银灰色，初级飞羽具黑色端斑，尾羽浅叉状；跗跖黑色。非繁殖羽喙端黑色，头顶黑色区域夹杂白色纵纹。幼鸟喙橙色，翼上覆羽沾褐色。

分布与生境 广泛分布于中国，于东部更为多见。栖息于沿海和内陆的各类湿地。北京为不常见旅鸟。

习性与生态 飞行较为缓慢，振翼不明显，略显僵硬。喜集小群在同一区域一起筑巢，营巢于地面，通常在湖心或河心的小岛或沙洲上；窝卵数 2~3 枚。

Morphological characteristics A medium-sized natatore (48-56 cm). Sexes alike. Rather large head and strong bill compared with other terns. Breeding adult has bright red bill; cap and nape glossy black with short crest; body largely white; mantle and wings pale grey; primaries tipped black; tail slightly forked; tarsi black. Non-breeding adult has black-tipped bill and crown streaked white. Juvenile has orange bill and pale brown on upperwing-coverts.

Distribution and habitat Widespread at both inland and coastal areas in China. More common in eastern China. Uncommon passage migrant at large rivers and reservoirs in Beijing.

Behaviour and ecology Flight slow with shallow wing beats. Usually seen in small flocks. Nests colonially on the ground near water, particularly on sand islands in lakes or rivers. Clutch size: 2-3.

白额燕鸥	*Sternula albifrons*	Little Tern	鸥科 Laridae

形态特征 小型游禽，体长 22~28 厘米。雌雄同色。成鸟繁殖羽喙黄色，喙端黑色；前额白色，具黑色的眼先、头顶和枕；背部、腰和两翼翼上浅灰色，尾羽浅叉状；下体白色；跗跖橙色。非繁殖羽喙黑色；头顶白色区域扩大，仅枕部和头顶后部为黑色，具白色细纹。幼鸟前额灰色，上体沾棕色。

分布与生境 广泛分布于中国西北和东部地区。多活动于沿海和内陆湿地的沙洲、河床附近。北京为不常见旅鸟和夏候鸟。

习性与生态 飞行轻盈，常悬停于空中，发现猎物后迅速俯冲而下。在沙滩的地面扒一浅凹即为巢；窝卵数 2~3 枚。

Morphological characteristics A small natatore (22-28 cm). Sexes alike. Breeding adult has yellow bill with black tip; forehead white; black from lores and crown to nape; mantle, rump and wings pale grey; tail forked; cheek, throat, neck and remaining underparts all-white; tarsi orange. Non-breeding adult has black bill and larger white patch on forehead; rear crown and nape black with some white streaks. Juvenile has grey forehead and brown-scaled upperwing-coverts.

Distribution and habitat Widespread in eastern and northwest China. Prefers coastal and inland sandy shores and rivers. Uncommon passage migrant and summer breeder in Beijing.

Behaviour and ecology Flight light and agile. When feeding, hovers above water, then dives for prey. Nests in scrapes on ground at sandy shores. Clutch size: 2-3.

IUCN 红色名录 无危 (LC) 2018.8.7

普通燕鸥	*Sterna hirundo*	Common Tern	鸥科 Laridae

繁殖羽 Breeding

繁殖羽 Breeding

幼鸟 Juvenile

IUCN 红色名录　无危 (LC) 2018.8.7

形态特征　小型游禽，体长 32~39 厘米。雌雄同色。喙较长而尖锐，红色或黑色。成鸟繁殖羽头顶至枕部黑色，背部及两翼浅灰色，尾羽白色，深叉形，下体白色或灰色。非繁殖羽喙黑色，前额白色。幼鸟似非繁殖羽个体，但上体具深褐色斑纹。

分布与生境　广泛分布于中国沿海和内陆湿地。北京为区域性常见旅鸟和夏候鸟。

习性与生态　典型的燕鸥类悬停、俯冲捕食习性。一般营巢于湿地水域岸边的沙滩、石滩和沼泽中；窝卵数 2~5 枚。

Morphological characteristics　A small natatore (32-39 cm). Sexes alike. Bill long and sharp, red or black. Breeding adult has black cap and nape; pale grey from mantle to wings; white tail deeply forked; underparts white or grey. Non-breeding adult has black bill and white forehead; black from rear crown to nape; upperparts paler. Juvenile resembles non-breeding adult but has dark fringes on upperparts.

Distribution and habitat　Widespread at both inland and coastal wetlands in China. Locally common summer breeder and passage migrant at reservoirs, lakes and rivers in Beijing.

Behaviour and ecology　Hovers and dives when feeding. Nests on sand, gravel or soil near water. Clutch size: 2-5.

灰翅浮鸥	*Chlidonias hybrid*	Whiskered Tern	鸥科 Laridae

形态特征 小型游禽，体长 23~28 厘米。雌雄相似。成鸟繁殖羽喙暗红色；头顶及枕黑色，颊、颏、喉、颈部白色，与黑色的头顶形成鲜明对比；上体及两翼大致为灰色，尾呈浅叉形，下体深灰色；跗跖暗红色。非繁殖期成鸟喙和跗跖黑色，前额白色，头顶后部和枕部为黑灰色，杂有白色纵纹；下体白色。幼鸟似成鸟非繁殖羽但上体沾棕色。

分布与生境 分布于中国东部和中部。北京见于水生植物丰富的较大河流、湖泊和水库。为常见夏候鸟和旅鸟。

习性与生态 一般集松散的群体在水面上放低空飞行。以水生植物为巢材建浮巢；窝卵数 2~4 枚。

Morphological characteristics A small natatore (23-28 cm). Sexes alike. Breeding adult has deep-red bill; cap and nape glossed-black, contrasting strongly with white cheek, chin, throat and neck; upperparts and wings largely grey; tail slightly forked; underparts blackish-grey; tarsi deep-red. Non-breeding adult has black bill and tarsi; forehead white; rear crown and nape blackish-grey with white streaks; underparts white. Juvenile similar to non-breeding adult with dark brown scales on upperparts.

Distribution and habitat Widespread in eastern and central China. Common summer breeder and passage migrant at shallow lakes, ponds, rivers and other well-vegetated wetlands in Beijing.

Behaviour and ecology Tends to form loose flocks. Flies above water to hunt. Nest is constructed with grass and reeds on top of floating vegetation. Clutch size: 2-4.

IUCN 红色名录 无危 (LC) 2016.10.1

繁殖羽 Breeding

幼鸟 Juvenile

繁殖羽 Breeding

白翅浮鸥 | *Chlidonias leucopterus* | White-winged Tern | 鸥科 Laridae

繁殖羽 Breeding

繁殖羽 Breeding

IUCN 红色名录 无危 (LC) 2016.10.1

形态特征 小型游禽，体长 20~27 厘米。雌雄同色。成鸟繁殖羽喙暗红色；头、颈、背和下体皆为黑色，两翼翼上覆羽白色，飞羽灰色，翼下覆羽黑色，腰、尾上覆羽、尾下覆羽和尾羽白色，尾羽浅叉形。非繁殖期成鸟喙黑色，头、颈、下体大致为白色，仅耳羽及头顶后部具黑斑。幼鸟似成鸟非繁殖羽，但背部褐色。

分布与生境 广布于中国。北京见于水生植物丰富且水深较浅的湖泊、水塘、沼泽等湿地，为不常见的夏候鸟和旅鸟。

习性与生态 飞行轻快而灵活，喜集小群，亦有时与灰翅浮鸥混群。一般在浮水植物上或水边地面筑巢；窝卵数 2~4 枚。

Morphological characteristics A small natatore (20-27 cm). Sexes alike. Breeding adult has deep-red bill; head, neck, mantle and underparts all-black, contrasting with white rump, uppertail-coverts and tail; tail slightly forked; upperwing-coverts white; flight feathers largely grey; underwing-coverts black; undertail-coverts white. Non-breeding adult has black bill; white from head and neck to underparts; black patch behind eyes and on rear crown. Juvenile resembles non-breeding adult but has brown fringes on upperparts.

Distribution and habitat Widespread in China. Uncommon summer breeder and passage migrant at well-vegetated shallow lakes, ponds or marshes.

Behaviour and ecology Flight light and agile, usually in small and loose flocks. Sometimes joins mixed flocks with Whiskered Tern. Nests on floating vegetation or on the ground near water. Clutch size: 2-4.

黑浮鸥	*Chlidonias niger*	Black Tern	鸥科 Laridae

形态特征 小型游禽，体长 23~28 厘米。雌雄相似。成鸟繁殖羽喙黑色；头、颈、胸至下体皆为黑色，背、两翼的翼上翼下、腰及尾羽灰色，尾羽为浅叉形，尾下覆羽白色。成鸟非繁殖羽前额、眼先和颊白色，眼后、耳羽至头顶后部黑色，颏、喉、颈、胸及下体白色，胸两侧各具一深灰色斑块。幼鸟似非繁殖羽成鸟，但上体具棕色斑。

分布与生境 分布于中国西北和东北地区，非繁殖期在东部和南部有零星记录，见于植被丰富的湖泊、池塘和沼泽。北京仅于 1933 年 8 月在通州有一次记载。

习性与生态 喜在水面上空优雅而灵活地飞行，捕捉飞行的昆虫。与白翅浮鸥偶有杂交现象。

Morphological characteristics A small natatore (23-28 cm). Sexes alike. Breeding adult has black bill; all-black from head and neck to breast and belly; mantle, wings (both upperwing and underwing), rump and tail mostly grey; tail forked; undertail-coverts white. Non-breeding adult has white on forehead, lores and cheeks, and black behind eyes and on rear crown; chin, throat, neck, breast and remaining underparts white; dark grey patch on breast-sides. Juvenile resembles non-breeding adult but upperparts fringed brown.

Distribution and habitat Mainly in northwest and northeast China. Scarce in southern and eastern China in non-breeding season. Seen at lakes, ponds or marshes with abundant vegetation. Very rare vagrant in Beijing, with only one histroical record from Tongzhou District in August 1933.

Behaviour and ecology Flies gracefully over water catching insects. Occasionally hybridises with White-winged Terns.

国家二级保护野生动物　IUCN 红色名录　无危 (LC) 2018.8.7

繁殖羽 Breeding

中贼鸥	*Stercorarius pomarinus*	Pomarine Skua	贼鸥科 Stercorariidae

幼鸟 Juvenile

IUCN 红色名录 无危 (LC) 2018.8.7

形态特征 中型游禽，体长 48~55 厘米。雌雄相似，有深浅二色型。喙基粉色，喙端黑色。浅色型成鸟头顶及眼先黑色，上体、两翼黑褐色，仅初级飞羽基部白色，尾羽亦为黑褐色，中央尾羽延长突出（但非繁殖期中央尾羽不延长），颏、喉、颈淡黄色，胸、腹部白色，胸侧和两胁具暗褐色横斑，尾下覆羽暗褐色。深色型成鸟较罕见，其通体黑色，仅初级飞羽基部白色。幼鸟全身褐色，上体具浅色羽缘，尾上、尾下覆羽密布白色横斑。

分布与生境 中国主要见于东南沿海，多在海面上活动。北京于郊区水库、河流等湿地有零星记载，为甚罕见迷鸟。

习性与生态 一般单独活动，喜追逐鸥类、燕鸥、鹱或其他海鸟，并抢夺它们捕到的鱼类作为食物。

Morphological characteristics A medium-sized natatore (48-55 cm). Sexes similar. Bill pink at base with black tip. Pale morph has black cap and lores; chin, throat and neck pale yellow; mantle, wings and tail mostly blackish-brown, with white at base of primaries; central rectrices long and spatulate in breeding season; breast and belly mostly white; breast-sides and flacks barred dark; undertail-coverts blackish-brown. The rarer dark morph is almost all-black with white at base of primaries. Juvenile is largely dark brown with light fringes on upperparts and white bars on uppertail- and undertail-coverts.

Distribution and habitat Along coast or offshore in east and south China. Scarce inland at large areas of open water. Very rare vagrant to rivers and reservoirs in Beijing.

Behaviour and ecology Usually solitary. Chases gulls, terns, shearwaters and other sea birds to pirate food.

长尾贼鸥 *Stercorarius longicaudus* Long-tailed Jaeger 贼鸥科 Stercorariidae

形态特征 中型游禽，体长 50~58 厘米。雌雄相似。有两种色型。浅色型成鸟头顶黑色，上体、两翼翼上、翼下及尾羽深灰色，中央一对尾羽极度延长；枕、颊在繁殖期淡黄色，非繁殖期深褐色；胸、腹部白色，臀至尾下覆羽黑褐色。深色型个体通体黑色，较浅色型个体罕见。幼鸟大致为黑褐色，上体和下体皆具浅色横斑。

分布与生境 中国东部和南部沿海有少量迁徙过境记录，偶有迷鸟至内陆。北京为甚罕见迷鸟。

习性与生态 喜单独活动。典型的贼鸥习性，在空中抢夺其他鸥类和燕鸥的食物。

Morphological characteristics A medium-sized natatore (50-58 cm). Sexes similar. The commoner pale morph has black on crown, lores and around eyes; upperparts, wings and tail dark grey; prominent tail projection; cheek and nape pale yellow in breeding season and dark in non-breeding season; white from breast to belly; blackish-brown from vent to undertail-coverts. Dark morph is all-black. Juvenile is largely blackish-brown, with extensive pale bars above and below.

Distribution and habitat Along coast or offshore in east and south China, scarce inland. Very rare vagrant in Beijing with only one record at Shahe Reservoir, Changping District in June 2013.

Behaviour and ecology Solitary. Pursues seabirds, forcing them to drop food, which it takes in flight.

IUCN 红色名录 无危 (LC) 2018.8.7

成鸟 Adult

幼鸟 Juvenile

III

北京鸟类图谱

涉禽
WADING BIRDS

BIRDS OF BEIJING

十、鸻形目（二）CHARADRIIFORMES

鹮嘴鹬	*Ibidorhyncha struthersii*	Ibisbill	鹮嘴鹬科 Ibidorhychidae

形态特征 小型涉禽，体长 37~41 厘米。雌雄同色。喙暗红色，细长而下弯。头顶、眼先、颊、颏、喉黑色，具白色边缘，其余头、颈、胸灰色，下胸具宽阔的黑色横带。上体及两翼灰色，但内侧数枚初级飞羽和外侧次级飞羽基部为白色；形成翼上之白色斑块，飞行时可见。尾羽灰色，具黑色横斑和较宽的黑色次端斑。下体白色；跗跖粉色。

分布与生境 分布于中国青藏高原、中部至华北地区。北京见于西部和北部郊区具砾石滩的河流，为不常见留鸟。

习性与生态 单只或成对活动。喜在浅水和岸边的石滩捕捉小型无脊椎动物，亦捕食小鱼。营巢于河岸砾石间；窝卵数 3~4 枚。

Morphological characteristics A small wading bird (37-41 cm). Sexes alike. Long, slender and decurved red bill; black from crown and face to chin and throat, with white borders; remainder of head, neck and breast grey; broad black band on lower breast; upperparts and wings largely grey; inner primaries and outer secondaries white at base, forming white patch on wing; tail grey with black bars and distinctive black subterminal band; underparts mostly white; tarsi pink.

Distribution and habitat From the Tibetan Plateau to central and north China. Uncommon resident along fast-flowing, stony rivers in rural areas of Beijing.

Behaviour and ecology Alone or in pairs. Feeds mainly on invertebrates that it finds under stones in shallow water or at river banks. Nests on stony ground near rivers. Clutch size: 3-4.

北京市一级保护野生动物　　国家二级保护野生动物　　IUCN 红色名录　无危 (LC) 2016.10.1

黑翅长脚鹬	*Himantopus himantopus*	Black-winged Stilt	反嘴鹬科 Recurvirostridae

北京市二级保护野生动物

IUCN 红色名录 无危 (LC) 2016.10.1

形态特征 小型涉禽，体长 29~41 厘米。喙和跗跖皆甚为细长，极易辨识。雄鸟喙黑色；头顶、枕部至后颈黑色，前额、头侧、颈部大致为白色，背部及两翼黑色，腰、尾上覆羽和尾羽白色；下体白色；跗跖粉色。雌鸟似雄鸟但上体羽色较淡。幼鸟上体具褐色羽缘。

分布与生境 中国广泛分布，生境以淡水湿地为主。北京见于开阔的湖泊、河流、沼泽等湿地的浅水处，为常见旅鸟和夏候鸟。

习性与生态 繁殖期和非繁殖期皆喜集群活动。繁殖期常发出连续的鸣叫声，似“piu piu piu”。营巢于露出水面的沼泽、浅滩和草地上；窝卵数 4 枚。

Morphological characteristics A small wading bird (29-41 cm). Black bill slender and straight. Male is black from crown, nape to hindneck; forehead, head-sides and neck white; mantle and wings black; rump, uppertail-coverts, tail and underparts all-white; tarsi long, slender and pink. Female is paler on upperparts. Juvenile has pale brown fringes on mantle and upperwing-coverts.

Distribution and habitat Widespread in China. Mainly seen at freshwater wetlands. In Beijing: common passage migrant and summer breeder at shallow lakes, rivers, marshes and other well-vegetated wetlands.

Behaviour and ecology Forms small flocks in both breeding and non-breeding seasons. Noisy in the breeding colony. Calls soft, like "piu piu piu". Nests on the ground in grassland near water, or on exposed mud in marshes. Clutch size: 4.

反嘴鹬	*Recurvirostra avosetta*	Pied Avocet	反嘴鹬科 Recurvirostridae

形态特征 中型涉禽，体长 40~45 厘米。雌雄相似。喙黑色，细长且上翘。成鸟头顶、枕、后颈黑色，颊、颏、喉和前颈白色，其余上体、两翼和下体亦大致为白色，仅初级飞羽、外侧翼上覆羽黑色；跗跖蓝灰色，较为粗壮。幼鸟似成鸟，但以头顶及上体灰褐色。

分布与生境 中国广泛分布。北京见于湖泊和河流的河滩及浅水处，为不常见旅鸟。

习性与生态 迁徙时喜单独活动或集成松散的小群，在水边或浅水处边走动边以喙在地面或水面扫动，亦擅游水。营巢于裸露的泥地上或沙滩上；窝卵数一般为 4 枚。

Morphological characteristics A medium-sized wading bird (40-45 cm). Sexes alike. Black bill is slender and upcurved. Adult is black from crown to nape and hindneck; remaining plumage largely white, but with black primaries and outer wing-coverts; tarsi strong, greyish-blue. Juvenile resembles adult but has greyish-brown crown and upperparts.

Distribution and habitat Widespread in China. Uncommon passage migrant to shallow water along rivers and at lakes in Beijing.

Behaviour and ecology Usually alone or in loose flocks on migration. Feeds by sweeping its bill in shallow water; often swims. Nests on sandy shores or exposed mud in marshes. Clutch size: 4.

IUCN 红色名录 无危 (LC) 2016.10.1

凤头麦鸡	*Vanellus vanellus*	Northern Lapwing	鸻科 Charadriidae

IUCN 红色名录 近危 (NT) 2016.10.1

形态特征 小型涉禽，体长 29~34 厘米。雌雄相似。成鸟繁殖羽喙黑色，较短；眼先、额、头顶及枕部黑色，具上翘的黑色冠羽，颊和耳羽白色，后颈褐色；背、腰和翼上覆羽绿色，具金属光泽，飞羽黑色，外侧初级飞羽端部白色，尾上、尾下覆羽橙棕色，尾羽基部白色，端部黑色；颏、喉、颈、胸大部分呈黑色，腹部白色；跗跖红色。非繁殖羽与繁殖羽相似，但颏、喉为白色。

分布与生境 中国全境皆有分布，繁殖于北部，越冬于长江以南。北京见于沼泽、湖泊和河流的浅水区域、近水草地和农田，为区域性常见旅鸟。

习性与生态 非繁殖期多集松散的小群活动。飞行轻盈而缓慢。一般营巢于近水草地；窝卵数 3~5 枚。

Morphological characteristics A small wading bird (29-34 cm). Sexes alike. Breeding adult has short, black bill; black from lores and crown to nape; prominent black crest; cheeks and ear-coverts white; hindneck brown; mantle, rump and upperwing-coverts glossy green; primaries largely black, with white tips on outer primaries; uppertail- and undertail-coverts orange; tail white at base and black at tip; chin, throat, neck and breast black; belly white; tarsi deep red. Non-breeding adult has more white and less black on face; chin and throat white.

Distribution and habitat Widespread in China. Breeds in northern China and winters in southern China. In Beijing: locally common passage migrant to marshes, shallow lakes and rivers, grassland near water and farmland.

Behaviour and ecology Forms loose flocks in non-breeding season. Flight light and slow. Usually nests on grassland near water. Clutch size: 3-5.

灰头麦鸡 | *Vanellus cinereus* | Grey-headed Lapwing | 鸻科 Charadriidae

形态特征 小型涉禽，体长 32~36 厘米。雌雄相似。成鸟繁殖羽喙较短，基部黄色，端部黑色；虹膜红色；头、颈、胸灰色，下胸具黑色横带；背、腰及翼上小覆羽褐色或灰褐色，中覆羽、大覆羽和次级飞羽白色，初级飞羽和初级覆羽黑色；尾上覆羽白色，尾羽白色，具宽阔的黑色次端斑；下体白色；跗跖黄色。非繁殖羽似繁殖羽，但下胸的黑色带不清晰。幼鸟无黑色胸带。

分布与生境 中国除西部外广泛分布。在北京为不常见旅鸟。主要栖息于平原沼泽、草地和农田，有时离水域甚远。

习性与生态 多单只、成对或集松散的小群活动，有时与凤头麦鸡混群，飞行时，振翼速度亦较慢。营巢于开阔的裸地；窝卵数 4 枚。

Morphological characteristics A small wading bird (32-36 cm). Sexes alike. Bill short with yellow base and black tip; eyes red; plain grey from head to neck and breast; lower breast has distinctive black band; mantle, rump and lesser coverts brown or greyish brown; median coverts, greater coverts and secondaries white; primaries and primary coverts black; uppertail-coverts black; tail white with broad black subterminal band; upperparts mostly white; tarsi yellow. Non-breeding adult resembles breeding adult but has indistinct breast band. Juvenile lacks breast band altogether.

Distribution and habitat Widespread in China except for western China. In Beijing: uncommon passage migrant to marshes, grassland and farmland, sometimes far from water.

Behaviour and ecology Often alone, in pairs or in loose flocks. Sometimes joins flocks of Northern Lapwing. Wingbeats slow. Nests on grassland or exposed mud. Clutch size: 4

IUCN 红色名录 无危 (LC) 2016.10.1

金鸻	*Pluvialis fulva*	Pacific Golden Plover	鸻科 Charadriidae

雄鸟 Male/ 繁殖羽 Breeding

非繁殖羽 Non-breeding

IUCN 红色名录　无危 (LC) 2016.10.1

形态特征　小型涉禽，体长 23~26 厘米。雌雄同色。成鸟繁殖羽喙黑色；头顶、后颈、上体及两翼翼上覆羽皆为黑色，密布金黄色和白色斑点，飞羽深褐色，有一条宽阔的白色带自前额经头侧、颈侧、胸侧延伸至两胁。眼先、颊、颏、喉、前颈、胸及下体黑色，尾下覆羽具白色横斑。成鸟非繁殖羽颊皮黄色，颏、喉及下体白色，体侧无白色宽带。幼鸟似成鸟非繁殖羽。

分布与生境　中国广泛分布。北京见于河滩浅水区、农田、草地、沼泽等地，为不常见旅鸟。

习性与生态　非繁殖期喜集小群或大群活动，常与其他鸻鹬混群，飞行快速。

Morphological characteristics A small wading bird (23-26 cm). Sexes alike. Breeding adult has black bill; crown, hindneck and upperparts black, with yellow and white spots; flight feathers blackish-brown; white band from forehead and supercilium to neck-sides, breast-sides and flanks; all-black from face, chin, throat and foreneck to breast and belly; vent and undertail-coverts barred black and white. Non-breeding adult and juvenile lack white band; face is buff; white from chin and throat to underparts.

Distribution and habitat Widespread in China. In Beijing: uncommon passage migrant to shallow lakes, marshes, grassland and farmland near water.

Behaviour and ecology Forms flocks in non-breeding season. Often joins mixed flocks with other plovers and sandpipers. Flight fast.

灰鸻	*Pluvialis squatarola*	Grey Plover	鸻科 Charadriidae

形态特征 小型涉禽，体长 27~32 厘米。雌雄同色。成鸟繁殖羽头顶、后颈至上体和两翼黑色，密布白色斑点；腰、尾上覆羽和尾羽白色，尾羽具黑色横斑；一宽阔的白色带自前额起，经眉、颈侧延伸至胸侧；头两侧、前颈、胸及下体纯黑色，仅臀至尾下覆羽白色；翼下覆羽白色，腋羽黑色。成鸟非繁殖羽头侧浅褐色，下体白色。

分布与生境 中国除西部外广泛分布。非繁殖期多活动于沿海地区，北京一般见于农田、草地、沼泽及其他浅水区域，为不常见旅鸟。

习性与生态 集群活动，非繁殖期喜集大群休息，觅食时则散开。有时与金鸻等其他鸻鹬类混群。

Morphological characteristics A small wading bird (27-32 cm). Sexes alike. Breeding adult has black crown, hindneck, mantle and wings, extensively spotted white; rump and uppertail-coverts white; tail also white but barred black; white from forehead to supercilium, which extends to neck and breast-sides; head-sides, foreneck, breast and underparts all-black; white from vent to undertail-coverts; underwing-coverts white with prominent black axillaries. Non-breeding adult is pale brown on head-sides; underparts largely white.

Distribution and habitat Widespread in China except for western areas. Usually seen along coast in non-breeding season. In Beijing: uncommon passage migrant to grassland, lakeshores, farmland and marshes.

Behaviour and ecology Forms loose flocks when feeding and dense flocks when roosting. Sometimes joins mixed flocks with Pacific Golden Plover and other shorebirds.

IUCN 红色名录 无危 (LC) 2019.8.14

雄鸟 Male/ 繁殖羽 Breeding

非繁殖羽 Non-breeding

剑鸻	*Charadrius hiaticula*	Common Ringed Plover	鸻科 Charadriidae

繁殖羽 Breeding

非繁殖羽 Non-breeding

IUCN 红色名录　无危 (LC) 2016.10.1

形态特征　小型涉禽，体长 18~24 厘米。雌雄相似。喙较短，基部橙黄色，端部黑色。成鸟繁殖羽前额白色，眼后方具白色眉纹；头顶前部、眼先、眼周及耳羽黑色，头顶后部灰褐色；上体及两翼大致为褐色，各级飞羽中部具白斑，尾羽端部深褐色；颏、喉、颈白色，胸部具黑色横带（雌鸟胸带较窄）；下体余部白色；跗跖橙色。成鸟非繁殖羽及幼鸟头部及胸部的黑色区域被深褐色取代。

分布与生境　中国主要于北部和东部有零星记载。栖息于河流、湖泊的岸边或沼泽。在北京为甚罕见迷鸟。

习性与生态　多单只或集松散的小群活动，飞行迅速敏捷。

Morphological characteristics　A small wading bird (18-24 cm). Sexes similar. Bill short, orange at base and black at tip. Breeding male has white on forehead, black from forecrown to lores, around eyes and ear-coverts; supercilium behind eye is white; rear crown greyish-brown; upperparts and wings largely brown; flight feathers white at centre, forming white bars on wings; tail has dark brown tip; chin and throat white; distinctive black band on breast (narrower in females); remainder of of underparts all-white; tarsi orange. Non-breeding adult and juvenile are dark brown, instead of black, on head-sides and breast.

Distribution and habitat　Scarce in northern and eastern China. Seen at sandy shores and marshes. Very rare vagrant in Beijing.

Behaviour and ecology　Alone or in small flocks. Flight fast and agile.

长嘴剑鸻 *Charadrius placidus* Long-billed Plover 鸻科 Charadriidae

形态特征 小型涉禽，体长 18~24 厘米。雌雄相似。成鸟繁殖羽喙黑色；前额白色，头顶前部、眼先至耳羽黑色，此黑色区域具白色边缘，眼圈较窄，呈黄色，头顶后部褐色，上体及两翼褐色；颏、喉、颈白色，具显著的黑色胸带；下体白色；跗跖浅黄色。成鸟非繁殖羽似繁殖羽，但头侧羽色较淡，胸带变为褐色，仅上缘黑色。

分布与生境 除西部地区外广布于中国。北京多活动于有砾石滩的河流附近，为区域性常见旅鸟和留鸟。

习性与生态 多单只或成对活动。营巢于沙滩或砾石滩的地面；窝卵数 3~4 枚。

Morphological characteristics A small wading bird (18-24 cm). Sexes similar. Breeding adult has black bill and narrow yellow eye-ring; forehead white; forecrown, lores and ear-coverts black, bordered white; eye-stripe dark brown; rear crown, upperparts and wings mostly brown; chin, throat and neck white; narrow black breast band; underparts white; tarsi pale yellow. Non-breeding adult resembles breeding adult but eye-stripe paler and breast band is brown, bordered black.

Distribution and habitat Widespread in China except for western areas. Locally common resident and passage migrant in Beijing. Mainly seen along riverbanks with pebbles and rocks.

Behaviour and ecology Often alone or in pairs. Nests at rocky riverbanks or pebble spits. Clutch size: 3-4.

IUCN 红色名录 无危 (LC) 2016.10.1

金眶鸻 | *Charadrius dubius* | Little Ringed Plover | 鸻科 Charadriidae

繁殖羽 Breeding

非繁殖羽 Non-breeding

幼鸟 Juvenile

IUCN 红色名录 无危 (LC) 2016.10.1

形态特征 小型涉禽，体长 15~18 厘米。雌雄相似。成鸟繁殖羽喙黑色；具显著的金黄色眼圈；前额白色，头顶前部至头两侧黑色，眼后方眉纹白色，头顶后部、上体、两翼及尾羽褐色，尾羽端部羽色较深；颏、喉、颈白色，具黑色胸带，其余下体白色，跗跖橙色。成鸟非繁殖羽头部及胸部以褐色取代黑色。幼鸟似非繁殖羽成鸟，但胸带断开。

分布与生境 广泛分布于中国大部分地区。见于沿海和内陆各种湿地的沙滩、泥滩和石滩以及近水草地。在北京为常见旅鸟和夏候鸟。

习性与生态 非繁殖期喜集松散的小群在水边快速走动。于近水地面的浅坑营巢；窝卵数 3~5 枚。

Morphological characteristics A small wading bird (15-18 cm). Sexes similar. Breeding adult has short, black bill and prominent yellow eye-ring; forehead white; black patch from forecrown to eye-stripe; white supercilium behind eye; rear crown, upperparts and wings largely brown; tail also brown but darker at tip; chin, throat and neck white; collar black; underparts white; tarsi orange. Non-breeding adult has brown head-sides and collar. Juveniles resembles non-breeding adult but has broken collar.

Distribution and habitat Widespread in China. Seen on grassland, sandy shores, rocky shores and mudflats. Common summer breeder and passage migrant at both urban and rural wetlands in Beijing.

Behaviour and ecology Forms small and loose flocks in non-breeding season. Runs rapidly. Nest is a small depression in the ground near water. Clutch size: 3-5.

环颈鸻 *Charadrius alexandrinus* Kentish Plover

鸻科 Charadriidae

形态特征 小型涉禽，体长 17~21 厘米。雄鸟繁殖羽喙黑色；前额及眉纹白色，头顶前部及贯眼纹黑色，头顶余部栗棕色，颊白色；上体及两翼沙褐色，尾羽暗褐色，外侧尾羽白色，尾端深色；颏、喉、胸白色，具断开的黑色窄胸带；下体余部白色；跗跖黑色。雌鸟及非繁殖羽雄鸟，头侧及胸带以褐色取代黑色，上体羽色较淡。

分布与生境 中国广泛分布。见于沿海和内陆的各种湿地。在北京为不常见旅鸟。

习性与生态 成对或集小群活动，行走快速。在沙地上刨一浅坑营巢；窝卵数 3~5 枚。

Morphological characteristics A small wading bird (17-21 cm). Breeding male has black bill; white on forehead, extending to supercilium; forecrown and eye-stripe black; rest of crown to nape rufous brown; cheek white; upperparts and wings sandy brown; tail dark brown, with darker tip and white sides; chin, throat and breast largely white; prominent broken narrow black collar; underparts whit; tarsi black. Female and non-breeding male are brown on head and collar instead of black.

Distribution and habitat Widespread in China. Seen at inland and coastal beaches, riverbanks and marshes. Uncommon passage migrant in Beijing.

Behaviour and ecology In pairs or small flocks. Runs fast then stops to feed. Nest is a shallow scrape on the ground. Clutch size: 3-5.

IUCN 红色名录 无危 (LC) 2016.10.1

雄鸟 Male

雌鸟 Female

蒙古沙鸻 | *Charadrius mongolus* | Lesser Sand Plover | 鸻科 Charadriidae

IUCN 红色名录 无危 (LC) 2016.10.1

形态特征 小型涉禽，体长 15~20 厘米。雌雄相似。成鸟繁殖羽喙黑色；头顶前部及两侧有一甚宽的黑色带延伸至贯眼纹和耳羽，头顶、头后、颈至胸赤褐色，胸带上缘黑色；上体及两翼褐色，尾羽深褐色，外侧尾羽白色。颏、喉白色；下体白色；跗跖灰绿色。成鸟非繁殖羽头部褐色取代黑色横带，前额及眉纹白色，上体羽色更为暗淡。

分布与生境 中国广泛分布。常活动于沿海沙滩、河口、农田和草甸等地。在北京为不常见旅鸟，多见于湖边滩地及河岸。

习性与生态 非繁殖季节喜单独或集小群活动，常与其他鸻鹬类混群。营巢于海滨沙滩或内陆湖滩上；窝卵数 2~5 枚。

Morphological characteristics A small wading bird (15-20 cm). Sexes similar. Breeding adult has black bill; black from forecrown to eye-stripe and ear-coverts; rear crown, nape, neck and breast band rufous-brown; breast band has black upper border; upperparts and wings brown; tail dark brown and tail-sides white; chin, throat and underparts white; tarsi greyish-green. Non-breeding adult has brown on head instead of black; supercilium white; upperparts duller than breeding adult.

Distribution and habitat Widespread in China. Mainly seen at coastal wetlands, farmland and grassland. Uncommon passage migrant to lakeshores and riverbanks in Beijing.

Behaviour and ecology Alone or in small flocks in non-breeding season. Often joins mixed flocks of plovers. Nests at both inland and coastal shores. Clutch size: 2-5.

铁嘴沙鸻	*Charadrius leschenaultii*	Greater Sand Plover	鸻科 Charadriidae

形态特征 小型涉禽，体长 19~23 厘米。雌雄相似。成鸟繁殖羽喙黑色；额基白色，前额、贯眼纹及耳羽黑色，形成与蒙古沙鸻相似的黑色斑纹；头顶后部、枕部、上体及两翼褐色，颏、喉白色，胸赤褐色；下体白色；跗跖黄绿色。成鸟非繁殖羽头顶及头侧褐色，无黑色横带，具白色眉纹，胸带浅褐色。似蒙古沙鸻，但本种体型较大，且喙较为长而粗壮。

分布与生境 除青藏高原外广布于中国各地。活动于海滩、河口及内陆湿地。在北京为不常见旅鸟，一般见于湖泊和河流的岸边。

习性与生态 喜集小群活动，有时与蒙古沙鸻或其他鸻鹬类混群。主食昆虫和软体动物。营巢于草甸、苔原或海滨沙滩地面；窝卵数 2~4 枚。

Morphological characteristics A small wading bird (19-23 cm). Sexes similar. Breeding adult has black bill and small white patch on forehead; forecrown, eye-stripe and ear-coverts black, forming prominent black mask on head as with Lesser Sand Plover; rear head, nape, upperparts and wings brown; chin and throat white; breast rufous-brown; underparts white; tarsi yellowish-green. Non-breeding adult has brown on forecrown and head-sides; forehead and supercilium white; breast pale brown. Resembles Lesser Sand Plover but is larger with a longer and thicker bill.

Distribution and habitat Widespread in China except for the Tibetan Plateau. Seen at both coastal and inland wetlands. Uncommon passage migrant to lakeshores and riverbanks in Beijing.

Behaviour and ecology Forms small flocks or joins mixed flocks with Lesser Sand Plover and other shorebirds. Nests on the ground in grassland, tundra or on beaches. Clutch size: 2-4.

IUCN 红色名录 无危 (LC) 2019.8.14

成鸟 Adult

幼鸟 Juvenile

东方鸻	*Charadrius veredus*	Oriental Plover	鸻科 Charadriidae

繁殖羽 Breeding

非繁殖羽 Non-breeding

繁殖羽 Breeding

IUCN 红色名录 无危 (LC) 2016.10.1

形态特征 小型涉禽，体长 22~26 厘米。雄鸟繁殖羽头、颈白色；上体、两翼及尾羽褐色，颏、喉白色，具显著的胸带，上胸栗褐色，下胸黑色，腹部及尾下覆羽白色；跗跖橙黄色。雌鸟头部灰褐色。成鸟非繁殖羽头褐色，胸部淡褐色，无栗色及黑色胸带。幼鸟似非繁殖羽成鸟，但上体羽缘淡黄色。

分布与生境 广布于中国东部、北部及中部。夏季见于内陆草原、荒漠和湿地，冬季迁徙至海滨。北京通常见于官厅水库、密云水库开阔的湿地浅滩区域和近水草地与农田，为不常见旅鸟。

习性与生态 在地面上来回奔跑，喜集松散的小群。飞行速度较快。

Morphological characteristics A small wading bird (22-26 cm). Sexes similar. Breeding male has white head and neck; upperparts, wings and tail brown; some individuals have variable pale brown patch on rear head and nape; chin and throat white; upper breast has prominent chestnut band; lower breast has a prominent black band; remainder of underparts white; tarsi orange-yellow. Non-breeding adult has brown head and pale brown breast, lacking the prominent chestnut-and-black breast band. Juvenile resembles non-breeding adult but has buff-fringed upperparts.

Distribution and habitat Widespread in northern, eastern and central China. Seen at inland grassland, semi-desert and wetland habitats in summer but prefers coastal areas in non-breeding season. In Beijing: uncommon passage migrant to large wetlands such as Guanting and Miyun Reservoirs and grassland or farmland near water.

Behaviour and ecology Forms small and loose flocks and runs on the ground. Flight fast.

彩鹬	*Rostratula benghalensis*	Greater Painted Snipe	彩鹬科 Rostratulidae

形态特征 小型涉禽，体长 24~28 厘米，体型紧凑，喙肉粉色，喙端向下弯曲。雌鸟头顶暗褐色，具淡黄色顶冠纹，眼周至眼后方白色，具黑色外缘，头侧至颈部栗红色；上体橄榄褐色，具黑色横纹，背部具两条黄色纵纹，两翼暗褐色，飞羽具皮黄色斑点；尾羽褐色，具皮黄色横斑；胸黑色，下体余部白色，白色区域经胸侧向上延伸；跗跖黄绿色。雄鸟羽色较雌鸟暗淡，头、颈、胸及上体大致为褐色。

分布与生境 除西部地区外广布于中国，于北方为夏候鸟，南方为留鸟。北京见于植被丰盛的湿地，为不常见夏候鸟。

习性与生态 多单独或成对活动，受到惊扰后多隐藏在草丛中不动。繁殖为一雌多雄制，由雄鸟孵卵、育雏，窝卵数 3~6 枚。

Morphological characteristics A small wading bird (24-28 cm) with plump body and pink bill with slightly decurved tip. Female has dark brown crown with pale yellow median crown-stripe; prominent white patch around and behind eye, which is bordered black; chestnut from head-sides to neck; upperparts olive-brown with barring on the back and two buff stripes; wings dark brown, with buff spots on remiges; tail brown, barred buff; breast black; remainder of underparts white, extending to breast-sides, forming braces-like stripes; tarsi yellowish-green. Male is duller than female; head and neck largely olive-brown instead of chestnut and black in female.

Distribution and habitat Widespread in China except for western areas. Summer breeder in northern China and resident in southern China. In Beijing, uncommon summer breeder to wetlands with extensive vegetation.

Behaviour and ecology Usually alone or in pairs. Freezes in reeds or grass when disturbed. In breeding season, male is courted by female. Polyandrous. Only male incubates the eggs and raises the young. Clutch size: 3-6.

IUCN 红色名录 无危 (LC) 2016.10.1

雌鸟 Female

雄鸟 Male

幼鸟 Juvenile

水雉	*Hydrophasianus chirurgus*	Pheasant-tailed Jacana	水雉科 Jacanidae

国家二级保护野生动物　IUCN 红色名录　无危 (LC) 2016.10.1

形态特征　中型涉禽，体型似雉类，体长 31~58 厘米。雌鸟繁殖羽喙蓝灰色，头、颈白色，后颈具金黄色饰羽，边缘黑色；背褐色，两翼大致为白色，仅初级飞羽端部黑色，但外侧初级飞羽全为黑色，腰、尾上覆羽及尾羽黑色，两对中央尾羽甚为延长，尾长通常占体长的 50% 或以上；下体黑色；跗跖黄绿色，趾甚长。雄鸟繁殖羽似雌鸟，但体型较小、尾羽较短。成鸟非繁殖羽上体褐色，下体白色，无延长的中央尾羽。

分布与生境　中国主要见于华北以南地区。主要活动于多浮水植物（菱角、莲等）的淡水湖泊和水塘。在北京为甚罕见夏候鸟。

习性与生态　在浮水叶片上快速行走觅食。一雌多雄制，一只雌鸟与多只雄鸟交配后产数窝卵，窝卵数 4 枚，由不同的雄鸟孵化并完成育雏。

Morphological characteristics A medium-sized pheasant-like wading bird (31-58 cm). Breeding female has greyish-blue bill; head and neck largely white; nape and hindneck bright yellow with black border; mantle brown; wings mostly white, but outer primaries black and other primaries tipped black; black from rump to tail; central rectrices very long; underparts largely black; tarsi yellowish-green; long toes. Breeding male has similar plumage but it is considerably smaller with shorter tail. Non-breeding adult is brown above and white below, retaining the yellow hindneck but lacking the long central rectrices.

Distribution and habitat North China, as well as in central and southern China. Mainly seen on ponds, lakes and rivers with extensive floating vegetation (*e.g. Trapa natans*). Very rare summer breeder in Beijing.

Behaviour and ecology Walks fast on surface vegetation in lakes. Polyandrous. A female court several males. Clutch size: 4. Male incubates and raises young.

丘鹬	*Scolopax rusticola*	Eurasian Woodcock	鹬科 Scolopacidae

形态特征 小型涉禽，体长 32~42 厘米，雌雄同色，羽色似沙锥。成鸟喙较长，呈黄褐色，喙端黑色。头、颈及上体黄褐色，头顶具黑色横斑，贯眼纹黑色，上体及两翼具黑色和淡黄色斑点，腰、尾上覆羽和尾羽具锈红色和黑色斑。颏、喉白色或淡皮黄色。下体淡灰褐色或灰白色，具黑褐色横斑。跗跖粉色。

分布与生境 中国除青藏高原外广泛分布，于东部更多见。北京地区为不常见旅鸟，一般活动于森林（林下）、灌丛和农田，偶尔亦见于城区公园和居民区绿化带中。

习性与生态 多单只或成对活动，行为隐秘，受惊时常隐藏在植物中不动。占域行为极具特征性：雄鸟在晨昏时于森林上空边来回飞行，边发出响亮而多变的鸣声。营巢于森林近水处；窝卵数 3~5 枚。

Morphological characteristics A small snipe-like wading bird (32-42 cm). Sexes alike. Adult has long, yellowish-brown bill, tipped black; head, neck and upperparts largely brown; crown barred black; eye-stripe black; upperparts and wings spotted black and buff; rump, uppertail-coverts and tail barred black and rufous; chin and throat white or buff; remainder of underparts mostly off-white or pale greyish-brown with dark bars; tarsi pink.

Distribution and habitat Widespread in China except for the Tibetan Plateau; more common in the east. Uncommon passage migrant in farmland, scrub or forest in Beijing. Sometimes also seen in urban parks, gardens and residential areas.

Behaviour and ecology Usually solitary or in pairs. Shy. Prefers to freeze in vegetation than fly when disturbed. In breeding season, male performs twilight display flights ("roding") over the forest, with varied loud calls. Nests in forest near water. Clutch size: 3-5.

IUCN 红色名录 无危 (LC) 2016.10.1

姬鹬	*Lymnocryptes minimus*	Jack Snipe	鹬科 Scolopacidae

IUCN 红色名录 无危 (LC) 2016.10.1

形态特征 小型涉禽，体长 17~20 厘米，雌雄同色。喙基浅褐色，喙端黑色。头顶黑色，眉纹甚宽阔，皮黄色，其间有一条黑色细纹将眉纹分割为上、下两部分，贯眼纹黑色，颊淡褐色，颊纹黑色。上体深褐色，具 4 条显著的淡黄白色纵纹。颏、喉白色，胸灰白色，密布褐色纵纹。下体白色。跗跖黄绿色。

分布与生境 中国见于西北、华北至东部沿海。活动于植被丰富的沼泽、湖泊、河流岸边。在北京为罕见旅鸟。

习性与生态 行为隐秘，受惊扰时常伏于地面不动，非繁殖期单独活动，不集群。营巢于植被丰富的湿地中；窝卵数 4 枚。

Morphological characteristics A small wading bird (17-20 cm). Sexes alike. Bill is pale brown at base, black at tip; crown black; supercilium broad and buff, divided into upper and lower parts by a black line; eye-stripe black; cheek pale brown with a black moustachial stripe; upperparts largely dark brown with four distinctive pale yellow stripes; chin and throat white; breast greyish white, with extensive dark brown streaks; remaining underpants white; tarsi yellowish-green.

Distribution and habitat Northwest and eastern China. Seen in well-vegetated marshes, lakeshores and riverbanks. Rare passage migrant in Beijing.

Behaviour and ecology Very shy. Freezes in vegetation if disturbed. Nests in wetlands with abundant vegetation. Clutch size: 4.

孤沙锥 *Gallinago solitaria* Solitary Snipe 鹬科 Scolopacidae

形态特征 小型涉禽，体长 26~32 厘米。雌雄同色。喙基部黄绿色，端部黑色，长而直。头顶褐色，具白色的较细顶冠纹，眉纹白色，贯眼纹深褐色，颊灰白色。上体褐色，具黑色横斑和白色纵纹。尾羽共 18 枚，3 对中央尾羽基部黑色，端部白色，具栗色次端斑，其余尾羽无栗色斑。颏、喉白色，胸部褐色。下体白色，具褐色横斑。跗跖黄绿色。

分布与生境 中国广泛分布。栖息于山区（特别是峡谷）林地及其附近的沼泽、水塘和河流附近。在北京为罕见旅鸟和冬候鸟。

习性与生态 性隐秘，单独活动，飞行缓慢。雄鸟在繁殖期会进行复杂的来回飞行炫耀行为。营巢于山区湿地；窝卵数 4~5 枚。

Morphological characteristics A small wading bird (26-32 cm). Sexes alike. Long, straight bill with yellowish-green base and black tip; crown brown with white median crown-stripe; supercilium white; eye-stripe dark brown; upperparts mostly brown, with black bars and white stripes; tail has 18 feathers, with the central six having a black base, white tips and a rufous subterminal band; remaining tail feathers lack rufous; chin and throat white; breast brown; underparts white, heavily barred brown; tarsi yellowish-green.

Distribution and habitat Widespread in China. Rare passage migrant and winter visitor in Beijing. Usually seen in fast-moving streams and rivers in forest or valleys.

Behaviour and ecology Solitary and shy. Flight slow. Male performs complicated switchback flight as courtship displays. Nests in mountainous wetlands. Clutch size: 4-5.

IUCN 红色名录 无危 (LC) 2016.10.1

针尾沙锥 | *Gallinago stenura* | Pin-tailed Snipe | 鹬科 Scolopacidae

IUCN 红色名录　无危 (LC) 2016.10.1

形态特征　小型涉禽，体长 21~29 厘米。雌雄同色。喙较长，喙基黄褐色，喙端黑色。头顶深褐色，顶冠纹淡黄色并延伸至枕部，眉纹、颊纹皆为淡黄色，贯眼纹深褐色。上体深褐色，具锈红色斑纹和 4 条土黄色纵纹，内侧翼上覆羽褐色，外侧翼上覆羽和各级飞羽黑褐色。尾羽 24~28 枚，中央 5 对尾羽基部黑色，近端部栗色，端部白色，其余外侧尾羽甚窄。颏、喉、胸近白色或淡褐色，两胁具褐色横斑，其余下体白色。跗跖黄绿色。

分布与生境　广布于中国。北京为区域性常见旅鸟，见于植被丰富的沼泽、湖泊和河流等生境。

习性与生态　觅食时会用喙在地面或泥土中探寻小型无脊椎动物。飞行时常突然改变高度，飞行路线似“Z”形。

Morphological characteristics　A small wading bird (21-29 cm). Sexes alike. Bill long and slender, pale brown at base and black at tip; crown dark brown, with distinctive pale yellow median crown-stripe, extending to nape; supercilium and moustachial stripe pale yellow; eye-stripe dark brown; upperparts dark brown with four buff stirpes as well as rufous scales; inner wing-coverts largely brown; outer wing-coverts and remiges blackish-brown; has 24-28 tail feathers, of which the 10 central tail feathers are black at base, white at tips and rufous near tips; outer tail feathers very narrow; chin and throat off-white or pale brown; breast streaked brown; breast-sides and flanks have dark brown bars; remainder of underparts white; tarsi yellowish-green.

Distribution and habitat　Widespread in China. Locally common passage migrant in Beijing. Seen at well-vegetated marshes, ponds, lakes and rivers.

Behaviour and ecology　When feeding, probes for invertebrates in the ground using its long bill. Flies off in zigzags.

大沙锥	*Gallinago megala*	Swinhoe's Snipe	鹬科 Scolopacidae

形态特征 小型涉禽，体长 26~29 厘米。雌雄同色。喙基浅褐色或粉褐色，喙端黑色。头褐色，顶冠纹和眉纹淡皮黄色，贯眼纹深褐色。上体褐色，具 4 条皮黄色纵纹和褐色横斑，次级飞羽端部白色，形成白色的翼后缘，飞行时可见，尾羽 18~26 枚，中央 4~6 对尾羽基部黑色，次端栗色，端部白色，其余外侧尾羽较窄。颏、喉近白色，颈至胸具褐色斑纹，下体余部白色。跗跖黄绿色。

分布与生境 分布于中国东部及中部。北京地区见于水田以及湿地的草丛和苇丛中，为不常见旅鸟。

习性与生态 迁徙时多单独活动，偶尔集成松散的小群。受惊时多选择在原地蹲伏，隐匿于植被中。

Morphological characteristics A small wading bird (26-29 cm). Sexes alike. Bill pale brown or brownish-pink at base, black at tip; head brown, with pale yellow median crown-stripe and supercilium; eye-stripe dark brown; upperparts largely rufous-brown with dark bars and four buff stripes; wings largely dark brown, but secondaries have white tips, which can be seen in flight; tail has 18-26 feathers, of which the central 4-6 pairs of have black base, white tips and rufous bars near tips; outer tail feathers narrow; chin and throat off-white; neck and breast have brown streaks; flanks barred dark; remainder of underparts white; tarsi yellowish-green.

Distribution and habitat Widespread in eastern and central China. In Beijing: uncommon passage migrant in grass and reeds near water.

Behaviour and ecology Solitary but can form loose flocks on migration. Freezes when disturbed.

IUCN 红色名录 无危 (LC) 2016.10.1

扇尾沙锥	*Gallinago gallinago*	Common Snipe	鹬科 Scolopacidae

IUCN 红色名录 无危 (LC) 2016.10.1

形态特征 小型涉禽，体长 24~30 厘米。雌雄同色。喙较长，基部淡褐色，端部黑褐色。头顶深褐色，具淡黄色的顶冠纹和眉纹，头两侧淡褐色。上体褐色，具 4 条淡黄色纵纹，其中外侧一对纵纹有时不清晰。次级飞羽的白色端部形成和大沙锥相似的白色翼后缘。尾羽仅 14 枚左右，除两枚最外侧尾羽外，其余尾羽皆具深褐色基部、白色端部和栗红色次端斑。胸部和两胁具褐色斑纹，其余下体白色。跗跖黄绿色。

分布与生境 广布于中国沿海和内陆湿地。非繁殖期的生境和大沙锥相似，迁徙过境时皆活动于植被丰富的湿地。在北京为区域性常见旅鸟。

习性与生态 过境时多集松散的小群活动。飞行时忽上忽下，似“Z”形。营巢于森林、草原、沼泽等地的地面；窝卵数 3~5 枚。

Morphological characteristics A small wading bird (24-30 cm). Sexes alike. Bill long and straight, with pale brown base and black tip; crown dark brown; median crown-stripe and supercilium pale yellow; head-sides pale brown; upperparts largely brown, with two pairs of pale yellow stripes, but outer pair sometimes indistinct; secondaries have white tips, resembling Swinhoe's Snipe; has 14 tail feathers, most of which have dark base, white tip and rufous subterminal band; breast and flanks heavily steaked and barred dark brown; remainder of underparts white; tarsi yellowish-green.

Distribution and habitat Widespread in both inland and coastal areas of China. Prefers similar habitat to Swinhoe's Snipe in non-breeding season. Locally common passage migrant at well-vegetated wetlands in Beijing.

Behaviour and ecology Usually forms loose flocks on migration. Flies off in zigzags. Nests on the ground in forests, grassland and marshes. Clutch size: 3-5.

长嘴半蹼鹬 *Limnodromus scolopaceus* Long-billed Dowitcher 鹬科 Scolopacidae

形态特征 小型涉禽，体长 27~30 厘米。雌雄相似。喙黑色，基部黄绿色，喙长且直。繁殖羽头顶红褐色，密布黑色短纵纹，头侧、颈、胸及下体亦为红褐色，眉纹白色或淡褐色，贯眼纹深褐色。背部及翼上覆羽黑色，羽缘褐色，下背至腰白色，尾羽褐色，具黑色横斑；跗跖黄色。非繁殖羽头部及上体大致为灰褐色，胸灰色，两胁具深色横斑，其余下体白色。

分布与生境 中国仅零星分布，主要见于东部沿海滩涂。在北京为甚罕见迷鸟，仅于 2009 年 10 月在延庆野鸭湖有过一次记载。

习性与生态 在其主要迁徙路线（北美）喜集群活动，但因中国并非位于其迁徙路线上，因此多为单只记载，或与其他鸻鹬尤其是塍鹬混群。

Morphological characteristics A small wading bird (27-30 cm). Sexes alike. Black bill long and straight. Breeding adult is largely rufous-brown from head, neck and breast to underparts; crown has fine black streaks; supercilium white or pale brown; eye-stripe dark brown; mantle and wing-coverts black with brown fringes; rump white; tail brown, barred black; tarsi yellow. Non-breeding adult has greyish-brown head and upperparts; breast grey; flanks with dark barring; remaining underparts white.

Distribution and habitat Scarce in eastern China. Usually seen along the coast. Very rare vagrant in Beijing with only one record at Yeyahu, Yanqing District in October 2009.

Behaviour and ecology Forms flocks on migration in North America, but in China it is usually alone or in mixed flocks with other shorebirds, particularly godwits.

IUCN 红色名录 无危 (LC) 2016.10.1

繁殖羽 Breeding

非繁殖羽 Non-breeding

半蹼鹬	*Limnodromus semipalmatus*	Asian Dowitcher	鹬科 Scolopacidae

繁殖羽 Breeding

亚成鸟 Immature

国家二级保护野生动物

IUCN 红色名录　近危 (NT)　2016.10.1

形态特征　小型涉禽，体长 31~36 厘米。雌雄相似。喙较长，呈黑色，喙端略膨大。繁殖羽头顶和眼先黑褐色，背部及翼上覆羽黑色，具棕红色和白色羽缘，腰、尾上覆羽及尾羽白色，具黑褐色横斑；头两侧、颈、胸至上腹棕红色，下腹及尾下覆羽白色，具黑色横斑；跗跖黑色。非繁殖羽上体深褐色，羽缘白色，颏、喉白色，颈及胸淡褐色，两胁具深灰色横斑，其余下体白色。

分布与生境　繁殖于中国东北地区，迁徙时过境东部及南部沿海。繁殖期喜草地，过境时多见于泥滩。北京为甚罕见旅鸟，于顺义潮白河、延庆官厅水库、密云水库等湿地的滩涂偶有记载。

习性与生态　迁徙时喜集群或与塍鹬混群。在泥滩及浅水中觅食时常一边向前移动，一边将喙垂直伸入水中或泥中探查食物。

Morphological characteristics　A small wading bird (31-36 cm). Sexes alike. All-black bill long and straight, swollen at tip. Breeding adult has dark crown and lores; mantle and wing-coverts black, fringed white and rufous; rump, uppertail-coverts and tail white, barred blackish-brown; rufous from head-sides, neck and breast to upper belly; vent and undertail-coverts white, with black bars; tarsi black. Non-breeding adult has blackish-brown upperparts with white fringes; chin and throat white; foreneck and breast greyish-brown; flanks barred dark grey. Rest of underparts white.

Distribution and habitat　Breeds in northeast China. Migrates along the coastal areas in eastern and southern China. Prefers grassland in breeding season and coastal mudflats in non-breeding season. In Beijing: rare passage migrant at rural wetlands such as Chaobai River, Shunyi District, Guanting Reservoir, Yanqing District and Miyun Reservoir, Miyun District.

Behaviour and ecology　Usually in flocks and joins mixed flocks with godwits on migration. When feeding at mudflats, vertically probes using bill as it walks forward.

黑尾塍鹬	*Limosa limosa*	Black-tailed Godwit	鹬科 Scolopacidae

形态特征 中型涉禽，体长 36~44 厘米。雌雄相似，但雌鸟繁殖羽略显暗淡。具较长而直的喙。喙基粉色，喙端黑色。繁殖羽头、颈、胸棕红色，头顶密布黑色短纵纹，眉纹白色，贯眼纹黑褐色；背部和翼上覆羽具黑色和棕红色斑点，飞羽黑褐色，内侧初级飞羽和次级飞羽基部白色，形成翼上宽阔的白色带；腰至尾上覆羽白色，尾羽基部白色，端部黑色，下体白色，下胸至上腹具黑色横斑；跗跖黑色。非繁殖羽头部及上体灰褐色，下体近白色，无黑色横斑。

分布与生境 中国除西藏外广泛分布。繁殖于草地，迁徙时见于沿海和内陆各种湿地的滩涂及浅水。在北京为区域性常见旅鸟。

习性与生态 在繁殖期与非繁殖期皆喜集群。营巢于湿地附近的草丛中；窝卵数 3~5 枚。

Morphological characteristics A medium-sized wading bird (36-44 cm). Sexes similar but female has duller breeding plumage. Long and straight bill, pink at base and black at tip. Breeding adult has rufous-brown head, neck and breast; crown has fine black streaks; supercilium white; eye-stripe black; mantle and upperwing-coverts spotted black and rufous; flight feathers blackish-brown; inner primaries and secondaries white at base, forming distinctive white band on the wings; rump and uppertail-coverts white; tail white at base and black at tip; underparts largely white, with extensive black bars on lower breast and upper belly; tarsi black. Non-breeding adult has greyish-brown head and upperparts; underparts largely white, lacking black bars.

Distribution and habitat Widespread in China except for the Tibetan Plateau. Breeds in grassland. Seen at both inland and coastal wetlands, especially mudflats. Locally common passage migrant in Beijing.

Behaviour and ecology Forms flocks in both breeding and non-breeding seasons. Nests on the ground in grass near water. Clutch size: 3-5.

IUCN 红色名录 近危 (NT) 2016.10.1

繁殖羽 Breeding

非繁殖羽 Non-breeding

斑尾塍鹬	*Limosa lapponica*	Bar-tailed Godwit	鹬科 Scolopacidae

繁殖羽 Breeding

非繁殖羽 Non-breeding

IUCN 红色名录 近危 (NT) 2016.10.1

形态特征 小型涉禽，体长 33~41 厘米。喙较长，呈黑色，基部粉色，喙端略上翘。雄鸟繁殖羽头、颈、胸、下体皆为棕红色，仅头顶具黑色细纵纹，贯眼纹黑色；背部黑褐色，具淡棕色斑点，两翼深褐色，翼上覆羽具灰白色羽缘；腰及尾上覆羽大致为白色，尾羽具深褐色与白色相间的横斑；跗跖黑色。雌鸟繁殖羽似雄鸟繁殖羽但棕红色较淡。成鸟非繁殖羽头、颈、胸及上体灰褐色，下体白色。

分布与生境 分布于中国除青藏高原外的大部分地区，但多见于东部及南部沿海滩涂有软泥的区域。在北京为郊区较大湿地的罕见旅鸟。

习性与生态 迁徙经过中国东部沿海时常形成大群，但于北京的记录多为单只或小群。常在低潮时于潮间带寻找小型无脊椎动物。飞行时振翼快速有力。

Morphological characteristics A small wading bird (33-41 cm). Bill long, slightly upcurved, pink at base, black at tip. Breeding male has prominent rufous plumage from head and neck to underparts; crown streaked black; eye-stripe also black; mantle blackish-brown, with pale rufous spots; wings mostly dark brown; upperwing-coverts fringed pale grey; rump and uppertail-coverts white; tail barred blackish-brown and white; tarsi black. Breeding female resembles breeding male but less rufous. Non-breeding adult has grey head, neck, breast and upperparts; underparts largely white.

Distribution and habitat Widespread in China except for the Tibetan Plateau; more common at coastal mudflats. Rare passage migrant at large wetlands in rural Beijing.

Behaviour and ecology Forms large flocks when migrating through coastal areas in China. But rare inland and records in Beijing are of singles. Prefers to feed for invertebrates at the intertidal zone at low tide. Flight fast and powerful.

小杓鹬 | *Numenius minutus* | Little Curlew | 鹬科 Scolopacidae

形态特征 小型涉禽，体长 29~32 厘米。雌雄同色。具较短而下弯的喙，下喙基部粉色，余部黑色。头顶黑褐色，顶冠纹土黄色，眉纹较宽，浅褐色，贯眼纹深褐色，但有时不清晰。上体深褐色，各羽具淡褐色羽缘。尾羽褐色，具深色横斑。颈、胸淡褐色，具稀疏的深褐色纵纹，其余下体近白色。跗跖灰褐色。本种虽然与大杓鹬、白腰杓鹬和中杓鹬羽色相似，但喙显著较短，亦为体型最小者，因此不难辨识。

分布与生境 分布于中国西北、东北至东南沿海。喜开阔的草地、农田、沼泽等，有时距水源较远。在北京为罕见旅鸟。

习性与生态 多单独或成小群活动，飞行时常发出三声一度的短促哨声。营巢于林缘或开阔林地；窝卵数 3~4 枚。

Morphological characteristics A small wading bird (29-32 cm). Sexes alike. Bill short and decurved, mostly black but pink at base of lower mandible; crown blackish-brown with buff median crown-stripe; supercilium broad, pale brown; eye-stripe dark brown, sometimes indistinct; upperparts dark brown with pale fringes; tail brown with dark bars; neck and breast pale brown with dark streaks; remaining underparts largely white; tarsi greyish-brown. Plumage resembles other curlew species, but smaller body and shorter bill.

Distribution and habitat Northern and eastern China, and coastal areas in southern China. Prefers grassland, farmland and marshes, sometimes far from water. Rare passage migrant in Beijing.

Behaviour and ecology Alone or in small flocks. Often gives three-note soft calls in flight. Nests on the ground in or near woodlands. Clutch size: 3-4.

国家二级保护野生动物 | IUCN 红色名录 无危 (LC) 2016.10.1

中杓鹬	*Numenius phaeopus*	Whimbrel	鹬科 Scolopacidae

IUCN 红色名录 无危 (LC) 2016.10.1

形态特征 中型涉禽，体长 40~46 厘米。雌雄同色。喙较长且略微下弯，喙基粉色，其余黑色。头顶深褐色，顶冠纹和眉纹淡褐色，贯眼纹深褐色，颊近白色或淡褐色。背和两翼深褐色，具浅褐色白点。腰白色，尾上覆羽白色，具深色斑点，尾羽褐色，密布黑褐色横斑。颏、喉白色，颈、胸皮黄色，具暗褐色纵纹，腹部白色，两胁和尾下覆羽具深色横斑。跗跖蓝灰色。

分布与生境 除青藏高原外广布于中国。主要见于滨海沙滩、泥滩、河口，以及内陆河流、湖泊沿岸的草地、滩涂、沼泽等地。在北京为不常见旅鸟，多见于郊区较大湿地。

习性与生态 非繁殖期常集小群或与其他鸻鹬混群。飞行时振翼有力。

Morphological characteristics A medium-sized wading bird (40-46 cm). Sexes alike. Bill long and slightly decurved, mostly black with pink base; crown dark brown; median crown-stripe and supercilium buff; eye-stripe dark; cheek off-white or pale brown; mantle and wings largely dark brown, with white-spotted fringes; rump and uppertail-coverts mostly white, but the latter spotted dark; tail brown, heavily barred blackish-brown; chin and throat white; neck and breast buff, with dark streaks; belly white; flanks and undertail-coverts have dark bars; tarsi greyish-blue.

Distribution and habitat Widespread in China except for the Tibetan Plateau. Prefers coastal mudflats, sandy shores, estuaries, as well as inland marshes, grassland and shores. Uncommon passage migrant to large wetlands in Beijing.

Behaviour and ecology Forms small flocks or joins mixed flocks with other curlews and sandpipers. Wingbeats strong in flight.

白腰杓鹬 | *Numenius arquata* | Eurasian Curlew | 鹬科 Scolopacidae

形态特征 中型涉禽，体长 57~63 厘米。雌雄同色。喙甚长（雌鸟喙长于雄鸟），端部向下弯曲，下喙基部粉色，余部黑色。头部灰褐色，头顶缀有黑色纵纹。背及翼上覆羽大致为淡褐色，具密布黑褐色纵纹，外侧数枚初级飞羽和初级覆羽几乎全为黑色，内侧初级飞羽和次级飞羽深褐色，具浅色横斑，腰白色，尾羽亦为白色，密布黑褐色横斑。颏、喉白色，颈及胸灰白色，具深色纵纹，翼下覆羽及下体余部白色。跗跖蓝灰色。

分布与生境 广布于中国全境。栖息于海滩、河口以及内陆水域沿岸的滩涂、草地、沼泽等各种生境。在北京为不常见旅鸟。

习性与生态 迁徙过境时常集成大群或与大杓鹬等其他鸻鹬混群，但于北京的记载仅为零星单只或小群。飞行时振翼缓慢。

Morphological characteristics A medium-sized wading bird (57-63 cm). Sexes alike. Bill very long (longest in female), decurved at tip, mostly black, but pink at base of lower mandible; head greyish-brown, with black streaks on crown; mantle and upperwing-coverts largely peat-brown, with extensive black streaks; primary coverts and outer primaries mostly black; remainder of flight feathers dark brown, with pale bars; rump white; tail white with dark brown bar; chin and throat white; neck and breast pale greyish-white, with dark streaks; remainder of underparts and underwing-coverts largely white; tarsi greyish-blue.

Distribution and habitat Widespread in China. Usually seen along coasts, estuaries, as well as inland wetlands. Uncommon passage migrant at large wetlands in rural areas of Beijing.

Behaviour and ecology On migration, prefers to form large flocks or join mixed flocks with Eastern Curlews or other shorebirds along coastal areas. But usually alone or in small flocks in Beijing. Flight heavy with slow wingbeats.

国家二级保护野生动物 | IUCN 红色名录 近危 (NT) 2017.10.1

大杓鹬 | *Numenius madagascariensis* | Eastern Curlew | 鹬科 Scolopacidae

国家二级保护野生动物　IUCN 红色名录　濒危 (EN) 2016.10.1

形态特征　中型涉禽，体长 54~65 厘米。雌雄同色。喙甚长（雌鸟喙较雄鸟长，喙长为头长的 3~5 倍）且下弯，基部粉色，端部黑色。头、颈、胸淡褐色，具黑色细纵纹。背部和两翼深褐色，各羽具淡褐色羽缘，腰、尾上覆羽及尾羽皆为褐色，具深褐色横斑。下体淡灰褐色，翼下覆羽淡褐色，密布黑褐色斑点。

分布与生境　于中国东部及中部迁徙过境时喜沿海泥滩或近海沼泽、草地、浅池塘等飞行，于内陆较为少见。北京见于郊区水库、湖滩和河岸，为罕见旅鸟。

习性与生态　非繁殖期在沿海喜集大群活动，主要以泥滩中的蟹类和软体动物为食。飞行时振翼缓慢。于北京的记录则多为单只。

Morphological characteristics A medium-sized wading bird (54-65 cm). Sexes alike. Bill very long (longest in female), decurved; mostly blackish-brown but pink at base of lower mandible; length of its bill is equivalent to 3-5 times the length of head; pale brown from head and neck to beast, streaked black; mantle and wings dark brown, with pale fringes; rump, uppertail-coverts and tail brown, with extensive black bars; underparts largely pale greyish-brown; underwing-coverts pale brown, heavily spotted blackish-brown.

Distribution and habitat Eastern and central China. Usually seen at coastal mudflats, swamps, wet grassland and shallow ponds. Scarce inland. Rare passage migrant at rural reservoirs, lake shores and riverbanks in Beijing.

Behaviour and ecology Forms large flocks in non-breeding season and feeds on crabs and molluscs in mud. Wingbeats slow in flight. Records in Beijing usually of singles.

鹤鹬	*Tringa erythropus*	Spotted Redshank	鹬科 Scolopacidae

形态特征 小型涉禽，体长 26~33 厘米。雌雄同色。喙较细长，呈黑色，下喙基部红色。繁殖羽头、颈、胸及下体黑色，具断开的白色眼圈；背和两翼深褐色，具白色斑点，初级飞羽和初级覆羽几全为黑色；下背及腰白色，尾上覆羽和尾羽黑褐色，密布白色横斑；跗跖暗红色。非繁殖羽头、颈及上体灰色，具白色眉纹和深灰色眼先；下体白色或灰白色；跗跖橙红色。

分布与生境 广布于中国。见于沿海和内陆各种湿地的浅水区域和滩涂、农田等地。在北京为区域性常见旅鸟。

习性与生态 喜单只或形成分散的群体活动。相对其他鹬鹬而言，鹤鹬喜在较深的水中行走觅食，有时甚至会游水。

Morphological characteristics A small wading bird (26-33 cm). Sexes alike. Breeding adult has fine, black bill, with red basal half of lower mandible; all-black from head and neck to underparts; broken white eye-ring; mantle and wings dark brown, spotted white; primaries and primary coverts black; rump white; uppertail-coverts and tail blackish-brown, with extensive white bars; tarsi deep red. Non-breeding adult has plain grey head, neck and upperparts; supercilium white; lores dark grey; underparts white or pale greyish-white; tarsi red-orange.

Distribution and habitat Widespread in China. Seen at inland and coastal wetlands, shores and farmland. Locally common passage migrant at wetlands in Beijing.

Behaviour and ecology Usually alone or in loose flocks in non-breeding season. Prefers to feed in deep water. Also swims.

IUCN 红色名录 无危 (LC) 2016.10.1

繁殖羽 Breeding

非繁殖羽 Non-breeding

红脚鹬	*Tringa totanus*	Common Redshank	鹬科 Scolopacidae

繁殖羽 Breeding

IUCN 红色名录　无危 (LC) 2016.10.1

形态特征　小型涉禽，体长 26~29 厘米。雌雄同色。喙基橙红色，端部黑色，喙较鹤鹬短。繁殖羽头褐色，眼先深褐色，眉纹较短，白色；上背及翼上覆羽褐色，具黑色斑点，初级覆羽及初级飞羽黑色，次级飞羽白色，形成飞行时可见的翼后缘大块白斑；下背及腰白色，尾上覆羽和尾羽亦为白色，但具细密的黑色横斑；颏、喉、胸白色，具黑色纵纹，两胁和尾下覆羽具黑色横斑，下体余部白色；跗跖橙红色。非繁殖羽头及上体灰褐色，几无深色斑纹。

分布与生境　广布于中国全境，繁殖期喜植被丰富的各种湿地，非繁殖期主要见于沿海。在北京为不常见旅鸟，喜河流、湖泊及其附近的草地和滩涂。

习性与生态　非繁殖期多形成松散的小群活动。营巢于湿地附近的草丛中；窝卵数 3~5 枚。

Morphological characteristics A small wading bird (26-29 cm). Sexes alike. Bill red-orange at base, black at tip, shorter than Spotted Redshank. Breeding adult has brown head, with dark lores and a short white supercilium; mantle and upperwing-coverts brown, spotted black; primaries and primary coverts black; secondaries white, forming a distinctive white patch on the wing; lower back and rump white; uppertail-coverts and tail also white but barred black; chin, throat and breast white, with black streaks; flanks and undertail-coverts barred black; remainder of underparts white; tarsi red-orange. Non-breeding largely plain greyish-brown on head; upperparts with fewer dark spots and streaks.

Distribution and habitat Widespread in China. Seen at well-vegetated wetlands in breeding season; prefers coastal wetlands in non-breeding season. Uncommon passage migrant at riverbanks, lake shores and grassland near water.

Behaviour and ecology Forms small and loose flocks in non-breeding season. Nests on the ground in grass near water. Clutch size: 3-5.

泽鹬	*Tringa stagnatilis*	Marsh Sandpiper	鹬科 Scolopacidae

形态特征 小型涉禽，体长 19~26 厘米。雌雄同色。喙黑色，基部有时呈黄绿色，喙较细长。成鸟繁殖羽头、颈淡灰褐色，具较短的白色眉纹；上背和两翼灰褐色，具显著的黑色斑点，下背、腰及尾上覆羽白色，尾羽亦为白色，但具深褐色横斑；颏、喉、胸白色，胸侧具褐色纵纹，下体余部白色；跗跖黄绿色。成鸟非繁殖羽上体为较淡的灰色，几无黑色斑点。幼鸟似非繁殖羽成鸟但上体偏褐色。

分布与生境 中国除西南地区外广泛分布。北京多见于沼泽、河流、湖泊等湿地的浅水区或水边，为区域性常见旅鸟。

习性与生态 非繁殖期多单只或集成较为分散的小群活动，常与林鹬、青脚鹬等混群。营巢于近水草地；窝卵数 3~5 枚。

Morphological characteristics A small wading bird (19-26 cm). Sexes alike. Bill fine, mostly black; some have yellowish-green bill base. Breeding adult has pale greyish-brown head and neck; supercilium white; mantle and wings greyish-brown with black spots; lower back, rump, and uppertail-coverts white; tail also white with dark bars; chin, throat and breast white; breast-sides streaked brown; remainder of underparts white; tarsi greenish-yellow. Non-breeding adult has pale grey upperparts lacking black spots. Juvenile resembles non-breeding adult but with more brown on upperparts.

Distribution and habitat Widespread in China except for the southwest. In Beijing, locally common passage migrant in shallow water or shores at marshes, lakes and rivers.

Behaviour and ecology Alone or in loose flocks in non-breeding season. Often joins mixed flocks with Common Greenshank and Wood Sandpiper. Nests on the ground in grass near water. Clutch size: 3-5.

IUCN 红色名录 无危 (LC) 2016.10.1

繁殖羽 Breeding

非繁殖羽 Non-breeding

青脚鹬 | *Tringa nebularia* | Common Greenshank | 鹬科 Scolopacidae

繁殖羽 Breeding

非繁殖羽 Non-breeding

非繁殖羽 Non-breeding

IUCN 红色名录　无危 (LC) 2016.10.1

形态特征　小型涉禽，体长 30~35 厘米。雌雄同色。喙较粗壮，略上翘；喙基灰色，端部黑色。成鸟繁殖羽头、颈、胸灰色，密布深褐色纵纹；上背和两翼灰褐色，具黑色斑点，下背、腰和尾上覆羽白色，尾羽白色但具细密的深色横斑；下体白色；跗跖黄绿色。成鸟非繁殖羽头、颈羽色较淡，纵纹稀疏，更偏白色，上体无黑色斑点。幼鸟似成鸟非繁殖羽，但上体偏褐色，且各羽具浅色羽缘。

分布与生境　中国广泛分布。北京多见于河流、湖泊、池塘、沼泽的浅水区域，为区域性常见旅鸟。

习性与生态　喜在浅水区域觅食，亦偶尔游至深水。非繁殖期多单独或集成 10 只以内的小群活动。

Morphological characteristics A small wading bird (30~35 cm). Sexes alike. Bill thick, slightly upcurved at tip, grey at base and black at tip. Breeding adult has pale grey head, neck and breast, with extensive dark streaks; upper back and wings greyish-brown with black spots; all-white from lower back and rump to uppertail-coverts; tail also white with fine dark bars; underparts white; tarsi yellowish-green. Non-breeding adult has pale greyish-white head and neck with fewer streaks; upperparts lack black spots. Juvenile resembles non-breeding adult but has brown upperparts with pale fringes.

Distribution and habitat Widespread in China. Locally common passage migrant in Beijing. Usually seen in shallow water at lakes, rivers, ponds and marshes.

Behaviour and ecology Mainly feeds in shallow water but occasionally swims in deeper water. Alone or in small flocks in non-breeding season.

白腰草鹬	*Tringa ochropus*	Green Sandpiper	鹬科 Scolopacidae

形态特征 小型涉禽，体长 20~24 厘米。雌雄同色。喙较短，基部灰绿色或黄绿色，端部黑色。繁殖羽头、颈及上体灰褐色，具较短的白色眉纹和较窄的白色眼圈，背和翼上覆羽具稀疏的白色斑点，飞羽黑色，无斑点，腰和尾上覆羽白色，尾羽亦为白色，具 2~3 道宽阔的黑色横带；颏、喉白色，胸部具褐色纵纹；下体白色；跗跖黄绿色。非繁殖羽似繁殖羽，但上体羽色较淡，偏灰色。

分布与生境 广布于中国。北京见于植被丰富的河流、湖泊、沼泽和农田，为区域性常见旅鸟和冬候鸟。

习性与生态 非繁殖期常单独活动，偶尔集小群或与其他鹬鹬混群。飞行时振翼甚快。营巢于森林中的河流、溪流和沼泽附近；窝卵数 3~4 枚。

Morphological characteristics A small wading bird (20-24 cm). Sexes alike. Bill short, straight, greyish-green or yellowish-green at base, black at tip. Breeding adult has largely greyish-brown head, neck and upperparts; short white supercilium and narrow white eye-ring; mantle and upperwing-coverts spotted white; flight feathers black, spotless; rump and uppertail-coverts white; tail also white with two or three broad black bands; chin and throat white; breast streaked brown; rest of underparts white; tarsi yellowish-green. Non-breeding adult resembles breeding adult with paler and greyish upperparts.

Distribution and habitat Widespread in China. Locally common passage migrant and winter visitor in Beijing. Usually seen at well-vegetated lakes, rivers, marshes and farmland.

Behaviour and ecology Usually alone in non-breeding season. Sometimes joins mixed flocks with other Tringa species. Wingbeats fast in flight. Nests at marshes, rivers and streams in forests. Clutch size: 3-4.

IUCN 红色名录　无危 (LC) 2016.10.1

繁殖羽 Breeding

非繁殖羽 Non-breeding

幼鸟 Juvenile

林鹬 *Tringa glareola* Wood Sandpiper

鹬科 Scolopacidae

繁殖羽 Breeding

非繁殖羽 Non-breeding

IUCN 红色名录 无危 (LC) 2016.10.1

形态特征 小型涉禽，体长 19~23 厘米。雌雄同色。喙短而直，基部黄绿色。成鸟繁殖羽头顶及眼先深褐色，眉纹白色，颊灰白色；上体深褐色，具白色斑点，尾上覆羽白色，尾羽白色，具深褐色横斑；颏、喉白色，胸灰褐色；下体白色；跗跖黄色、黄绿色或黄褐色。非繁殖羽与繁殖羽相似，但上体羽色较淡，且具较少的浅色斑点。

分布与生境 繁殖于中国西北和东北，迁徙时旅居中国全境。北京见于植被丰富的平原湿地，为常见旅鸟。

习性与生态 非繁殖期多集松散的小群活动。繁殖期营巢于近水林地的树上，有时亦于地面筑巢；窝卵数约 4 枚。

Morphological characteristics A small wading bird (19-23 cm). Sexes alike. Bill short and straight, mostly black, but greenish-yellow at base. Breeding adult has dark brown crown and lores; supercilium white; cheek pale greyish-white; upperparts largely dark brown, spotted white; uppertail-coverts white; tail also white with dark brown bars; chin and throat white; breast greyish-brown; underparts white; tarsi yellow, yellowish-green or yellowish-brown. Non-breeding adult similar to breeding adult but with paler and less spotted upperparts.

Distribution and habitat Breeds in northeast and northwest China. Migrates from northern China to southern China. In Beijing, a common passage migrant at lowland wetlands with abundant vegetation.

Behaviour and ecology Tends to form small and loose flocks in non-breeding season. Nests in trees or on the ground near water in forests. Clutch size: 4.

灰尾漂鹬 | *Tringa brevipes* | Grey-tailed Tattler | 鹬科 Scolopacidae

形态特征 小型涉禽，体长 25~28 厘米。雌雄同色。体型紧凑，喙和跗跖都较短。喙基黄绿色，喙端黑色。成鸟繁殖羽头顶灰色，眉纹白色，贯眼纹深灰色，颊和颈部白色，具深灰色细纵纹；上体、两翼和尾羽皆为灰色，无斑点；颏、喉白色，胸及两胁具灰褐色横斑；下体白色；跗跖黄色。成鸟非繁殖羽与繁殖羽相似，但颈、胸及两胁皆为灰色，无深色斑纹。

分布与生境 分布于中国东部及中部。非繁殖期一般活动于沿海泥滩和礁石区域。北京于温榆河等湿地有零星记载，为罕见旅鸟。

习性与生态 常单独活动，在水边或浅水处缓慢行走觅食。

Morphological characteristics A small wading bird (25-28 cm). Sexes alike. A compact *Tringa* species with relatively short bill and tarsi. Bill is yellowish-green at base, black at tip. Breeding adult has grey crown, white supercilium and dark eye-stripe; cheek and neck white with fine grey streaks; upperparts, wings and tail are plain grey; chin and throat white; breast and flanks barred greyish-brown; remaining underparts white; tarsi yellow. Non-breeding adult resembles breeding adult but has grey neck, breast and flanks, lacking dark streaks.

Distribution and habitat Eastern and central China. Prefers coastal mudflats and rocky areas. Rare passage migrant in Beijing. Seen at lake shores and riverbanks in rural areas, such as the Wenyu River.

Behaviour and ecology Solitary in non-breeding season. Walks slowly along water edge when feeding.

IUCN 红色名录 近危 (NT) 2016.10.1

翘嘴鹬	*Xenus cinereus*	Terek Sandpiper	鹬科 Scolopacidae

繁殖羽 Breeding

非繁殖羽 Non-breeding

IUCN 红色名录　无危 (LC) 2016.10.1

形态特征　小型涉禽，体长 22~25 厘米。雌雄同色。喙略上翘，基部橙色，端部黑色。繁殖羽头顶灰褐色，眉纹白色，贯眼纹黑色，上体灰褐色，两侧各具一条黑色纵纹自肩羽延伸至翼上覆羽；次级飞羽端部白色，飞行时甚为清晰；颏、喉白色，胸浅灰色，腹部及尾下覆羽白色；跗跖橙黄色。非繁殖羽头及上体羽色较淡，上体无黑色带。

分布与生境　广布于中国全境。多见于沿海滩涂和礁石区域。北京见于河流、湖泊、沼泽等湿地，为罕见旅鸟。

习性与生态　迁徙时多单独或集小群活动。觅食时常将头压低，快速跑动，追逐螃蟹等无脊椎动物，或将喙于水下左右扫动觅食。

Morphological characteristics　A small wading bird (22-25 cm). Sexes alike. Bill upcurved, with orange base and black tip. Breeding adult has greyish-brown crown; white supercilium and black eye-stripe; upperparts largely greyish-brown, with two distinctive stripes from scapulars to upperwing-coverts; secondaries have white tips, distinctive in flight; white from chin to throat; breast pale grey; belly and undertail-coverts white; tarsi orange-yellow. Non-breeding adult has paler upperparts and lacks the two black stripes.

Distribution and habitat　Widespread in China. Mainly seen at coastal mudflats and rocky shores. Rare passage migrant in shallow waters of rivers, lakes and marshes in Beijing.

Behaviour and ecology　Alone or in small flocks on migration. When feeding, usually chases prey (e.g. crabs) actively with head held low. Sometimes also moves bill from side to side in water when feeding.

矶鹬	*Actitis hypoleucos*	Common Sandpiper	鹬科 Scolopacidae

形态特征 小型涉禽，体长 16~22 厘米。雌雄同色。身体紧凑。喙短而直，呈深灰色。繁殖羽头顶及后颈灰褐色，具较短的白色眉纹和深褐色贯眼纹，眼圈白色；上体和翼上覆羽褐色，具黑褐色横斑，飞羽黑色，中段具白色斑，形成一道清晰的白色带；颏、喉白色，胸侧灰褐色，胸侧外缘白色，下体余部白色；跗跖黄绿色。非繁殖羽似繁殖羽，但上体羽色较淡。

分布与生境 中国见于全境。北京见于城区和郊区水库河流、湖泊、沼泽岸边，为常见旅鸟。

习性与生态 喜单独活动，有时也与其他鸻鹬混群，停歇时尾连续上下摆动，叫声似“叽－叽”。

Morphological characteristics A small, compact wading bird (16-22 cm). Sexes alike. Dark grey bill short and straight. Breeding adult is greyish-brown from crown to hindneck; supercilium short and white; eye-stripe dark brown, with white eye-ring; upperparts and upperwing-coverts brown, with blackish-brown bars; remiges black, with white centres, forming a white band on wings; chin and throat white; breast-sides brown, with distinctive white outer border; remainder of underparts white; tarsi greenish-yellow. Non-breeding adult resembles breeding adult but upperparts paler.

Distribution and habitat Widespread in China. Common passage migrant in both rural and urban reservoirs, lakes, rivers and marshes in Beijing.

Behaviour and ecology Usually alone or in small and loose flocks; sometimes also joins mixed-flocks of other sandpipers. Calls are high-pitched "ji-ji".

IUCN 红色名录 无危 (LC) 2016.10.1

翻石鹬 *Arenaria interpres* Ruddy Turnstone

鹬科 Scolopacidae

繁殖羽 Breeding

非繁殖羽 Non-breeding

国家二级保护野生动物

IUCN 红色名录 无危 (LC) 2019.8.14

形态特征 小型涉禽，体长 18~25 厘米。雌雄相似。体型甚为紧凑。喙黑色，楔形，较短。成鸟繁殖羽头大致为白色，头顶具黑色纵纹，前额基部和眼周具黑色带，黑色带自眼下延伸至颈部和胸部，因个体不同黑色区域的形状有所差异；上体橙棕色（雌鸟偏褐），具黑色斑，下背及尾上覆羽白色，腰黑色，尾羽基部和端部白色，具宽阔的黑色次端斑；下体白色；跗跖橙色。成鸟非繁殖羽头、胸及上体深褐色。

分布与生境 中国广泛分布，但主要见于东部沿海，特别是有石滩的地区。北京于延庆野鸭湖、昌平沙河水库等湿地有零星记录，为罕见旅鸟。

习性与生态 常单独或形成松散的小群活动。觅食时一般使用强健的喙翻开水边的小石，寻找藏在其下的小型无脊椎动物。

Morphological characteristics A small and compact wading bird (18-25 cm). Black bill is short and wedge-shaped. Breeding male has white head with black streaks; variable black bands from forehead and around eyes across neck-sides to breast; upperparts largely orange-brown with black patches and spots; lower back and uppertail-coverts white; rump black; tail white with broad black subterminal band; underparts white; tarsi orange. Breeding female is browner on upperparts. Non-breeding adult has dark brown on head, breast and upperparts.

Distribution and habitat Widespread in China, but mainly seen in coastal areas, particularly rocky shores. In Beijing, a rare passage migrant at rural wetlands such as Yeyahu and Shahe Reservoir.

Behaviour and ecology Alone or in small, loose flocks. When feeding, often flips over rocks and pebbles to look for small invertebrates.

红腹滨鹬	*Calidris canutus*	Red Knot	鹬科 Scolopacidae

形态特征 小型涉禽，体长 23~25 厘米。雌雄相似。体型紧凑，喙和跗跖相对较短。繁殖羽喙黑色；头顶至后颈淡棕色，具黑色纵纹，眉纹橙红色；上体灰褐色，具黑色斑点，腰和尾上覆羽白色，具深色短横斑，尾羽灰褐色；颊、颈、胸、腹皆为橙红色，下腹至尾下覆羽白色；跗跖黄绿色。非繁殖羽头及上体灰色，眉纹白色，下体白色。

分布与生境 在中国主要分布于东部沿海地区，于内陆有零星记载。多见于沿海滩涂、河口、潮间带等地。在北京为罕见旅鸟。

习性与生态 迁徙时常集成非常密集的大群活动，常在沿海泥滩上快速走动觅食。

Morphological characteristics A small and compact wading bird (23-25 cm) with relatively short bill and tarsi. Sexes alike. Breeding adult has black bill; crown, nape and hindneck pale brown, with fine black streaks; supercilium orange-red; upperparts greyish-brown, spotted black; rump and uppertail-coverts white with dark bars; tail greyish-brown; reddish from cheek, chin, throat and breast to belly; vent and undertail-coverts white; tarsi greenish-yellow. Non-breeding adult has grey head and upperparts with white supercilium; underparts white.

Distribution and habitat Along coastal areas in eastern China; scarce inland. Mainly seen at intertidal mudflats and estuaries. Rare passage migrant in Beijing.

Behaviour and ecology Forms dense flocks on migration. Walks rapidly on mudflats when feeding.

IUCN 红色名录 近危 (NT) 2018.8.10

繁殖羽 Breeding

幼鸟 Juvenile

繁殖羽 Breeding

三趾滨鹬	*Calidris alba*	Sanderling	鹬科 Scolopacidae

非繁殖羽 Non-breeding

繁殖羽 Breeding

IUCN 红色名录　无危 (LC) 2016.10.1

形态特征　小型涉禽，体长 20~21 厘米。雌雄相似。喙较短，黑色。成鸟繁殖羽头顶和颈部灰褐色，具深色细纵纹，头侧棕红色，上体及翼上覆羽棕褐色，具黑色和砖红色斑点，飞羽黑褐色，具显著的白色带，尾羽灰褐色；颏、喉白色，胸棕红色，具黑色纵纹；下体白色；跗跖黑色，后趾缺失，仅有前三趾。成鸟非繁殖羽头顶和上体浅灰色。

分布与生境　中国广泛分布。非繁殖期多见于沿海。北京见于河流、湖泊的滩涂和岸边浅水处，为罕见旅鸟。

习性与生态　非繁殖期集群活动，有时也与其他滨鹬混群。

Morphological characteristics A small wading bird (20-21 cm). Sexes alike. Bill relatively short and black. Breeding adult has greyish-brown head and neck, with fine dark streaks; head-sides rufous; upperparts and wings largely brown, with black and rufous spots; flight feathers blackish, with white bases, forming white band on the wings; tail greyish-brown; chin and throat white; breast rufous, streaked black; remainder of underparts white; tarsi black; feet tridactyl, lacking hind toe. Non-breeding adult is pale grey on head and upperparts.

Distribution and habitat Widespread in China. Prefers coastal mudflats in non-breeding season. Rare passage migrant at lake shores and riverbanks in Beijing.

Behaviour and ecology Forms flocks in non-breeding season. Sometimes join mixed flocks with other sandpipers.

勺嘴鹬	*Calidris pygmeus*	Spoon-billed Sandpiper	鹬科 Scolopacidae

形态特征 小型涉禽，体长 14~16 厘米。雌雄相似。喙黑色，喙端扁平，扩展成铲状。成鸟繁殖羽头、颈棕红色，头顶具黑色纵纹；上体各羽黑色，具较宽的棕红色羽缘，两翼深褐色；尾羽深灰色，胸淡棕红色，具黑色斑点，下体余部白色；跗跖黑色。成鸟非繁殖羽头顶、头两侧至上体浅灰褐色，眉纹白色，下体全为白色。幼鸟具白色眉纹和深色耳羽，上体羽深色，具棕色羽缘。

分布与生境 中国见于东部及南部沿海滩涂、河口区域。北京为甚罕见旅鸟，仅多年前于通州有过一次记载（蔡其侃，1987）。

习性与生态 单独或集小群活动，常与其他滨鹬混群。种群数量极为稀少，目前全球成年个体数量仅为 240~456 只（IUCN, 2018）。

Morphological characteristics A small wading bird (14-16 cm). Sexes similar. Black bill short and spatula-shaped. Breeding adult has rufous head and neck, with a few fine black streaks; upperparts largely black, with distinctive rufous fringes; wings dark brown; tail dark grey; breast pale rufous, spotted black; remainder of underparts white; tarsi black. Non-breeding adult is greyish-brown with black streaks on head and upperparts; supercilium white; breast and belly white. Juvenile has brown ear-coverts and white supercilium with upperparts dark, fringed brown.

Distribution and habitat Coastal wetlands in eastern and southern China. Very rare passage migrant in Beijing, with only one historic undated record from Tongzhou District (Cai, 1987).

Behaviour and ecology Alone or in small flocks. Often joins mixed flocks with other shorebirds such as Red-necked Stint. Population has dropped dramatically in recent decades, now estimated to have a population of 240-456 mature individuals (IUCN, 2018).

国家一级保护野生动物　IUCN 红色名录 极危 (CR) 2018.8.19

繁殖羽 Breeding

非繁殖羽 Non-breeding

红颈滨鹬	*Calidris ruficollis*	Red-necked Stint	鹬科 Scolopacidae

繁殖羽 Breeding
幼鸟 Juvenile

IUCN 红色名录　近危 (NT)　2016.10.1

形态特征　小型涉禽，体长 13~17 厘米。雌雄相似。体型紧凑，喙和跗跖皆较短，呈黑色。成鸟繁殖羽头、颈砖红色，头顶具黑色纵纹，眼先黑色；上体各羽黑色，羽缘砖红色，腰和尾上覆羽中部黑色，两侧白色；胸棕红色或橙红色，具深褐色斑点，腹部和尾下覆羽白色。成鸟非繁殖羽上体灰色，具黑色羽干纹，下体白色。幼鸟似成鸟非繁殖羽，但上体偏褐色。

分布与生境　除西部外见于中国各地，但主要分布于沿海区域。北京见于水库、湖泊、河流和沿岸的泥滩和浅水区域，为区域性常见旅鸟。

习性与生态　于沿海迁徙过境时常集大群。于内陆过境时则单独活动或形成分散的小群。常在水边快速走动并将喙插入泥中觅食。

Morphological characteristics　A small and compact wading bird (13-17 cm). Sexes alike. Relatively short, black bill and tarsi. Breeding adult has rufous head and neck; crown has fine black streaks; lores black; upperparts largely black, with distinctive red fringes; rump and uppertail-coverts have black centres and white sides; breast rufous or reddish-orange, with dark spots; belly and undertail-coverts white. Non-breeding adult has grey upperparts with narrow black shafts; underparts white. Juvenile resembles non-breeding adult but with brown ear-coverts and dark upperparts with brown or buff fringes.

Distribution and habitat　Widespread in China except for western areas. More common along coastal wetlands, especially mudflats. Locally common passage migrant in Beijing, usually seen in shallow water at shores of reservoirs, lakes and rivers, or mudflats near water.

Behaviour and ecology　Forms large flocks at coastal wetlands on migration. Inland, most often seen alone or in loose flocks. Walks rapidly and makes rapid stabbing actions when feeding.

小滨鹬	*Calidris minuta*	Little Stint	鹬科 Scolopacidae

形态特征 小型涉禽，体长 12~16 厘米。雌雄相似。喙黑色。成鸟繁殖羽头顶淡栗色，具黑色细纵纹，头侧栗褐色，眉纹白色，但不甚清晰；上体各羽中央黑色，具宽阔的栗色羽缘，腰和尾上覆羽中央黑色，两侧白色，中央尾羽黑褐色，两侧灰色；颏、喉、胸及下体余部皆为白色，仅胸两侧栗褐色，具黑色斑点。非繁殖羽上体灰褐色，下体白色。幼鸟似成鸟非繁殖羽，但上体具深色斑点。幼鸟甚似红颈滨鹬幼鸟，但本种头顶羽色略深，且背部具更为明显的白色条纹。

分布与生境 分布于中国东部沿海和西部。北京见于大型水库、河流的岸边，为罕见旅鸟。

习性与生态 在中国东部大多单只活动或与红颈滨鹬混群。

Morphological characteristics A small wading bird (12-16 cm). Sexes alike. Bill black. Breeding adult has pale rusty-red crown, with fine black streaks; head-sides rufous; supercilium whitish and indistinct; upperparts largely black, with distinctive chestnut fringes; rump and uppertail-coverts mostly white with black centres; central rectrices black; tail-sides grey; chin, throat, breast and remaining underparts all white; breast-sides rufous with dark spots. Non-breeding adult is greyish-brown on upperparts and white on underparts. Juvenile has dark crown and ear-coverts; upperparts dark with brown fringes; closely resembles juvenile Red-necked Stint but has darker crown and more distinctive white stripes on mantle.

Distribution and habitat Western China and coastal wetlands in eastern China. Rare passage migrant in Beijing.

Behaviour and ecology Records in eastern China usually of singles or small groups; often joins mixed flocks of Red-necked Stint and other shorebirds.

IUCN 红色名录 无危 (LC) 2018.11.21

繁殖羽 Breeding

非繁殖羽 Non-breeding

青脚滨鹬	*Calidris temminckii*	Temminck's Stint	鹬科 Scolopacidae

繁殖羽 Breeding
非繁殖羽 Non-breeding

IUCN 红色名录 无危 (LC) 2018.11.21

形态特征 小型涉禽，体长 12~17 厘米。雌雄同色。喙黑色，喙基暗绿色。成鸟繁殖羽头至颈褐色，头顶具不清晰的深色纵纹，眼先深色；上体和翼上中、小覆羽黑褐色，各羽具棕褐色羽缘，初级飞羽深褐色，大覆羽和次级飞羽黑褐色，具白色端斑；中央尾羽深褐色，外侧尾羽白色；颏、喉近白色，胸灰褐色，其余下体白色；跗跖绿色或黄绿色。成鸟非繁殖羽头、胸、上体皆为灰色。幼鸟上体各羽具淡棕色羽缘。

分布与生境 广布于中国全境。在内陆湿地和沿海滩涂皆可见。在北京为郊区开阔的水库、湖泊、河流、沼泽等地的区域性常见旅鸟。

习性与生态 喜在水边集松散的群体活动，常与其他滨鹬混群。营巢于湿地附近的草丛或灌丛下；窝卵数 3~4 枚。

Morphological characteristics A small wading bird (12-17 cm). Sexes alike. Bill mostly black with dull dark green base. Breeding adult has brown head and neck; crown has indistinct dark streaks; lores dark; mantle, lesser coverts and median coverts blackish-brown with rufous-brown fringes; primaries blackish-brown; secondaries and greater coverts also blackish-brown with white tips; chin and throat whitish; breast greyish-brown; remaining underparts white; tarsi green or yellowish-green. Non-breeding adult has grey head, upperparts and breast. Juvenile has pale brown-fringed upperparts.

Distribution and habitat Widespread in China. Seen at both inland wetlands and coastal mudflats. In Beijing, locally common passage migrant at rivers, marshes, lakes and reservoirs in rural areas.

Behaviour and ecology Feeds on mudflats near water. Tends to form loose flocks or joins mixed flocks with other shorebirds. Nests on the ground in grass or underneath shrubs. Clutch size: 3-4.

长趾滨鹬	*Calidris subminuta*	Long-toed Stint	鹬科 Scolopacidae

形态特征 小型涉禽，体长 13~15 厘米。雌雄相似。成鸟繁殖羽头顶栗棕色，具黑褐色纵纹，眉纹白色，眼先及耳羽深褐色；上体各羽中部黑色，边缘栗棕色，腰和尾上覆羽中部黑色，两侧白色，中央尾羽黑色，外侧尾羽灰色；颏、喉、胸、腹、尾下覆羽皆为白色，但胸部具黑褐色纵纹；跗跖黄绿色。成鸟非繁殖羽上体灰褐色取代栗棕色。

分布与生境 广布于中国沿海和内陆各种湿地。北京多见于水库、河流、湖泊的浅水区域和岸边的滩涂与草地，为区域性常见旅鸟。

习性与生态 非繁殖期常集小群活动，或与其他滨鹬混群。飞行速度快而灵活。

Morphological characteristics A small wading bird (13-15 cm). Sexes alike. Breeding adult has rufous crown with blackish-brown streaks; supercilium white; lores and ear-coverts dark brown; mantle and upperwing-coverts have black centres and rufous fringes; rump and uppertail-coverts black, with white sides; central rectrices black; tail-sides white; chin, throat, breast and remaining underparts largely white; breast-sides streaked black; tarsi greenish-yellow. Non-breeding adult has greyish-brown-fringed upperparts instead of rufous.

Distribution and habitat Widespread at both inland and coastal wetlands in China. In Beijing, locally common passage migrant in shallow water, such as riverbanks, edges of marshes, lakes and reservoirs, as well as mudflats and grassland near water.

Behaviour and ecology Forms small flocks or in mixed flocks with other shorebirds in non-breeding season. Flight fast and agile.

IUCN 红色名录 无危 (LC) 2016.10.1

繁殖羽 Breeding

非繁殖羽 Non-breeding

斑胸滨鹬 | *Calidris melanotos* | Pectoral Sandpiper | 鹬科 Scolopacidae

IUCN 红色名录 无危 (LC) 2016.10.1

形态特征 小型涉禽，体长 19~24 厘米，属滨鹬中体型较大的种类，雌雄相似，但雌鸟体型略小。喙略下弯，呈黑色，喙基褐色。繁殖羽头顶褐色，密布黑色细纵纹，眉纹白色，头侧淡褐色；上体各羽黑色，具较宽的棕色羽缘，腰至尾上覆羽中部黑色，两侧白色，中央尾羽黑色，外侧尾羽褐色；颏、喉近白色，颈和胸淡黄色，密布黑色纵纹，腹部白色，与胸形成清晰的分界；跗跖黄绿色。成鸟非繁殖羽与繁殖羽相似，但上体偏灰色而非棕色。

分布与生境 中国偶见于东部和南部沿海。北京甚罕见于密云水库、昌平沙河、延庆野鸭湖等郊区开阔湿地，为迷鸟。

习性与生态 非繁殖期多集群活动，但中国并不属于其主要分布区，因此多为单只零星活动，亦与尖尾滨鹬等其他鸻鹬混群。

Morphological characteristics A small wading bird (19-24 cm). Relatively large Calidris species. Sexes similar but female is smaller than male. Bill slightly curved, black at tip and brown at base. Breeding adult has brown, dark-streaked crown; supercilium is whitish and indistinct; head-sides pale brown; upperparts black, with broad brown fringes; rump and uppertail-coverts have black centres and white sides; central rectrices black; outer tail feathers brown; chin and throat off-white; neck and breast pale buff, with extensive black streaks, forming distinctive division from white belly; tarsi greenish-yellow. Non-breeding adult resembles breeding adult but with more greyish upperparts.

Distribution and habitat Scarce along the coast in east and south China. In Beijing, very rare passage migrant at grassland or mudflat near water.

Behaviour and ecology Forms flocks in North America. But in China usually alone or in mixed flocks with the similar Sharp-tailed Sandpiper and other shorebirds.

尖尾滨鹬	*Calidris acuminata*	Sharp-tailed Sandpiper	鹬科 Scolopacidae

形态特征 小型涉禽，体长 16~23 厘米。雌雄相似，但雌鸟体型较小。成鸟繁殖羽头顶栗褐色，具较细的黑色纵纹，眉纹白色，头两侧淡褐色；上体黑色，具棕色羽缘；两翼暗褐色，大覆羽端部白色，形成一道翼斑，飞行时可见；腰至尾上覆羽中部黑色，两侧白色；颏、喉白色，颈、胸棕黄色，其余白色，自颏、喉向下至上腹和两胁皆密布黑色斑点；跗跖黄绿色。非繁殖羽头顶和上体栗棕色较淡，耳羽黑褐色，下体黑色点斑亦显著少于繁殖羽个体。

分布与生境 在中国主要分布于东部和南部，偶见于西部。在沿海区域更为易见。北京见于河流、湖泊等湿地的浅滩，为不常见旅鸟。

习性与生态 迁徙过境时常形成大群或与其他滨鹬混群。

Morphological characteristics A small wading bird (16-23 cm). Sexes alike, female smaller. Breeding adult has rufous crown with fine black streaks and distinctive whitish supercilium; cheek and ear-coverts pale brown; upperparts black, fringed brown; wings mostly dark brown; greater coverts tipped white; rump and uppertail-coverts have black centre and white sides; chin and throat white; neck and breast pale brown; remainder of underparts white; extensive black spots and streaks from breast to belly; tarsi greenish-yellow. Non-breeding adult has paler cap and upperparts, dark ear-coverts and underparts with fewer spots and streaks.

Distribution and habitat Mainly eastern and southern China, scarce in the west. More common at coastal wetlands such as intertidal mudflats. Uncommon passage migrant at lake shores and riverbanks in Beijing.

Behaviour and ecology In large flocks or joins mixed flocks with other shorebirds on migration.

IUCN 红色名录 无危 (LC) 2016.10.1

阔嘴鹬	*Calidris falcinellus*	Broad-billed Sandpiper	鹬科 Scolopacidae

	国家二级保护野生动物	IUCN 红色名录 无危 (LC) 2016.10.1

形态特征 小型涉禽，体长 16~18 厘米。雌雄相似。体型紧凑，但喙较长、直且粗壮，繁殖羽头顶深褐色，白色眉纹自眼上方起被一道深色线分叉为上下两道，耳羽栗褐色，颊灰白色；上体各羽黑色，具白色和栗褐色羽缘，翼上大覆羽具白色端斑，形成一道白色翼斑，仅飞行时可见；颏、喉白色或灰白色，颈、胸灰褐色，腹部及尾下覆羽白色；跗跖暗绿色。非繁殖羽头部及上体偏灰褐色。

分布与生境 分布于中国东部沿海及西部地区。活动于沿海沙滩、泥滩等地。在北京为不常见旅鸟，主要见于湖泊、河流附近的草地和泥滩等。

习性与生态 常形成分散的小群活动，或与其他滨鹬（特别是红颈滨鹬和黑腹滨鹬）混群。

Morphological characteristics A small and compact wading bird (16-18 cm). Sexes alike. Bill long, straight and thick. Breeding adult has dark brown cap; a dark line splits the white supercilium into upper and lower parts; ear-coverts rufous; cheek pale grey; upperparts black with white and rufous-brown fringes; greater coverts have white tips, forming a narrow white band on the wing; chin and throat off-white; neck and breast greyish-brown; remaining underparts white; tarsi deep green. Non-breeding adult has paler, greyish-brown upperparts.

Distribution and habitat Western and eastern China. Prefers coastal beaches and mudflats. In Beijing, an uncommon passage migrant at lake shores, mudflats and grassland near water.

Behaviour and ecology In small and loose flocks, often feeds and roosts with other stints and sandpipers, particularly Dunlin and Red-necked Stint.

流苏鹬	*Calidris pugnax*	Ruff	鹬科 Scolopacidae

形态特征 小型涉禽，体长 26~32 厘米。体型较其他滨鹬粗壮，雄鸟较雌鸟大。喙下弯，褐色、橙色或黄色。雄鸟繁殖羽眼先和前额具裸皮，头部和颈部具蓬松而上翘的饰羽，饰羽羽色可能为黑色、白色、栗色、棕色等；上体黑色，具橙色鳞状斑，尾上覆羽较长，中央黑色，两侧白色，尾羽灰色；胸黑色，下体白色；跗跖较长，橙色。雌鸟繁殖羽无饰羽，头及上体褐色，具黑色短纵纹和横斑。成鸟非繁殖羽上体灰褐色，下体近白色。

分布与生境 在中国主要分布于东部及西部地区。见于沿海、内陆各类湿地和草原。在北京为不常见旅鸟。

习性与生态 于中国多为单只或集小群活动。繁殖期雄鸟会展开饰羽进行复杂的求偶表演和争斗。雌鸟营巢于草地地面；窝卵数 4 枚。

Morphological characteristics A small wading bird (26-32 cm). Larger than most Calidris stints and sandpipers. Male is larger than female. Bill relatively short, slightly curved, brown, orange or yellow. Breeding male has variable bare skin around lores and forehead; variable, prominent erectile ear-tufts and neck-ruff, which may be black, white, rufous or brown; upperparts largely black, with orange scales; uppertail-coverts relatively long, with black centres and white sides; tail grey; breast black; remaining underparts mostly white; tarsi relatively long, orange. Breeding female lacks ear-tufts and neck-ruff; head and upperparts brown, with dark streaks and bars. Non-breeding adult is greyish-brown above and whitish below.

Distribution and habitat Eastern and western China. Seen in inland and coastal wetlands as well as grassland. Uncommon passage migrant in Beijing.

Behaviour and ecology Alone or in small flocks. Male makes complicated courtship displays using its ruff. Nests on the ground in grass. Clutch size: 4.

IUCN 红色名录 无危 (LC) 2016.10.1

雌鸟 (左) 和幼鸟 (右) Female (left) and juvenile (right)

雌鸟 Female

雄鸟 Male

雄鸟 Male

雄鸟 Male

雄鸟 Male

弯嘴滨鹬	*Calidris ferruginea*	Curlew Sandpiper	鹬科 Scolopacidae

形态特征 小型涉禽，体长 19~23 厘米。雌雄相似。喙黑色，较长且明显下弯。繁殖羽头、颈、胸、腹皆为栗褐色，甚为鲜艳；上体黑褐色，具栗色斑点，两翼深褐色，初级飞羽基部白色，大覆羽端部白色，腰和尾上覆羽白色，尾羽褐色，臀及尾下覆羽白色；跗跖黑色。非繁殖羽头顶、胸及上体淡灰褐色，具白色眉纹，下体白色。

分布与生境 中国除西南地区外广泛分布。多见于沿海滩涂和内陆近水滩涂、沼泽、水田区域。在北京为不常见旅鸟。

习性与生态 非繁殖期喜集松散的群体或与其他滨鹬混群觅食、休息，飞行快速，飞行时常发出轻柔的哨音。

Morphological characteristics A small wading bird (19-23 cm). Sexes alike. The bill is relatively long, curved and black. Breeding adult has prominent rusty head, neck, breast and belly; upperparts dark brown with rusty-orange and white spots; primaries blackish-brown, white at the base; greater coverts tipped white; rump and uppertail-coverts white; tail brown; vent and undertail-coverts white, contrasting strongly with the rusty belly; tarsi black. Non-breeding adult has pale greyish-brown head and upperparts; supercilium white; underparts also white.

Distribution and habitat Widespread in China except for the southwest. Usually seen at coastal mudflats as well as inland lake shores, riverbanks, paddy fields and marshes. Uncommon passage migrant in Beijing.

Behaviour and ecology Often in loose flocks or in mixed flocks with other shorebirds when feeding and roosting on migration. Flight fast with soft flight calls.

IUCN 红色名录　近危 (NT)　2016.10.1

繁殖羽 Breeding

非繁殖羽 Non-breeding

黑腹滨鹬	*Calidris alpina*	Dunlin	鹬科 Scolopacidae

繁殖羽 Breeding

非繁殖羽 Non-breeding

IUCN 红色名录　无危 (LC) 2016.10.1

形态特征　小型涉禽，体长 16~22 厘米。雌雄同色。喙较长，端部略下弯，呈黑色。繁殖羽头顶栗棕色，具黑色细纵纹，头侧和颈部灰白色；背部羽毛中心黑色，羽缘栗棕色，两翼大致为深褐色，内侧初级飞羽基部和大覆羽端部白色；腰及尾上覆羽黑褐色，中央尾羽黑褐色，外侧尾羽白色；颏、喉白色，胸具深色细纵纹，下体余部白色，但腹部中央具一块显著的黑色斑；跗跖黑色。非繁殖羽上体灰褐色，胸淡褐色，腹部白色。

分布与生境　中国见于东部地区及新疆。活动于沿海滩涂和内陆沼泽、河滩等地。在北京为不常见旅鸟，一般见于郊区较大的水库、湖泊和河流。

习性与生态　非繁殖期常集大群活动，在北京则多为单只记载或与其他滨鹬混群。常在水边快速奔跑，频繁将喙插入泥沙中取食小型无脊椎动物。

Morphological characteristics A small wading bird (16-22 cm). Sexes alike. Long, slightly curved black bill. Breeding adult has rufous-brown crown with fine black streaks; head-sides and neck pale greyish-white; black-centred mantle has rufous fringes; wings largely dark brown; inner primaries have white bases; greater coverts have white tips; rump and uppertail-coverts blackish-brown; central rectrices black; tail-sides white; chin and throat white; breast has fine dark streaks; distinctive black patch on belly; remainder of underparts white; tarsi black. Non-breeding adult is greyish above and white below with a pale grey breast.

Distribution and habitat Widespread in eastern and western China (mainly Xinjiang Uygur Autonomous Region). Found along coastal wetlands and inland beaches, shores, paddy fields and marshes. Uncommon passage migrant at large reservoirs, lakes and rivers in Beijing.

Behaviour and ecology Forms large flocks on migration along coastal wetlands. Records in Beijing usually of singles or small flocks. Runs fast on sandy breaches or mudflats, frequently stopping to catch small invertebrates.

红颈瓣蹼鹬 | *Phalaropus lobatus* | Red-necked Phalarope | 鹬科 Scolopacidae

形态特征 小型涉禽，体长 18~21 厘米。喙较细，黑色。雌鸟繁殖羽头深灰色，眼上方有一小块白色眼斑，颈红色，甚为鲜艳；上体和两翼大致为深灰色，背部具浅黄色纵纹，大覆羽具白色端斑；颏、喉白色，胸及两胁灰色，腹部至尾下覆羽白色；跗跖较短，灰色，趾具明显的瓣蹼。雄鸟繁殖羽似雌鸟，但羽色较淡。非繁殖羽头、颈至下体大致为白色，头顶黑色，眼后至耳羽有一黑色斑块，上体灰色。

分布与生境 分布于中国东部和西部，于东部地区多见于沿海的浅海及近海池塘中，于内陆则见于淡水池塘、湖泊、河流和沼泽。在北京为罕见旅鸟。

习性与生态 喜在浅水处游水，在水面不断转圈捕食小型无脊椎动物。营巢于草丛或苇丛中；窝卵数 4 枚；孵卵、育雏主要由雄鸟完成。

Morphological characteristics A small wading bird (18-21 cm). Bill fine, black. Breeding female has dark grey head, with a small white patch above eye; neck is red; upperparts and wings largely dark grey; mantle striped pale yellow; greater coverts tipped white; chin and throat white; breast and flanks grey; belly and undertail-coverts white; tarsi short, grey; toes lobed. Breeding male resembles breeding female but with duller plumage. Non-breeding adult has largely white head, neck and underparts; crown black; black patch from eyes to ear-coverts; upperparts greyish.

Distribution and habitat Widespread in eastern and western China. In eastern areas, mainly seen along coastal wetlands. In land prefers freshwater ponds, lakes, rivers and swamps. Rare passage migrant in Beijing.

Behaviour and ecology When feeding, often swims in a tight circle in shallow water. May form large flocks offshore in non-breeding season. Clutch size: 4. Male incubates eggs and raises the chicks.

IUCN 红色名录 无危 (LC) 2018.8.8

非繁殖羽 Non-breeding

繁殖羽 Breeding

灰瓣蹼鹬	*Phalaropus fulicarius*	Red Phalarope	鹬科 Scolopacidae

繁殖羽 Breeding

非繁殖羽 Non-breeding

IUCN 红色名录 无危 (LC) 2018.8.8

形态特征 小型涉禽，体长 20~24 厘米。喙黄色，端部黑色。雌鸟繁殖羽头顶及眼先黑色，头两侧白色；上体各羽黑色，羽缘棕黄色，中央尾羽黑色，其余尾羽灰色；颏、喉、颈、胸及其余下体皆为栗红色。雄鸟繁殖羽似雌鸟，但体型较小，且腹部杂有白色区域。非繁殖羽成鸟头顶后部、眼后及上体浅灰色，其余皆为白色，似红颈瓣蹼鹬非繁殖羽但本种喙较短粗。

分布与生境 在中国主要分布于东部沿海，于内陆湿地偶有过境个体记载。北京于昌平沙河、密云水库有零星记录，为甚罕见旅鸟。

习性与生态 在其主要迁徙路线上（美洲）多集群活动，但北京及中国其他地区的记录多为单只，习性似红颈瓣蹼鹬，在浅水处游泳打转觅食。

Morphological characteristics A small wading bird (20-24 cm). Bill short, black, with pale yellow base. Breeding female has black crown and lores; head-sides white; upperparts black with brown fringes; central rectrices black; outer rectrices grey; chestnut from chin, throat, neck and breast to remaining underparts. Breeding male similar to female but smaller with white belly. Non-breeding adult has pale grey hindcrown and upperparts and a blackish-grey patch from eyes to ear-coverts. Non-breeding adult closely resembles Red-necked Phalarope but bill is shorter and thicker at tip.

Distribution and habitat Along coastal wetlands in eastern China. Scarce inland. Very rare passage migrant in Beijing; recorded at Shahe and Miyun Reservoir.

Behaviour and ecology Forms flocks on migration in the Americas. Records in Beijing are of singles. Behaviour similar to Red-necked Phalarope: swims in a tight circle when feeding.

黄脚三趾鹑	*Turnix tanki*	Yellow-legged Buttonquail	三趾鹑科 Turnicidae

IUCN 红色名录 无危 (LC) 2016.10.1

形态特征 小型涉禽，体长 12~18 厘米。体型紧凑似鹌鹑，仅有三趾。喙黄色，虹膜灰白色。雌鸟头顶至后颈深褐色，具淡黄色斑纹，上体灰褐色，具黑色和橙色斑点，尾羽甚短；颏、喉白色或黄色，颈侧橙红色，前颈、胸、两胁淡黄色，胸两侧具黑色斑点，腹部为白色或皮黄色；跗跖黄色。雄鸟较雌鸟略小，羽色与雌鸟相似，但较暗淡。

分布与生境 分布于中国东部和南部地区。在北京地区活动于平原或低山的灌丛、草地、农田等地，有时见于城市公园或高校，为不常见旅鸟和夏候鸟。

习性与生态 受惊时在地面快速奔跑或就地隐蔽，较少起飞。雌鸟与数只雄鸟先后交配并产下数窝卵，由雄鸟单独孵卵、育雏，窝卵数 3~4 枚。

Morphological characteristics A small wading bird (12-18 cm). Compact body shape similar to a quail but has only three toes. Yellowish bill and greyish-white eyes. Female is dark brown from crown to hindneck, with pale yellow bars; upperparts largely greyish-brown, with black and orange spots; tail very short; chin and throat whitish or buff; neck-sides orange-red; foreneck, breast and flanks pale yellow; breast-sides spotted black; belly white or buff; tarsi yellow. Male is similar to female but smaller and duller.

Distribution and habitat Eastern and southern China. Uncommon passage migrant and summer breeder in Beijing. Seen in scrub, grassland and farmland in the lowlands and low hills. On migration, sometimes also seen in urban parks and university campuses.

Behaviour and ecology Freezes or runs away rapidly rather than flying when disturbed. Female mates with several males and lays several clutches of eggs. Male incubates the eggs and raises the young. Clutch size: 3-4.

普通燕鸻	*Glareola maldivarum*	Oriental Pratincole	燕鸻科 Glareolidae

形态特征 小型涉禽，体长 20~28 厘米。雌雄同色。成鸟喙黑色，喙基红色；头、颈及上体大致为灰褐色，尾上覆羽白色，尾羽黑色，呈叉形；颏、喉黄色，具显著的白色（内侧）和黑色（外侧）边缘；胸淡棕色，翼下覆羽红色，腹部尾下覆羽白色；跗跖黑色。幼鸟喙全为黑色，上体褐色，下体白色。

分布与生境 中国除西部外广泛分布。在北京活动于开阔草地、农田及水库、湖泊、河流等湿地，为不常见旅鸟和夏候鸟。

习性与生态 喜集群活动，飞行迅速而灵活，常边飞边发出尖锐的鸣叫声。在湿地附近的土地上扒一浅坑营巢；窝卵数 2~4 枚。

Morphological characteristics A small wading bird (20-28 cm). Sexes alike. Adult has black bill with red base; head, neck and upperparts largely greyish-brown; uppertail-coverts white; forked tail black; chin and throat yellow with distinctive white (outer) and black (inner) borders; breast pale brown; underwing-coverts red; belly and undertail-coverts white; tarsi black. Juvenile has all-black bill, brown upperparts and white underparts.

Distribution and habitat Widespread in China except for western China. In Beijing: an uncommon passage migrant and summer breeder at grassland, farmland, lakes, rivers and reservoirs.

Behaviour and ecology In small flocks. Flight fast and agile with harsh flight calls. Nests on the ground near water. Clutch size: 2-4.

IUCN 红色名录 无危 (LC) 2016.10.1

成鸟 Adult

幼鸟 Juvenile

十一、鹈形目（二）PELECANIFORMES

白琵鹭	*Platalea leucorodia*	Eurasian Spoonbill	鹮科 Threskiornithidae

非繁殖羽 Non-breeding

繁殖羽 Breeding

国家二级保护野生动物　IUCN 红色名录　无危 (LC) 2016.10.1

形态特征　大型涉禽，体长 80~95 厘米。雌雄相似。黑色的喙长直而平扁，中段狭窄，先端扩展成匙状，形似琵琶，先端黄色，上喙具皱纹。全身羽毛白色。繁殖羽枕部具发丝状冠羽，颈下部具浅黄色环带。非繁殖羽无冠羽和环带。跗跖黑色。

分布与生境　在中国繁殖于北方，越冬于东南沿海。北京见于郊区湿地的开阔水面，为常见旅鸟。

习性与生态　多集群，觅食时在浅水处左右摆动琵琶形长喙搜索实物。主要以鱼、虾、蛙、昆虫等小动物为食。5~7 月营巢于干旱苇丛或树上，有时在地面；窝卵数 3~4 枚。

Morphological characteristics　A large wading bird (80-95 cm). Sexes alike. Spoon-shaped bill black, long, straight and flat, with yellow tip and wrinkled upper mandible. All-white plumage. Breeding plumage has crest on nape and an ochre chest band, absent in non-breeding plumage. Tarsi black.

Distribution and habitat　Breeds in northern China, winters on the southeast coast. In Beijing, fairly common in spring and autumn in areas of open water. Residential status: migrant.

Behaviour and ecology　Often found in flocks, feeds in shallow water by sweeping long spatulate bill side to side. Feeds on fish, shrimps, frogs, insects and other small creatures. Nests on dry reeds or trees, occasionally on the ground, from May to July. Clutch size: 3-4.

黑脸琵鹭	*Platalea minor*	Black-faced Spoonbill	鹮科 Threskiornithidae

形态特征 中型涉禽，体长 60~78 厘米。雌雄相似。似白琵鹭而体型较小，脸部黑色区域达眼后，黑色喙长直而平扁，中段狭窄，先端扩展成匙状，形似琵琶，上喙具皱纹。全身大致白色，繁殖期头后枕部有发丝状淡黄色冠羽，颈下部具浅黄色环带。跗跖黑色。

分布与生境 中国见于沿海地区。栖息于海滨、河湖的浅水区。北京见于远郊区大型开阔水域（如密云水库，2017 年 6 月），为甚罕见旅鸟。

习性与生态 单独或成小群。觅食活动主要在白天，多在水边浅水处。以琵琶形长喙插进水中，半张着喙在浅水一边涉水前进一边左右晃动头部扫荡，通过触觉捕捉水底层的鱼、虾、蟹、软体动物等。

Morphological characteristics A medium-sized wading bird (60-78 cm). Sexes alike. Similar to Eurasian Spoonbill but on average slightly smaller and black patch on face extends to the back of the eye. Black, long spoon-shaped bill straight and flat, with wrinkled upper mandible. White overall, with yellow crest on nape and an ochre chest band in breeding plumage. Tarsi black.

Distribution and habitat Coastal China. Found in shallow water of seashore, rivers and lakes. In Beijing, very rare passage migrant in in areas of open water (e.g. Miyun Reservoir in June 2017).

Behaviour and ecology Feeds alone or in small groups. Forages for food in shallow water by sweeping long, spatulate bill from side to side. Feeds on fish, shrimps, crabs, and molluscs, etc.

国家一级保护野生动物　IUCN 红色名录 濒危 (EN) 2017.10.1

繁殖羽 Breeding

非繁殖羽 Non-breeding

大麻鳽	*Botaurus stellaris*	Eurasian Bittern	鹭科 Ardeidae

形态特征 中型涉禽，体长 64~80 厘米。雌雄相似。喙黄褐色，喙峰暗褐色。眼先黄绿色，眉纹淡黄白色。额、头顶和枕部黑色。身体较粗胖，颈和跗跖较粗短。背黄褐色，具较粗的黑褐色斑点，下体淡黄褐色，具黑褐色较粗纵纹。跗跖黄绿色。

分布与生境 中国除青藏高原外广泛分布。北京多见于山地丘陵和山脚下平原地带的河湖苇丛、菖蒲中，为不常见的旅鸟、夏候鸟和冬候鸟。

习性与生态 多单独活动，喜苇丛生境，常立于有干枯芦苇的水边，头、颈向上伸直凝神不动，秋冬季隐蔽性极强，身体颜色与干枯芦苇浑然一体，不易发觉。以鱼、虾、蛙和昆虫等动物为食。单独营巢在苇丛、草丛之中；窝卵数 4~6 枚，卵橄榄褐色。

Morphological characteristics A medium-sized wading bird (64-80 cm). Sexes alike. Yellow-brown bill with dark brown culmen; yellow green lores, buff supercilium; black forehead, crown and nape; body is stocky with thick neck and short tarsi; brown back mottled with dark brown markings and buff brown underparts with bold black-brown streak; tarsi short, yellowish-green.

Distribution and habitat Widespread in China, excluding the Tibetan Plateau. In Beijing, mostly found in reed beds around lakes and along rivers. Residential status: uncommon migrant, summer breeder and winter visitor.

Behaviour and ecology Usually seen alone at the base of reeds. When startled, freezes with head and neck pointing up, resembling reeds, demonstrating excellent camouflage. Feeds on fish, shrimps, frogs and insects. Nests in reeds and grass. Clutch size: 4-6, olive-brown eggs.

北京市二级保护野生动物 | IUCN 红色名录 无危 (LC) 2016.10.1

黄斑苇鳽	*Ixobrychus sinensis*	Yellow Bittern	鹭科 Ardeidae

成鸟 Adult

成鸟 Adult

幼鸟 Juvenile

北京市二级保护野生动物　　IUCN 红色名录　无危 (LC) 2016.10.1

形态特征　小型涉禽，体长 30~37 厘米。雌雄相似。喙黄绿色，喙峰暗褐色；眼橙黄色。头顶和枕部蓝黑色，头侧、颈侧黄白色，微微沾粉。颈后部和背部黄褐色，腹部和翅覆羽土黄色，飞羽和尾羽黑色。跗跖黄绿色。

分布与生境　中国除西藏、青海、新疆外，见于各省。北京见于平原和低山丘陵地带的湿地苇丛、菖蒲中，为常见夏候鸟和旅鸟。

习性与生态　常单独活动，性安静，以取食鱼类为主，常弯折芦苇茎、叶在离水面不高的苇杆上筑巢；窝卵数 5~7 枚，卵白色。

Morphological characteristics　A small wading bird (30-37 cm). Sexes alike. Yellowish-green bill with dark brown culmen; orange-yellow eyes; bluish-black crown and nape; buff on the sides of head and neck washed with pink; yellow-brown mantle and back; ochre belly and wing-coverts with black primaries and tail; tarsi yellowish-green.

Distribution and habitat　Widespread in China except for the west. Common migrant and summer breeder in Beijing, mostly found in reed beds around lakes and along rivers.

Behaviour and ecology　Usually solitary and unobtrusive. Mainly feeds on fish. Nests in broken reeds above the water level. Clutch size: 5-7, white eggs.

紫背苇鳽	*Ixobrychus eurhythmus*	Von Schrenck's Bittern	鹭科 Ardeidae

形态特征 小型涉禽，体长 32~40 厘米。喙基黄色，喙峰黑褐色。虹膜黄色，瞳孔狭长近似“一”字。雄鸟头顶暗栗褐色，其余上体紫栗色。腹部淡土黄色，从喉至胸有一栗褐色纵纹。飞羽黑色，翼上覆羽灰黄色。雌鸟从头顶至背部紫栗色，背部有小的白色斑点。跗跖和趾黄绿色。

分布与生境 分布于中国东北至华南各地。北京见于开阔平原草地、岸边植物丰富的水域等处，为不常见夏候鸟及旅鸟。

习性与生态 性隐匿孤僻，常晨昏活动。主要以小鱼、虾、蛙、昆虫等小动物为食。营巢于植物茂密的草地上，甚为简陋；窝卵数 3~5 枚。

Morphological characteristics A small wading bird (32-40 cm). It has yellow bill base; blackish-brown culmen; yellow eyes with narrow long pupil. Male has maroon upperparts and a dark chestnut crown; buff underparts with a chestnut streak from throat to breast; black primaries and greyish-yellow upperwing-coverts. Female is maroon from crown to back with small white speckles on back. Both sexes have yellowish-green tarsi and toes.

Distribution and habitat Eastern and southern China. In Beijing, uncommon migrant and summer breeder in wet open grassland and well-vegetated wetlands.

Behaviour and ecology Solitary and secretive, active at dusk and dawn. Feeds on small fish, shrimps, frogs, insects and other small creatures. Nests in lush grass or reed bed. Clutch size: 3-5.

北京市二级保护野生动物

IUCN 红色名录 无危 (LC) 2016.10.1

雄鸟 Male

雌鸟 Female

栗苇鳽	*Ixobrychus cinnamomeus*	Cinnamon Bittern	鹭科 Ardeidae

雄鸟 Male

雄鸟 Male

北京市二级保护野生动物

IUCN 红色名录 无危 (LC) 2016.10.1

形态特征 小型涉禽，体长 32~37 厘米。喙黄褐色，喙峰黑褐色。虹膜黄色，瞳孔狭长近似“一”字，眼先黄绿色。雄鸟上体从头顶至尾部及飞羽、覆羽均为栗红色；下体淡红褐色，喉至胸部具一褐色纵纹；胸侧部缀有黑白两色斑点。雌鸟头顶暗栗红色，背部暗红褐色，杂有白色斑点，腹部土黄色，从颈至胸部有数条黑褐色纵纹。跗跖黄绿色。

分布与生境 在中国主要分布于除西北以外地区。北京见于平原和低山丘陵地带的湿地苇丛、菖蒲中，为不常见夏候鸟及旅鸟。

习性与生态 性隐匿孤僻，常单独活动。晨昏和夜间活动，捕食鱼、虾、昆虫等小动物。营巢于水边苇丛或灌丛中；窝卵数 3~6 枚，卵白色。

Morphological characteristics A small wading bird (32-37 cm). It has yellow-brown bill with dark brown culmen; yellow eyes with narrow long pupil; yellowish-green lores. Male has cinnamon rufous upperparts from crown to tail and wings; light rufous underparts with a brown streak from throat to breast, and black and white speckles on the sides of breast. Female has dark chestnut crown, dark rufous back with white speckles, buff underparts with blackish-brown streaks from throat to breast. Both sexes have yellow-green tarsi.

Distribution and habitat Most areas in China except for the northwest. In Beijing, uncommon summer breeder and migrant in wetlands with reeds or calamus.

Behaviour and ecology Solitary and secretive, most active at dusk and night. Feeds on fish, shrimps, insects and other small creatures. Nests in reeds or bushes near water. Clutch size: 3-6 white eggs.

黑苇鳽 | *Ixobrychus flavicollis* | Black Bittern | 鹭科 Ardeidae

形态特征　中型涉禽，体长 49~59 厘米。雄鸟周身大部分为蓝黑色，虹膜红色，头、后颈缀有蓝色，颏白色，颈侧和喉部有黄色和黑色杂斑，前胸具黑白相间条纹。雌鸟上体羽色暗褐色，头两侧、眼下栗色，颏、喉和前颈白色，并具棕色或黑褐色羽端斑。腹部淡褐色而具黄白色羽缘。

分布与生境　在中国主要分布于长江以南地区。北京为罕见迷鸟。见于河湖、溪流湿地生境或水边灌木林中，有时也进到村庄附近的竹林和小树林。

习性与生态　常单个或成对在开阔的多挺水植物的水域活动，主夜行性。窝卵数 4 ~6 枚。

Morphological characteristics A medium-sized wading bird (49-59 cm). Male is mostly bluish-black; head and nape tinged with blue; white chin, yellow and black speckles on neck and throat; black and white streaks on breast; iris red. Female has dark brown upperparts, chestnut on the sides of head and below eyes; white on chin, throat and front of neck with brown or dark feather tips, and light brown underparts with buff feather fringes

Distribution and habitat South of Yangtze River in China. Rare vagrant in Beijing. Seen along rivers, streams or in shrubs near water.

Behaviour and ecology Feeds individually or in pairs in open water with emergent plants, sometimes in bamboo forest and groves, often near villages. Mostly nocturnal. Clutch size: 4-6.

IUCN 红色名录　无危 (LC) 2016.10.1

雄鸟 Male

雌鸟 Female

栗头鳽	*Gorsachius goisagi*	Japanese Night Heron	鹭科 Ardeidae

成鸟 Adult

成鸟 Adult

成鸟 Adult

国家二级保护野生动物

IUCN 红色名录　易危 (VU) 2020.8.18

形态特征　中型涉禽，体长 43~49 厘米。雌雄相似。颈、背暗栗色，头顶黑褐色。翼上具黑白色肩斑。下体皮黄，有由深褐色纵纹形成的中线。飞行时灰黑色的飞羽与褐色覆羽成对比。

分布与生境　中国较罕见，仅东部和南部沿海地区有零星记录。北京曾见于天坛公园，为甚罕见迷鸟。

习性与生态　甚隐匿，性胆小而机警，常单独活动，很少飞行。主夜行性，有时白天也在林下阴暗处觅食。

Morphological characteristics　A medium-sized wading bird (43-49 cm). Sexes alike. Russet neck and back with a dark brown crown; black and white marking on scapulars; buff underparts with a middle line of dark-brown fine streaks; greyish primaries contrasting with brown coverts in flight.

Distribution and habitat　Scarce in east and south coastal areas in China. In Beijing, very rare vagrant with only one record at the Temple of Heaven.

Behaviour and ecology　Usually solitary, elusive, timid and vigilant. Usually nocturnal but occasionally forages in deep shadows in woodland during daylight hours.

夜鹭	*Nycticorax nycticorax*	Black-crowned Night Heron	鹭科 Ardeidae

形态特征 中型涉禽，体长 45~65 厘米。雌雄相似。喙黑色，虹膜深红色，颈短，头顶及背部蓝灰色，眼先黄绿色。颊、颈侧、胸和两胁淡灰色，其余下体白色。跗跖黄色。繁殖期枕部向后伸 2~3 枚辫状白色饰羽，下垂至背上。幼鸟上体暗褐色，具浅色斑点，下体白，满布浅色细纵纹。

分布与生境 广泛分布于中国东部和南部。北京常见于各种湿地。主要为夏候鸟和旅鸟。

习性与生态 常集小群，傍晚开始活跃。成群筑巢于高大乔木上；窝卵数 3~5 枚，卵蓝绿色。

Morphological characteristics A medium-sized wading bird (45-65 cm). Sexes alike. Bill black, iris red; short neck with bluish-grey crown and back, yellowish green lore. Pale grey cheek, neck sides, breast and flanks; white on the rest of underparts, yellow tarsi. Braid-like white nape plumes in breeding season. Juvenile has dark brown upperparts with pale speckles and finely sreaked white underparts.

Distribution and habitat Widely spread in eastern and southern China. In Beijing, common at all sorts of wetlands. Residential status: summer breeder and migrant.

Behaviour and ecology Often encountered in small groups; become most active at dusk; nests colonially on tall trees. Clutch size: 3-5. Eggs bluish-green.

北京市二级保护野生动物		IUCN 红色名录 无危 (LC) 2016.10.1

成鸟 Adult

亚成鸟 Immature

绿鹭	*Butorides striata*	Striated Heron	鹭科 Ardeidae

成鸟 Adult

亚成鸟 Immature

北京市二级保护野生动物

IUCN 红色名录　无危 (LC) 2016.10.1

形态特征　中型涉禽，体长 40~47 厘米。雌雄相似。喙黑色，下喙基和边缘黄绿色。虹膜黄色。颈短，体较粗胖。头顶和长冠羽墨绿色，颈和上体灰绿色。颏、喉白色，胸部、胁部灰色，其余下体污白色。跗跖黄绿色。

分布与生境　分布于中国东北东南部、华东、华南等地。北京不常见于山区溪流和平原湿地，为旅鸟和夏候鸟。

习性与生态　性独居，黄昏以后更为活跃。常守候在溪流边的石块或树枝上，捕食鱼、虾、蛙、蟹及其他水生动物。营巢于水附近的树杈上；窝卵数 5 枚，卵绿青色。

Morphological characteristics　A medium-sized wading bird (40-47 cm). Sexes alike. Black bill with yellow-green on the edge of lower mandible; yellow eyes. Appears short-necked and stocky. Crown and long crest blackish-green; neck and upperparts grayish-green; chin and throat white, breast and flanks gray, with the remaining underparts white. Yellowish-green tarsi.

Distribution and habitat　Eastern and southern China. In Beijing, uncommon near mountain streams and plain wetlands. Residential status: migrant and summer breeder.

Behaviour and ecology　Usually solitary and most active after dusk. Often waits in the middle of a shallow stream where it hunts for fish, shrimps, frogs, crabs and other aquatic creatures. Nests on branches near water. Clutch size: 5. Eggs pale blue.

池鹭	*Ardeola bacchus*	Chinese Pond Heron	鹭科 Ardeidae

形态特征 中型涉禽，体长 37~54 厘米。雌雄相似。繁殖期喙黄色，尖端黑色，基部蓝色；虹膜黄色，眼先和脸黄绿色；头、后颈、颈侧和胸均为棕红色；具显著冠羽，长至背部，肩部蓝黑色，具蓑羽并向后伸达尾羽末端；翼、尾、颏、喉部、前颈和腹部白色，飞翔时白翅白尾与黑色背非常明显。非繁殖期头、颈到胸部白色，具暗黄褐色纵纹，背暗褐色。

分布与生境 广泛分布于中国华北至长江以南地区。北京甚常见于各种湿地，为夏候鸟。

习性与生态 性大胆，不甚惧人。以鱼、蛙、小型爬行动物、昆虫等为食。集群营巢于高大乔木之上；窝卵数 2~5 枚，卵蓝绿色。

Morphological characteristics A medium-sized wading bird (37-54 cm). Sexes alike. In breeding season, head, nape, neck and breast are russet; long crown plumes reach to back and mantle; long cape feathers extend to tip of tail; wings, tail, chin, throat and belly are white; yellow eyes, yellow bill with black tip and blue base, yellowish-green lores and face. In flight, white wings, white tail and dark back conspicuous. In non-breeding plumage head, neck and breast white with dull yellowish-brown streaks and dark brown back.

Distribution and habitat Widespread from northern China to south of Yangtze River. In Beijing, abundant at wetland habitats. Residential status: summer breeder.

Behaviour and ecology Bold, careless of people. Feeds on fish, frog, and small reptiles. Nests colonially on tall trees. Clutch size: 2-5. Eggs bluish-green.

北京市二级保护野生动物

IUCN 红色名录 无危 (LC) 2016.10.1

繁殖羽 Breeding

繁殖羽 Breeding

亚成鸟 Immature

牛背鹭	*Bubulcus ibis*	Cattle Egret	鹭科 Ardeidae

繁殖羽 Breeding

繁殖羽 Breeding

非繁殖羽 Non-breeding

北京市二级保护野生动物 | IUCN 红色名录　无危 (LC) 2016.10.1

形态特征　中型涉禽，体长 46~55 厘米。雌雄相似。非繁殖羽甚似白鹭而头颈部短粗，喙黄色。繁殖羽头、颈、上胸及背部为橘黄色，余部纯白。喙亦变为橙红色，甚为鲜艳。跗跖暗黄至近黑。

分布与生境　中国除西北部均有分布。北京区域性常见于平原草地、农田之中，为旅鸟和少量夏候鸟。

习性与生态　成对或小群活动，常伴随家畜，喜捕食被家畜吸引或惊飞的昆虫，有时落于牛背之上。主要捕食昆虫等小型无脊椎动物，而非鱼类，为鹭类中所少有。集群营巢于树上；窝卵数 5~7 枚。

Morphological characteristics　A medium-sized wading bird (46-55 cm). Sexes alike. Non-breeding adult is very similar to Little Egret but with shorter and thicker head and neck region and yellow bill. Breeding adult has bright orange-red bill, orange-yellow head, neck, breast and back, white on the rest; greyish-yellow or black tarsi.

Distribution and habitat　Widespread in China except for the northwest. In Beijing, common in grassland, and cultivations. Residential status: migrant and summer breeder.

Behaviour and ecology　Usually feeds in pairs or small groups. Often accompanies livestock, catching insects disturbed by these animals. Sometimes perches on the back of livestock. Mostly feeds on insects and small invertebrates. Nests on trees. Clutch size: 5-7.

苍鹭	*Ardea cinerea*	Grey Heron	鹭科 Ardeidae

形态特征 大型涉禽，体长 90~98 厘米。雌雄相似。喙黄色。眼先裸皮黄绿色。繁殖期头侧、枕和两条长辫状冠羽黑色。喙、颈和跗跖皆甚长，身体细瘦。上体苍灰，头、颈和下体白色，前颈有 2~3 列纵行黑斑，体侧有大型黑色块斑。飞行时颈部缩成“S”形，跗跖伸于尾后，双翅缓慢扇动。

分布与生境 中国各省均有分布。北京甚常见于城区和郊区各种湿地。为夏候鸟、旅鸟和冬候鸟。

习性与生态 常集群分散开缩着脖子站立不动，民间素有“长脖儿老等”之称。以鱼、虾、蛙、等动物为食，亦见食小型哺乳动物。一般筑巢于岸边悬崖峭壁或高大乔木上；窝卵数 3~6 枚。

Morphological characteristics A large wading bird (90-98 cm). Sexes alike. Bill yellow with yellowish-green lores. In breeding season, head-sides and nape are black with two long braid-like black crown feathers. Bill, neck and tarsi are long; body is thin. Ashy-grey above, white on head, neck and below; two or three black streaks on front of neck and a large black patch on the side of the body. Flies with neck drawn back into S-shape; legs extend beyond tail, and bowed wings beat slowly.

Distribution and habitat Widespread in China. In Beijing, abundant in both urban and rural wetland. Residential status: summer breeder, migrant and winter visitor.

Behaviour and ecology Often spreads out into loose flocks; stands still with neck contracted. Feeds on fish, shrimps, frogs and sometimes young birds and small mammals. Nests on cliffs or tall trees near water, sometimes in reeds. Clutch size: 3-6.

北京市二级保护野生动物 | IUCN 红色名录 无危 (LC) 2019.8.13

成鸟 Adult

成鸟 Adult

幼鸟 Juvenile

草鹭	*Ardea purpurea*	Purple Heron	鹭科 Ardeidae

北京市二级保护野生动物		IUCN 红色名录　无危 (LC) 2019.8.14

形态特征　大型涉禽，体长 78~97 厘米。雌雄相似。喙黄褐色。虹膜黄色，眼先裸皮黄绿色。头顶蓝黑色，枕部具 2 条黑色辫状冠羽。颈栗褐色，细长，两侧有蓝黑色纵纹，上体蓝黑色，肩栗褐色，胸部、腹部中央铅灰黑色，两侧暗栗色。

分布与生境　中国除西北部外均有分布。北京不常见于开阔平原和低山丘陵的河湖、水库岸边，为夏候鸟和旅鸟。

习性与生态　行动迟缓。食性与苍鹭相似。营巢于有芦苇等挺水植物的水域岸边；窝卵数 4~5 枚，卵蓝灰色或灰绿色。

Morphological characteristics　A large wading bird (78-97 cm). Sexes alike. It has yellow-brown bill, yellow eyes, yellow-green lores, bluish-black crown, two long black stripes on nape; long narrow neck is chestnut-brown with bluish-black streaks on the sides; bluish black upperparts and brown-red mantle; breast and belly are lead grey-black with dark brown on flanks.

Distribution and habitat　Widespread in China except for the northwest. In Beijing, uncommon in vegetated wetlands. Residential status: summer breeder and migrant.

Behaviour and ecology　Moves slowly. Similar diet to Grey Heron. Nests close to open water in areas with emergent plants including reed beds. Clutch size: 4-5. Eggs bluish-grey or greyish-green.

大白鹭	*Ardea alba*	Great Egret	鹭科 Ardeidae

形态特征　大型涉禽，体长 82~100 厘米。雌雄相似。全身洁白。繁殖期背和前颈下部有长蓑羽；喙黑色，喙裂延至眼后；眼先蓝绿色；跗跖和趾黑色。非繁殖期无蓑羽，喙和眼先黄色。

分布与生境　中国除青藏高原外，几乎均有分布。北京常见于开阔平原和山地丘陵地区的河、湖、沼泽地，为夏候鸟、旅鸟及不常见越冬鸟。

习性与生态　通常白天成小群在岸边取食。以昆虫、小鱼、蛙、蜥蜴等小动物为食。营巢于高大乔木或苇丛中；窝卵数 3~6 枚，卵蓝色 。

Morphological characteristics　A large wading bird (82-100 cm). Sexes alike. All-white plumage with long plumes on breast and back in breeding season. Bill black with the gape extending behind the eye, lores bluish-green. Tarsi and toes black. In non-breeding plumage, plumes are absent and bill and lores are yellow.

Distribution and habitat　Widespread in China except for the Tibetan Plateau. In Beijing, common along rivers, lake and swamps in lowland and hills. Residential status: summer breeder and migrant, and scarce winter visitor.

Behaviour and ecology　Feeds at the shore or in shallow water in small flocks. Feeds on insects, small fish, frogs and reptiles. Nests in tall trees or reeds. Clutch size: 3-6. Eggs blue.

北京市一级保护野生动物　　IUCN 红色名录　无危 (LC)　2016.10.1

中白鹭	*Ardea intermedia*	Intermediate Egret	鹭科 Ardeidae

繁殖羽 Breeding

繁殖羽 Breeding

非繁殖羽 Non-breeding

北京市一级保护野生动物

IUCN 红色名录 无危 (LC) 2020.8.13

形态特征 中型涉禽，体长 62~70 厘米。雌雄相似。成鸟繁殖羽喙黑色；虹膜黄色，眼先绿色或黄绿色；全身羽毛白色，背和前颈下部有丝状长蓑羽；跗跖黑色。非繁殖期喙黄而先端黑色；眼先黄色；无蓑羽。

分布与生境 在中国主要分布于秦岭－淮河以南，亦见于华北。北京不常见于郊区的河湖、水库、池塘等浅水处和稻田，为旅鸟和夏候鸟。

习性与生态 常单独或成对活动，有时亦与其他鹭混群。以鱼、蛙、昆虫等小动物为食。营巢于树上；窝卵数 3~5 枚，卵蓝绿色。

Morphological characteristics A medium-sized wading bird (62-70 cm). Sexes alike. All-white plumage; tarsi black. Breeding adult has black bill and yellow eyes; lores green or yellowish-green; long plumes adorned at back and foreneck. Non-breeding adult has yellow bill with black tip and yellow lores, plumes absent.

Distribution and habitat Mostly found south of the Qinling–Huaihe Line and north China. In Beijing, seen in shallow wetlands including rivers, lakes, ponds and rice paddies. Residential status: uncommon summer breeder and migrant.

Behaviour and ecology Usually solitary or in pairs, sometimes in mixed flocks with other herons. Feeds on insects, small fish and frogs. Nests in trees. Clutch size: 3-5. Eggs blue-green.

白鹭 | *Egretta garzetta* | Little Egret | 鹭科 Ardeidae

北京市二级保护野生动物 | IUCN 红色名录 无危 (LC) 2016.10.1

形态特征 中型涉禽，体长 52~68 厘米。雌雄相似。体羽全白；喙黑色；眼先黄绿色；跗跖黑色，趾黄色。繁殖期枕部着生 2 根狭长而软的矛状饰羽；背和前颈具有蓑羽；面部裸皮黄色或黄绿色，繁殖期亦见粉色。非繁殖期全身白色，无饰羽。

分布与生境 中国除西藏外均有分布。北京甚常见于各湿地生境，为旅鸟和夏候鸟及不常见冬候鸟。

习性与生态 喜成小群活动于浅水处，以小鱼、虾、昆虫、蛙等小动物为食。集群筑巢在高大乔木上；窝卵数 3~6 枚，卵灰蓝色或蓝绿色 。

Morphological characteristics A medium-sized wading bird (52-68 cm). Sexes alike. All-white plumage with black bill; yellowish-green lore, black tarsi and yellow toes. In breeding season, two narrow and long spear-like plumes on nape; plumes adorned at back and foreneck. Bare face is yellow or yellowish-green, may turn pink in peak of breeding season. Non-breeding plumage is white without plumes.

Distribution and habitat Widespread in China except for Tibet. In Beijing, common in most types of wetlands. Residential status: summer breeder, migrant, and scarce winter visitor.

Behaviour and ecology Often found in small group in shallow water. Feeds on small fish, frogs, shrimps, insects and other invertebrates. Nests colonially in tall trees. Clutch size: 3-6. Eggs greyish-blue or bluish-green.

十二、鹳形目 CICONIIFORMES

黑鹳	*Ciconia nigra*	Black Stork	鹳科 Ciconiidae

形态特征 大型涉禽，体长 100~120 厘米。喙粗长呈红色。虹膜黑褐色，眼周红色。头、颈、上体和上胸黑色，颈背、肩、胸和翼具紫、绿色金属光泽。下胸、腹、两胁和尾下覆羽白色。跗跖红色。雌鸟体羽金属光泽较雄鸟稍暗。亚成鸟上体褐色，下体白色。

分布与生境 中国除青藏高原外均有分布。北京区域性常见于门头沟、房山、密云等有悬崖的河谷、湿地，为留鸟。

习性与生态 常单独或聚小群活动，以鱼、虾、蛙、蜥蜴、啮齿类、昆虫等动物为食。营巢于山谷峭壁之上；窝卵数 4~5 枚。

Morphological characteristics A large wading bird (100-120 cm). Adult male has black head, neck, upperparts and upper breast with metallic purple or green gloss on back of neck, mantle, breast and wings; lower breast, belly, flanks and undertail-coverts are white; long, thick bill, blackish-brown eye with red orbital skin; tarsi red. Female has duller metallic gloss than male. Immature has brown upperparts and white underparts.

Distribution and habitat Found in most of China except for the Tibetan Plateau. In Beijing, locally common in valleys or wetlands with cliffs in Mentougou, Fangshan and Miyun Districts. Residential status: resident.

Behaviour and ecology Often found alone or in small groups. Feeds on fish, shrimps, frogs, lizards, rodents or insects. Nests on mountain valley cliffs. Clutch size: 4-5.

国家一级保护野生动物 | IUCN 红色名录 无危 (LC) 2016.10.1

成鸟 Adult

成鸟 Adult

亚成鸟 Immature

东方白鹳 | *Ciconia boyciana* | Oriental Stork | 鹳科 Ciconiidae

国家一级保护野生动物 | IUCN 红色名录 濒危 (EN) 2018.8.9

形态特征 大型涉禽，体长 110~128 厘米。雌雄相似。虹膜乳白色，眼周裸皮、眼先朱红色。喙黑而粗厚。除飞羽、大覆羽、初级覆羽为黑色外，体羽和尾羽均为白色。跗跖朱红色。

分布与生境 分布于中国东北至华南地区。北京罕见于开阔而偏僻的水域，迁徙时通常在平原地区河流、沿岸的沼泽地和有水草的浅水处（如沙河、野鸭湖等）活动觅食，为旅鸟。

习性与生态 常单独或集群活动。筑巢于高大乔木顶端或高压电塔上；巢体甚大；窝卵数 3~5 枚。

Morphological characteristics A large wading bird (110-128 cm). Sexes alike. Milky-white iris with bright red lores and orbital skin, and thick black bill. Primaries, greater coverts and primary coverts black; the rest of the plumage is white. Bright red tarsi.

Distribution and habitat Northeast to southern China. In Beijing, rare passage migrant in open wetlands, such as rivers, marsh, or shallow water with aquatic plants (e.g. at Shahe and Yeyahu, etc.).

Behaviour and ecology Often found alone or in small flocks. small flocks. Builds a large nest, usually on top of a tall tree or tower. Clutch size: 3-5.

十三、红鹳目 PHOENICOPTERIFORMES

大红鹳	*Phoenicopterus roseus*	Greater Flamingo	红鹳科 Phoenicopteridae

形态特征 大型涉禽，体长 125~145 厘米。雌雄相似。成鸟喙粗厚且端部下弯，基部粉色，喙端黑色；虹膜白色；头、颈、上体及下体皆为白色，略沾粉色，粉色深度具个体差异，颈甚为细长；两翼翼上覆羽红色，飞羽黑色；跗跖亦极为细长，呈粉红色。幼鸟全身灰白色，喙基和跗跖皆为灰色。

分布与生境 中国主要见于西部地区，在东部仅有零星记录。喜在浅海和咸水湖活动。在北京为甚罕见迷鸟，仅于 2015 年 12 月有过一次记载，在昌平沙河、温榆河等地游荡数天后离去。

习性与生态 在主要分布区（欧亚大陆西部和非洲）集群活动，于中国的记载多为单只。觅食时将头倒置伸入水面下，用喙过滤水中的藻类、小型无脊椎动物和植物作为食物。

Morphological characteristics A large wader (125-145 cm). Sexes alike. Adult has shallow keeled bill with pink base and black tip; eyes white; head, neck, upperparts and underparts largely white or variable pale pinkish-white; neck slender and long; upperwing-covers red; flight feathers black; legs and tarsi very long, pink. Juvenile is mostly greyish-white with grey bill base and tarsi.

Distribution and habitat Mainly western China. Scarce in eastern China. Prefers coastal wetlands and salt lakes. Very rare vagrant in Beijing with only one record at Shahe Reservoir, Wenyu river (and areas nearby) in December 2015.

Behaviour and ecology In small and large flocks in Africa and western Eurasia. Records in Beijing and other areas in eastern China are usually of singles. When feeding, submerges its head upside-down and filters for food items (plants, aquatic invertebrates and small fish) with its bill.

IUCN 红色名录　无危 (LC) 2018.8.7

成鸟 Adult

亚成鸟 Immature

十四、鸨形目 OTIDIFORMES

大鸨	*Otis tarda*	Great Bustard	鸨科 Otididae

形态特征 大型涉禽，体长 75~105 厘米，体型甚为粗壮，雌鸟较雄鸟小。雄鸟繁殖羽头、颈灰色，颏、喉处具细长的白色须状饰羽，颈基部具白色和棕红色饰羽；上体及翼上小覆羽黄褐色，具黑色横斑，其余翼上覆羽灰白色，飞羽基部白色，端部黑色，其余下体白色。雄鸟非繁殖羽和雌鸟颏、喉部无须状饰羽，胸白色。

分布与生境 分布于中国西北和东部地区，活动于草原和半荒漠地区。在北京见于大型湿地（如密云水库、官厅水库等）附近的农田和草地，迁徙过境时亦飞经西、北部山区，为不常见旅鸟。

习性与生态 多成对或集小群活动。奔跑快速，飞行时振翼缓慢有力。营巢于草地浅坑中；窝卵数 2~4 枚。

Morphological characteristics A large wading bird (75-105 cm). The body is stout and heavy. Female smaller than male. Breeding male has grey head and neck; chin and throat have greyish whiskers; neck and breast have long white and chestnut-brown bristles; upperparts and lesser coverts brown, barred black; remaining upperwing-coverts greyish-white; remiges white at base and black at tips; underparts white. Female and non-breeding male lack whiskers; breast white.

Distribution and habitat Northwest and eastern China. Usually seen in steppe and semi-desert habitats. Uncommon passage migrant to grassland and farmland near large wetlands in Beijing. Also seen migrating over mountains in western and northern Beijing.

Behaviour and ecology Usually in pairs or small flocks. Runs fast on the ground. Flight slow and heavy. Nests on the ground in grassland. Clutch size: 2-4.

国家一级保护野生动物 | IUCN 红色名录 易危 (VU) 2017.10.1

非繁殖羽 Non-breeding

繁殖羽 Breeding

繁殖羽 Breeding

十五、鹤形目 GRUIFORMES

花田鸡	*Coturnicops exquisitus*	Swinhoe's Rail	秧鸡科 Rallidae

	国家二级保护野生动物	IUCN 红色名录 易危 (VU) 2016.10.1

形态特征 小型涉禽，体长 12~14 厘米，体型甚为纤小。雌雄同色。喙深褐色，基部稍淡。头褐色，头顶具黑色纵纹，头两侧羽色较淡。上体大致为深褐色，具黑色纵纹和甚细的白色横斑，两翼甚短，大致与上体同色，但次级飞羽为白色。颏、喉灰白色，胸具褐色横斑，其余下体白色。跗跖灰色或粉色。

分布与生境 分布于中国东部地区，多见于沿海的草地和植被丰富的沼泽等地，亦偶见于内陆湿地。北京为甚罕见旅鸟。

习性与生态 单独或成对活动，行为隐秘，于晨昏较活跃。营巢于湿地草丛中；窝卵数 6 枚左右。

Morphological characteristics A small wading bird (12-14 cm). Significantly smaller than other rails in Beijing. Sexes alike. Bill dark brown; crown brown, streaked black; head-sides paler; upperparts dark brown, with broad black streaks and fine white bars; wings very short, largely concolourous with back; secondaries white; chin and throat greyish-white; breast extensively barred brown; rest of underparts white; tarsi grey or pink.

Distribution and habitat Eastern China. Usually seen in coastal grassland and well-vegetated marshes. Scarce in inland. Very rare passage migrant in Beijing.

Behaviour and ecology Solitary or in pairs. Very shy, most active at dawn and dusk. Nests in grass at wetlands. Clutch size: usually 6.

灰胸秧鸡 *Lewinia striata* Slaty-breasted Rail 秧鸡科 Rallidae

形态特征 小型涉禽，体长 22~31 厘米。雌雄同色。喙红色。虹膜暗红色。头顶至后颈栗褐色，上体及两翼黑褐色，密布白色横斑和点斑。颏、喉近白色，颊、颈、胸蓝灰色，腹部和尾下覆羽深褐色，具白色横斑。跗跖灰绿色。

分布与生境 在中国主要分布于秦岭－淮河以南地区，偶见于华北。活动于沼泽、水田以及湖泊、水塘、河流岸边的草丛和苇丛中。甚罕见于北京，仅有两次记载，分别于 2000 年 5 月和 2018 年 6 月。居留型尚不明确。

习性与生态 单独或成对活动，擅奔跑，亦擅游水。营巢于植被丰富的湿地的苇丛或草丛中；窝卵数 5~9 枚。

Morphological characteristics A small wading bird (22-31 cm). Sexes alike. Red bill and deep-red eyes; distinctive rufous-brown from crown and nape to hindneck; upperparts and wings largely blackish-brown with extensive white bars and spots; chin and throat whitish; cheek, neck and breast blue-grey; belly and undertail-coverts dark brown, bared white; tarsi greyish-green.

Distribution and habitat Widespread in southern China (i.e. south of the Qinling–Huaihe Line), scarce in north China. Occurs in grass and reeds at marshes and paddy fields, also near lakes, ponds and rivers. Very rare in Beijing with only two records in May 2000 and June 2018. Residential status: unclear.

Behaviour and ecology Usually encountered alone or in pairs. Runs rapidly and also swims. Nests in reedbeds or grasses at wetlands with abundant vegetation. Clutch size: 5-9.

IUCN 红色名录 无危 (LC) 2016.10.1

成鸟 Adult

幼鸟 Juvenile

西秧鸡 | *Rallus aquaticus* | Water Rail | 秧鸡科 Rallidae

IUCN 红色名录 无危 (LC) 2016.10.1

形态特征 小型涉禽，体长 24~28 厘米。雌雄同色。喙较长，略下弯，喙峰至喙端黑色，余部红色。虹膜红色。头顶、后颈至上体褐色，具黑色纵纹。颏、喉灰白色，头两侧、颈、胸及腹部蓝灰色，两胁、臀至尾下覆羽为黑白相间的横斑。跗跖红褐色。与普通秧鸡相似，但本种眼周羽色较淡，且下体不沾棕色。

分布与生境 中国主要见于西部地区，于东部仅有零星记录。活动于湿地灌丛、高草丛和苇丛中。在北京见于奥林匹克森林公园和沙河水库等植被丰富的湿地，为罕见迷鸟（或为冬候鸟）。

习性与生态 性隐秘，非繁殖期多单独活动，主食鱼类、小型无脊椎动物及少量植物性食物。

Morphological characteristics A small wading bird (24-28 cm). Sexes alike. The red bill is long and slightly curved, with black culmen and tip; eyes red; brown from crown and hindneck to upperparts, streaked black; chin and throat pale greyish-white; head-sides, neck, breast and belly blue-grey; flanks, vent and undertail-coverts barred black and white; tarsi reddish-brown. Resembles Brown-cheeked Rail but lacks dark eye-stripe and underparts lack brownish wash.

Distribution and habitat Mainly in western China. Scarce in eastern China. Occurs in scrub, tall grass and reeds at wetlands. In Beijing, rare vagrant (mostly in winter) at well-vegetated wetlands such as Shahe Reservoir and the Olympic Forest Park.

Behaviour and ecology Shy. Usually solitary in non-breeding season. Mainly eats small fish, aquatic invertebrates and plant matter.

普通秧鸡	*Rallus indicus*	Brown-cheeked Rail	秧鸡科 Rallidae

形态特征 小型涉禽，体长 23~29 厘米。雌雄同色。喙峰和喙端黑色，余部红色。虹膜暗红色。头顶暗褐色，头侧淡灰褐色，具较宽的深褐色贯眼纹。上体棕褐色，具显著的黑色纵纹。颏、喉白色，前颈至胸蓝灰色沾棕色，腹部蓝灰色，两胁及尾下覆羽黑色，具白色横斑。跗跖红褐色。

分布与生境 分布于中国东部及中部，繁殖于东北及华北地区，越冬于南方。北京地区见于植被丰富的低海拔湿地，亦偶见于城市公园湿地，为不常见旅鸟、夏候鸟及冬候鸟。

习性与生态 繁殖期成对、非繁殖期多单独于晨昏或夜间活动，受惊扰时快速奔跑、隐藏至浓密的植被中。营巢于苇丛中；窝卵数 6~9 枚。

Morphological characteristics A small wading bird (23-29 cm). Sexes alike. Bill red with black culmen and tip; eyes deep-red; crown dark brown; head-sides pale greyish-brown with broad dark eye-stripe; upperparts mid-brown with distinctive black streaks; chin and throat white; foreneck and breast blue-grey with brown wash; belly grey-blue; prominent black and white bars from flanks to vent and undertail-coverts; tarsi reddish-brown.

Distribution and habitat Widespread in eastern and central China. Breeds in north and northeast China; winters in south China. Uncommon passage migrant, summer breeder and winter visitor in Beijing. Occurs in both rural and urban wetlands with rich vegetation.

Behaviour and ecology Usually in pairs in breeding season and alone in non-breeding season. Partly nocturnal. Runs rapidly into reeds or grass when disturbed. Nests in extensive reeds. Clutch size: 6-9.

IUCN 红色名录 无危 (LC) 2016.10.1

小田鸡	*Zapornia pusilla*	Baillon's Crake	秧鸡科 Rallidae

IUCN 红色名录 无危 (LC) 2019.8.15

形态特征 小型涉禽，体长 15~19 厘米。雌雄同色。成鸟喙黄绿色；虹膜红色；头顶、后颈及上体褐色，具黑色纵纹和白色斑点；颏、喉白色，头两侧、颈、胸至上腹蓝灰色，两胁、下腹和尾下覆羽为黑白相间的横斑；跗跖黄绿色。幼鸟颈、胸近白色，胸侧沾淡棕色，两胁具褐色横斑。

分布与生境 广泛分布于中国除青藏高原外的地区。多见于海拔较低之处的湖泊、河流、池塘、沼泽、水田等地。在北京为不常见旅鸟和夏候鸟。

习性与生态 性羞怯，多单独、成对或以家族群活动。营巢于浓密的近水苇丛、草丛中；窝卵数 6~9 枚。

Morphological characteristics A small wading bird (15-19 cm). Sexes alike. Adult has short, thick, yellowish-green bill; eyes red; crown, hindneck and upperparts are mid-brown with black streaks and white spots; chin and throat white; head-sides, neck, breast and upper belly plain blue-grey; distinctive black and white bars on flanks, vent and undertail-coverts; tarsi yellowish-green. Juvenile has off-white breast and belly; breast-sides have pale brown wash; flanks barred brown.

Distribution and habitat Widespread in China except for the Tibetan Plateau. Occurs at lowland well-vegetated shallow lakes, rivers, ponds, marshes and paddy fields. Uncommon passage migrant and summer breeder in Beijing.

Behaviour and ecology Shy. Usually solitary, in pairs or in family groups. Nests in dense reeds and tall grass at wetlands. Clutch size: 6-9.

红胸田鸡	*Zapornia fusca*	Red-breasted Crake	秧鸡科 Rallidae

形态特征 小型涉禽，体长 19~23 厘米。雌雄同色。喙深灰色，部分个体下喙基部灰绿色。虹膜红色。头顶、上体、两翼及尾羽深褐色。颏、喉近白色，头两侧、颈、胸至上腹栗红色。下腹褐色，两胁后部及尾下覆羽具黑白相间的横斑。跗跖红色。

分布与生境 广泛分布于中国东部及南部地区。活动于较低海拔且近岸植被丰富的湿地，亦见于城市湿地。在北京为区域性常见的旅鸟和夏候鸟。

习性与生态 性羞怯，多单独活动，于晨昏及夜间比较活跃，常在夜间发出“ko ko ko”的叫声。营巢于湿地附近的草丛或灌丛中；窝卵数 5~9 枚。

Morphological characteristics A small wading bird (19-23 cm). Sexes alike. Bill dark grey, sometimes greyish-green at base of lower mandible; eyes red; crown, upperparts, wings and tail dark brown; chin and throat whitish; plain chestnut from head-sides, neck and breast to upper belly; lower belly brown; rear flanks and undertail-coverts barred black and white; tarsi red.

Distribution and habitat Widespread in eastern and southern China. Usually seen in lowland wetlands with dense fringe vegetation, sometimes also in urban wetlands. Locally common passage migrant and summer breeder in Beijing.

Behaviour and ecology Shy; usually solitary and partly nocturnal. Call is a "ko ko ko" mostly given at night. Nests in dense shrubs or tall grass near water. Clutch size: 5-9.

IUCN 红色名录 无危 (LC) 2016.10.1

斑胁田鸡	*Zapornia paykullii*	Band-bellied Crake	秧鸡科 Rallidae

	国家二级保护野生动物	IUCN 红色名录 近危 (NT) 2017.10.1

形态特征 小型涉禽，体长 22~27 厘米。雌雄同色。喙灰绿色。虹膜红色。头顶、后颈及上体余部和两翼皆为褐色，部分个体翼上覆羽具数道较细的浅色横斑。颏、喉淡红色或白色，头两侧、颈、胸至上腹栗红色，下腹、两胁及尾下覆羽具黑白相间之横斑。跗跖橙红色。甚似红胸田鸡，但本种上体羽色较淡，且跗跖颜色亦较淡，为橙红色而非红色。

分布与生境 在中国主要分布于东部地区。多活动于海拔 800 米以下的湖泊、水塘和沼泽中的苇丛和草丛中，亦见于水田。在北京为罕见旅鸟。

习性与生态 多单独活动，在水边浓密的草丛和灌丛中潜行，于晨昏比较活跃。多营巢于草丛或灌丛中；窝卵数 6~9 枚。

Morphological characteristics A small wading bird (22-27 cm). Sexes alike. Bill short, thick, greyish-green; eyes red; crown, hindneck, upperparts and wings brown; some individuals have very fine pale bars on upperwing-coverts; chin and throat off-white or pale red; head-sides, neck, breast and upper belly chestnut; extensive black and white bars from flanks and lower belly to undertail-coverts; tarsi orange-red. Closely resembles Red-breasted Crake but has paler upperparts and tarsi.

Distribution and habitat Eastern China. Occurs in wet grassland and reeds at ponds, marshes and lakes below 800 m, also in paddy fields. Rare passage migrant in Beijing.

Behaviour and ecology Usually solitary. Skulks in dense fringe vegetation. Active at dawn and dusk. Nests in wet grass or shrubs near water. Clutch size: 6-9.

白胸苦恶鸟	*Amaurornis phoenicurus*	White-breasted Waterhen	秧鸡科 Rallidae

形态特征 小型涉禽，体长 26~35 厘米。雌雄同色。喙黄绿色，上喙基部红色。虹膜暗红色。头顶、后颈及上体和两翼皆为黑灰色或深青灰色，尾羽深褐色。前额、头两侧、颏、喉、颈、胸及上腹白色，下腹及尾下覆羽红色或橙红色。跗跖黄绿色。

分布与生境 在中国主要分布于华北及南方地区。一般活动于水边植被丰富的河流、沼泽、水塘和水田等湿地。北京为区域性常见夏候鸟和旅鸟。

习性与生态 喜单独或成对在苇丛和草丛中穿行，捕食小型无脊椎动物，偶尔亦在水中游泳。鸣叫声似重复的“ku-e ku-e”。营巢于水边的草丛或灌丛中；窝卵数 4~8 枚。

Morphological characteristics A small wading bird (26-35 cm). Sexes alike. Bill yellowish-green with a red spot at base of upper mandible; eyes deep-red; plain blackish-grey or dark slate-grey from crown and hindneck to upperparts and wings; tail dark brown; forehead, head-sides, chin, throat, neck, breast and upper belly all-white; lower belly and undertail-coverts red or orange-red; tarsi yellowish-green.

Distribution and habitat Widespread in north and southern China. Usually seen at well-vegetated ponds, rivers, marshes and paddy fileds. Locally common summer breeder and passage migrant in Beijing.

Behaviour and ecology Often alone or in pairs. Feeds mainly on small invertebrates in reeds and grass. Occasionally swims. Gives repeated calls like "ku-e ku-e". Nests in grass or shrubs near water. Clutch size: 4-8.

IUCN 红色名录 无危 (LC) 2016.10.1

董鸡 | *Gallicrex cinerea* | Watercock | 秧鸡科 Rallidae

IUCN 红色名录 无危 (LC) 2016.10.1

形态特征 小型涉禽，体长 35~43 厘米。繁殖期雄鸟喙黄色，上喙基部红色，具上翘的红色额甲；虹膜暗红色；全身大致呈深蓝黑色或近黑色，但背、腰及尾上覆羽具较窄的棕色羽缘，两翼黑褐色，尾下覆羽皮黄色；跗跖黄绿色。雌鸟及非繁殖期雄鸟额甲较小，不上翘且为黄绿色；上体深褐色，下体皮黄色。

分布与生境 分布于中国东部及南部地区。栖息于植被丰富的水塘、河流、沼泽或水田中。北京为罕见旅鸟和夏候鸟。

习性与生态 多单独或成对，有时集 3~5 只左右的小群活动。一般为夜行性。鸣叫声清脆而响亮，似“dong dong”。一般营巢于苇丛中；窝卵数 3~8 枚。

Morphological characteristics A small wading bird (35-43 cm). Male larger than female. Breeding male has yellow bill with red-based upper mandible and a prominent red shield on forehead, extending to an erectile horn; eyes deep-red; head and body largely dark slate-blue or blackish; back, rump and uppertail-coverts have narrow brown fringes; wings mostly blackish-brown; undertail-coverts buff; tarsi yellowish-green. Female and non-breeding male have much smaller yellowish-green shield, not raised on forehead; dark brown above and buff below.

Distribution and habitat Eastern and southern China. Occurs at well-vegetated ponds, rivers, marshes and paddy fileds. Rare passage migrant and summer breeder in Beijing.

Behaviour and ecology Usually alone or in pairs. Sometimes forms small flocks (three to five individuals). Nocturnal. Gives a "dong dong" call at night. Nests in dense reeds. Clutch size: 3-8.

黑水鸡 | *Gallinula chloropus* | Common Moorhen | 秧鸡科 Rallidae

形态特征 小型涉禽，体长 24~35 厘米。雌雄同色。成鸟喙鲜红色，喙端黄色或黄绿色，前额具红色额甲；虹膜暗红色；头、颈、上背、胸、腹蓝黑色，下背、两翼、腰、尾上覆羽和尾羽深褐色，两胁具白色纵纹，尾下覆羽中央黑色，两侧各具一块显著白斑；跗跖黄绿色。幼鸟通体大致为褐色，颏、喉白色，无红色额甲。

分布与生境 广布于中国西北、东部及南部地区。活动于植物丰富的池塘、湖泊、河流、水库等地，亦见于城市公园水域。在北京为甚常见夏候鸟，于冬季不封冻水面有少量越冬。

习性与生态 喜集小群或家族群活动。常在岸边的植被间穿行，亦擅游水。不擅飞，起飞前需在水面助跑。营巢于苇丛中；窝卵数 6~10 枚。

Morphological characteristics A small wading bird (24-35 cm). Sexes alike. Adult has bright red bill with yellow or greenish-yellow tip; prominent red shield on forehead; eyes deep-red; dark slate-blue from head and neck to upper back and underparts; lower back, wings, rump, uppertail-coverts and tail dark brown; distinctive white streaks on flanks; undertail-coverts have black centres and white sides; tarsi greenish-yellow. Juvenile largely brown with white chin and throat, lacking red shield.

Distribution and habitat Widespread in northwest, eastern and southern China. Seen in well-vegetated ponds, lakes, rivers and reservoirs, as well as urban wetlands. Abundant summer breeder in Beijing with a few wintering, particularly in parks and areas of unfrozen water with reeds.

Behaviour and ecology Usually in small flocks or family groups. Skulks in dense fringe vegetation, also swims. Flight weak and heavy. Runs across water to take off. Nests in reeds. Clutch size: 6-10.

IUCN 红色名录 无危 (LC) 2016.10.1

成鸟 Adult

亚成鸟 Immature

亚成鸟 Immature

白骨顶 | *Fulica atra* | Common Coot | 秧鸡科 Rallidae

形态特征 小型涉禽，体长 35~43 厘米。雌雄同色。成鸟喙和额甲白色；虹膜暗红色；通体黑色，两翼黑褐色，内侧次级飞羽端部白色，但仅飞行时可见，翼下灰色；跗跖黄绿色，趾具瓣蹼。幼鸟虹膜深褐色；上体深灰色，颏、喉、前颈、胸白色，下体淡灰色。

分布与生境 广泛分布于中国各省。活动于各类淡水水域，尤喜静水或水流速较慢且岸边具植被之地。在北京为常见旅鸟和夏候鸟，于冬季不封冻水面有少量越冬。

习性与生态 常游于水面，擅潜水，较少似其他秧鸡般在近水植被中穿行。非繁殖期常集群活动于开阔水面。一般营巢于浓密的苇丛中；窝卵数 8~10 枚。

Morphological characteristics A small wading bird (35-43 cm). Sexes alike. Adult has white bill and shield; eyes deep-red; head and body mostly black; wings short, largely blackish-brown; inner secondaries have white tips, seen only in flight; underwing-coverts grey; tarsi yellowish-green; toes lobed. Juvenile has dark eyes and blackish-grey upperparts; chin, throat, foreneck and breast white; underparts pale greyish-white.

Distribution and habitat Widespread in China. Seen at freshwater lakes, reservoirs and ponds. Prefers slow flowing or still water bodies with fringe vegetation. Common passage migrant and summer breeder in Beijing, also some winters in unfrozen water.

Behaviour and ecology Prefers to swim in water than walk in reeds and grasses, dives well. Forms small to large flocks in non-breeding season. Usually nests in dense reeds. Clutch size: 8-10.

IUCN 红色名录 无危 (LC) 2019.8.14

成鸟 Adult

亚成鸟 Immature

白鹤 | *Grus leucogeranus* | Siberian Crane | 鹤科 Gruidae

成鸟 Adult

成鸟 Adult

幼鸟 Juvenile

国家一级保护野生动物

IUCN 红色名录 极危 (CR) 2018.8.17

形态特征 大型涉禽，体长 130~140 厘米，体态优雅。雌雄同色。成鸟喙红褐色；前额至眼先为红色裸皮；虹膜淡黄色；全身大致为白色，仅初级飞羽黑色，三级飞羽特别延长，停歇时覆盖住黑色的初级飞羽；跗跖甚长，呈暗红色。幼鸟前额被羽，头、颈及上体黄褐色取代白色。

分布与生境 分布于中国东部地区。北京为罕见旅鸟。活动于开阔的平原沼泽、草地和湖泊的浅水处。

习性与生态 杂食性，主要取食水生植物的根、茎、叶等，亦食少量鱼或鼠类等小型动物。非繁殖期集群活动，通常为家族群或更大的群体。于西部利亚苔原的近水土丘上营巢；窝卵数 2 枚。

Morphological characteristics A large and elegant wading bird (130-140 cm). Sexes alike. Adult has long, reddish-brown bill; bare red skin from forehead to lores; eyes pale yellow; head and body mostly white; white tertials are long, covering the black primaries at rest; tarsi very long and deep red. Juvenile lacks bare skin on head; brownish head, neck and upperparts replace white of adult.

Distribution and habitat Eastern China. Occurs at lowland marshes, wet grassland and shallow lakes. Rare passage migrant in Beijing, with a few records from shores and farmland near large reservoirs such as Guanting Reservoir and Miyun Reservoir.

Behaviour and ecology Omnivorous. Usually feeds on roots, stems and leaves of aquatic plants; also eats small animals such as fish or rodents. Found in flocks (from family groups to large flocks) in non-breeding season. Nests on boggy ground surrounded by water on Siberian tundra. Clutch size: 2.

沙丘鹤	*Grus canadensis*	Sandhill Crane	鹤科 Gruidae

形态特征　大型涉禽，体长 100~110 厘米。雌雄同色。成鸟喙黑灰色；虹膜黄色；前额至眼先具红色裸皮；全身大致为灰色，上体沾褐色，外侧数枚初级飞羽黑灰色，三级飞羽延长呈弓状，两翼收拢时覆盖于初级飞羽和尾羽之上；颏、喉白色；跗跖黑灰色。幼鸟头部皆被羽，无裸皮；头、颈及上体棕色。

分布与生境　主要见于北美洲。中国于东部地区有少量记录。在北京为罕见旅鸟。活动于开阔的湿地、农田、草原等地。

习性与生态　在主要分布区（北美洲）集大群活动，但在中国多为单只或小群的零星记录，并与其他鹤类混群。飞行缓慢，振翼动作显著。

Morphological characteristics A large wading bird (100-110 cm). Sexes alike. Adult has blackish-grey bill and yellow eyes; red bare skin from forehead to rear of eye; head and body mostly grey; upperparts variable brownish; outer primaries blackish-grey; tertials long and drooping, covering tail and dark primaries at rest; chin and throat white; tarsi blackish-grey. Juvenile has brownish head and neck, and lacks red bare skin on head.

Distribution and habitat Mainly North America. Scarce in eastern China. Rare passage migrant in Beijing. Occurs in open grassland, wetlands and farmland in rural areas.

Behaviour and ecology Forms large flocks in North America, but mostly singles or small flocks in China. In Beijing, tends to join mixed flocks with other cranes (e.g. Common Crane). Flight slow with deep wingbeats.

	国家二级保护野生动物	IUCN 红色名录　无危 (LC) 2016.10.1

白枕鹤 | *Grus vipio* | White-naped Crane | 鹤科 Gruidae

国家一级保护野生动物 | IUCN 红色名录 易危 (VU) 2018.8.9

形态特征 大型涉禽，体长 120~150 厘米。雌雄同色。成鸟喙较长，淡黄白色；虹膜黄色或橙色；喙基至前额及眼先具黑灰色绒羽，颊至眼周具红色裸皮；耳羽灰色，头顶、枕至后颈白色，颏、喉及前颈上部亦为白色；前颈下部、胸、上体及下体大致为灰色；飞羽及初级覆羽近黑色，三级飞羽灰白色，延长而弯曲；跗跖淡红色。幼鸟似成鸟，但头和上颈沾棕色。

分布与生境 中国见于东部地区。繁殖于东北，过境于华北，越冬于长江下游流域。在北京为不常见旅鸟。活动于大型水库的浅水区及开阔的湖泊、草地、农田等地。

习性与生态 非繁殖期多集大群或家族群活动。繁殖期攻击性较强；营巢于水生植物丰富的沼泽中；窝卵数 2 枚。

Morphological characteristics A large wading bird (120-150 cm). Sexes alike. Adult has long, pale yellowish-horn bill and yellow or orange-yellow eyes; blackish-grey from forehead to lores; distinctive bare facial skin on cheek and around eye; ear-coverts grey; white from crown and nape to hindneck; chin and throat also white; foreneck and neck-sides grey-and-white-striped; breast, upperparts and underparts largely grey; remiges and primary coverts blackish; tertials pale greyish-white, long, covering other flight feathers and tail at rest; tarsi very long, pale and red. Juvenile similar to adult but has brownish head and upper neck.

Distribution and habitat Eastern China. Breeds in northeast China; migrates across north China and winters in the lower Yangtze River valley. Uncommon passage migrant in Beijing. Seen at open shallow lakes, reservoirs, wet grassland and farmlands.

Behaviour and ecology Usually in large flocks or family groups in non-breeding season. Aggressive in breeding season. Nests in marshes with abundant vegetation. Clutch size: 2.

蓑羽鹤	*Grus virgo*	Demoiselle Crane	鹤科 Gruidae

形态特征 大型涉禽，体长 70~92 厘米。雌雄同色。成鸟喙较短，淡黄绿色；虹膜红色（雄鸟）或橙色（雌鸟）；耳羽白色，延长并下垂；头、前颈、胸黑色，其中颈及胸具显著延长并下垂的黑色饰羽；上体余部及下体大致为灰色，仅飞羽黑色；三级飞羽特别延长，两翼收拢时垂于初级飞羽和尾羽之上；跗跖深灰色。幼鸟虹膜褐色，耳、颈、胸无饰羽，头大致为淡灰色。

分布与生境 广泛分布于中国长江以北地区。繁殖期多栖息于草原、草甸沼泽或半荒漠地区，高可至海拔 5000 米。非繁殖期多见于各种开阔的湿地。在北京为罕见旅鸟。

习性与生态 喜集群活动，非繁殖期常与其他鹤类混群。营巢于草甸或沼泽的草丛中，窝卵数 1~3 枚。

Morphological characteristics A large wading bird (70-92 cm). Sexes alike. Adult has relatively short, pale greenish-yellow bill; eyes red (male) or orange (female); ear tufts white, extending to nape; head, foreneck and breast plain black, with prominent black chest plumes; body largely grey; primaries black; grey tertials long and covering the dark primaries; tarsi dark grey. Juvenile has brown eyes and lacks ear tufts and chest plumes; head and neck largely pale grey.

Distribution and habitat Widespread north of Yangtze River. Seen in steppe, well-vegetated marshes or semi-desert habitats up to 5,000 m in breeding season and at open wetlands in non-breeding season. Rare passage migrant in Beijing.

Behaviour and ecology Tends to form flocks. Often joins mixed flocks with other cranes. Nests on the ground in grassland. Clutch size: 1-3.

国家二级保护野生动物　　IUCN 红色名录　无危 (LC) 2018.8.9

丹顶鹤	*Grus japonensis*	Red-crowned Crane	鹤科 Gruidae

成鸟 Adult

成鸟 Adult

亚成鸟 Immature

亚成鸟 Immature

国家一级保护野生动物　IUCN 红色名录　濒危 (EN) 2016.10.1

形态特征　大型涉禽，体长 120~160 厘米，体态优雅。雌雄同色。成鸟喙淡黄角质色；虹膜暗红色；头顶红色，眼后至枕部白色，前额、眼先、颊、颏、喉及颈黑色，上体及下体大致为白色；两翼亦大致为白色，仅次级飞羽和三级飞羽黑色，延长而下垂的三级飞羽在停歇时覆盖住白色的初级飞羽和尾羽；跗跖黑褐色。幼鸟头、颈呈褐色，头顶无红色区域。

分布与生境　在中国繁殖于东北，迁徙季节旅经华北，越冬于长江中下游流域。栖息在草原以及开阔的河流、湖泊、水库附近的草地、浅滩和农田。在北京为罕见旅鸟。

习性与生态　甚擅鸣，常发出响亮的“ko-ko-ko”声。飞行缓慢。非繁殖期多集家族群或更大的群体活动。营巢于苇丛或水草丛中；窝卵数 2 枚。

Morphological characteristics A large and elegant wading bird (120-160 cm). Sexes alike. Adult has yellowish-horn bill and deep red eyes; red patch on crown and a white patch from eye to nape; black on forehead, lores, cheek, chin, throat and neck; body and wings largely white; secondaries and tertials black; tertials are long, covering the white primaries and tail at rest; tarsi blackish-brown. Juvenile has brownish head and neck and crown lacks red patch.

Distribution and habitat Eastern China. Breeds in northeast China; migrates across north China and winters in the lower Yangtze River valley. Occurs in steppe, open grassland and farmland near lakes, rivers and reservoirs. Rare passage migrant in Beijing.

Behaviour and ecology Flight slow and leisurely. Gives loud "ko ko ko" call. Usually in family groups or larger flocks in non-breeding season. In breeding season, nests in reeds or wet grass. Clutch size: 2.

III

涉禽

灰鹤	*Grus grus*	Common Crane	鹤科 Gruidae

形态特征 大型涉禽，体长 100~125 厘米。雌雄同色。成鸟喙角质色；虹膜橙黄色；头顶红色，前额、眼先、颏、喉至前颈黑色，眼后方有一白色的带向后延伸于后颈汇合，其余上体、下体大致为灰色；初级飞羽黑色，三级飞羽黑灰色，明显延长，停歇时松散下垂覆盖于初级飞羽和尾羽之上；跗跖黑色。幼鸟头、颈为淡皮黄色。

分布与生境 中国除青藏高原外广泛分布。北京为区域性常见冬候鸟及旅鸟。活动于开阔的大型水库、湖泊（如密云水库、官厅水库）周围的草地、浅滩、农田等地。

习性与生态 非繁殖期集大群活动。繁殖期营巢于草原、沼泽中的干燥地面上；窝卵数 2 枚。

Morphological characteristics A large wading bird (100-125 cm). Sexes alike. Adult has yellowish-horn bill and orange eyes; cap red; forehead, lores, chin, throat and foreneck all-black; a white band extends from eye to nape; remainder of upperparts and underparts largely grey; primaries black; tertials blackish-grey and long, covering the primaries and tail at rest; tarsi black. Juvenile has pale buff head and neck.

Distribution and habitat Widespread in China except for the Tibetan Plateau. In Beijing, a locally common winter visitor and passage migrant. Seen in open grassland and farmland near large reservoirs and lakes, such as Miyun Reservoir and Guanting Reservoir.

Behaviour and ecology Forms large flocks in non-breeding season. Nests on the ground in steppe or marshes. Clutch size: 2.

	国家二级保护野生动物	IUCN 红色名录 无危 (LC) 2016.10.1

白头鹤	*Grus monacha*	Hooded Crane	鹤科 Gruidae

国家一级保护野生动物　IUCN 红色名录　易危 (VU) 2016.10.1

形态特征　大型涉禽，体长 90~97 厘米。雌雄同色。成鸟喙淡黄绿色；虹膜橙红色或暗红色；头顶前部至眼先黑色，此间包围着面积较小的红色区域，头余部至颈上部皆为白色，颈下部及上体、两翼和下体皆为深灰色；跗跖灰褐色。幼鸟头、颈为淡褐色，头顶无黑、红相间之区域。

分布与生境　分布于中国东部。繁殖于内蒙古、黑龙江，迁徙经东北、华北，至长江中下游区域越冬。北京为不常见旅鸟。一般活动于开阔的泥滩、沼泽、草地、农田中。

习性与生态　非繁殖期集群活动或与灰鹤等其他鹤类混群。营巢于泰加林间及林缘的草地和沼泽地上，或开阔湖泊、河流附近的草地中；窝卵数 2 枚。

Morphological characteristics A large wading bird (90-97 cm). Sexes alike. Adult has pale greenish-yellow bill and orange or deep red eyes; forehead red, bordered black from forecrown to lores; remainder of head and upper neck white; lower neck, body and wings mostly dark grey; tarsi greyish-brown. Juvenile has pale brown head and neck; crown lacks black and red patches.

Distribution and habitat Eastern China. Breeds in Inner Mongolia and Heilongjiang Province. Migrates across northeast and north China to winter in the lower Yangtze River valley. Uncommon passage migrant in Beijing. Usually occurs in open marshes, grassland and farmland.

Behaviour and ecology In flocks or joins mixed flocks with other cranes in non-breeding season. Usually nests on the ground in mossy taiga swamps or in grass near lakes or rivers. Clutch size: 2.

IV

北京鸟类图谱

攀禽

SCANSORIAL BIRDS

BIRDS OF BEIJING

十六、夜鹰目 CAPRIMULGIFORMES

普通夜鹰	*Caprimulgus indicus*	Grey Nightjar	夜鹰科 Caprimulgidae

形态特征 中等体型的攀禽，体长 26~28 厘米。雄鸟整体为棕褐色，杂以灰白、黑褐色细斑；颏、喉部黑褐色，喉具一显著白斑；外侧初级飞羽具一白斑；中央尾羽黑色，外侧尾羽具白色次端斑。雌鸟似雄鸟，但尾羽和初级飞羽无白色斑块或不显著。

分布与生境 中国除西北地区外，几乎皆有分布。北京区域性常见于中低山阔叶林中，为旅鸟和夏候鸟。

习性与生态 夜行性，白天在林中，贴伏于较大树枝上，体色与树皮融为一体，极难发现，民间俗称“贴树皮”。飞行快而无声，飞行中捕捉蛾、蚊和甲虫等。鸣声为连续快速而单调的“qiu. qiu. qiu. qiu……”，彻夜不停，一般为九声一度。营巢于地面；窝卵数 2 枚。

Morphological characteristics A medium-sized scansorial bird (26-28 cm). Male is overall dark brown coarsely marked with greyish-white and blackish-brown; chin and throat black-brown with a large white patch under throat; one white bar on outer primaries; central tail feathers black and outer tail feather has a white subterminal band. Female is similar to male, but the bars on tail and primaries are rufous-brown.

Distribution and habitat Widespread in China except for the northwest. In Beijing, locally common in broadleaved forest in lower mountain areas. Residential status: migrant and summer breeder.

Behaviour and ecology Nocturnal. In the daytime, often rests by lying horizontal along a branch, camouflaged with the bark. Flies fast and silently, catching moths, mosquitos and beetles on the wing. Song is a continuous, fast and monotonous "qiu.qiu.qiu.qiu….". Nests on the ground. Clutch size: 2.

北京市一级保护野生动物 | IUCN 红色名录 无危 (LC) 2016.10.1

短嘴金丝燕	*Aerodramus brevirostris*	Himalayan Swiftlet	雨燕科 Apodidae

IUCN 红色名录　无危 (LC) 2016.10.1

形态特征　小型攀禽，体长 13~14 厘米。雌雄相似。喙黑色。虹膜褐色。整体烟黑色略具蓝色金属光泽，翅长而钝，折合时明显突出于尾端。腰部多为较浅淡灰褐色。下体偏烟灰色，尾呈浅叉状。跗跖黑色略被羽。

分布与生境　分布于中国中部和西南部。主要栖息于海拔 500~4000 米的山坡石灰岩溶洞中。北京罕见记录于野鸭湖湿地公园及温榆河附近，推断为迷鸟。

习性与生态　白天常成群在栖息地上空飞翔猎食。主要以各种昆虫为食。通常营巢于岩壁洞中岩壁之上；巢呈浅盘形，主要由苔藓和涎液构成。

Morphological characteristics A small scansorial bird (13-14 cm). Sexes alike. Bill black; iris brown. Overall greyish-black with a metallic blue gloss; lighter rump greyish-brown and underparts greyish. Wings long and blunt, extending well beyond the tail when folded; tail slightly forked. Tarsi black.

Distribution and habitat Central and southwest China. Breeding habitat is usually limestone caves between 500-4,000 m elevation. In Beijing, rarely recorded at Yeyahu wetland park in Yanqing District and near the Wenyu River in Shunyi District. Residential status: rare vagrant.

Behaviour and ecology Often encountered in flocks around the breeding habitat. Feeds on insects. Nests on cliffs inside caves with plate-like nest consisting of moss and saliva.

白喉针尾雨燕 *Hirundapus caudacutus* White-throated Needletail 雨燕科 Apodidae

形态特征 小型攀禽，体长 19~20 厘米，为中国北方体型最大的雨燕。雌雄相似。喙为短阔的三角形。全身大致为黑褐色，翅狭长。翼覆羽和尾羽具暗绿色金属光泽，尾羽轴末端延长呈针状。颏、喉白色。两肋和尾下覆羽白色，飞翔时从下面可见明显的白色马蹄形斑。

分布与生境 中国见于东北、西南和东部地区。主要栖息于山地森林、河谷等开阔地带。北京不常见于山区和郊区河流上空，为夏候鸟和旅鸟。

习性与生态 单只或集小群飞翔于山头盘旋，速度极快，捕食空中飞虫。筑巢于悬崖石缝和树洞中；窝卵数 2~6 枚，卵白色。

Morphological characteristics A small scansorial bird (19-20 cm), while the largest swift species in northern China. Sexes alike. Bill is short, wide and triangular; body mostly blackish-brown; wings long and narrow; metallic green gloss on upperwing-coverts and tail; tail feathers extend to a needle shape. Chin and throat white; flanks and undertail-coverts white, forming a distinct horseshoe-shaped white patch seen from underneath in flight.

Distribution and habitat Northeast, southwest and eastern China. Found in mountain forest and river valleys. In Beijing, uncommon over mountains or rivers, usually high in the sky but can descend during bad weather. Residential status: summer breeder and migrant.

Behaviour and ecology Powerful flier. Often seen circling at high speed over mountain tops, alone or in small groups, catching flying insects on the wing. Nests in rock crevices on cliffs or in tree hollows. Clutch size: 2-6 white eggs.

北京市一级保护野生动物 | IUCN 红色名录 无危 (LC) 2016.10.1

普通雨燕 | *Apus apus* | Common Swift | 雨燕科 Apodidae

北京市一级保护野生动物

IUCN 红色名录　无危 (LC) 2016.10.1

形态特征　小型攀禽，体长 17~18 厘米。雌雄相似。全身几乎纯深褐色。喙短阔而平扁，呈纯黑色。颏、喉部近白色。胸部、腹部和尾下覆羽近暗烟褐色。两翼极狭长呈镰刀状。具浅开叉的叉尾。

分布与生境　中国大部分地区均有分布，但于北方更为常见。在北京多见于城区，特别是古建筑较多的公园、高校等地。为常见夏候鸟和旅鸟。

习性与生态　晨昏常在巢区附近疾速飞翔，叫声尖锐，白天飞行高度较高。在疾飞中张口捕食飞虫。在北京，普通雨燕多营巢于古建斗拱结构的缝隙中，亦在现代建筑（包括楼宇、桥梁乃至机场等）适合的洞状空间中筑巢；巢由唾液或泥土混合空中收集的草茎、枯叶、棉絮等筑成，呈平碟状；窝卵数 2~4 枚。

Morphological characteristics A small scansorial bird (17-18 cm). Sexes alike. Overall dark brown plumage, with white chin and throat. Short, wide and flat bill all black. Breast, belly and undertail-coverts dull ashy-brown. Sickle-shaped wings long and narrow and tail forked.

Distribution and habitat Widespread in China, mainly northern China. Common summer breeder and migrant in urban Beijing, especially in urban parks and universities with ancient architecture.

Behaviour and ecology Often found in fast-flying screaming parties, especially early morning and evening; higher at daytime. Feeds on the wing by catching flying insects. Breeds in the roof beams of traditional-style buildings such as old city gates or temples; more recently has adapted to modern structures such as road bridges and airport terminals. Nest is plate-like, made of mixed with grass stems, dry leaves, cotton and other materials collected on the wing attached using saliva or mud. Clutch size: 2-4.

白腰雨燕	*Apus pacificus*	Fork-tailed Swift	雨燕科 Apodidae

形态特征 小型攀禽，体长 17~19 厘米。雌雄相似。似普通雨燕，喙黑色，颏喉部白色具细黑褐色羽干纹。上体羽、两翼和尾大部分为黑褐色，腰部白色显著。腹部和尾下覆羽均为黑褐色。跗跖紫黑色。

分布与生境 中国除西北地区外均有分布。从山地至沿海皆可见，特别是近河边的悬崖峭壁，迁徙时不常见于北京，为旅鸟；少量个体繁殖于东灵山。

习性与生态 主要以飞虫等为食。营巢于岩壁上，用枯草、枯叶、细须根、残羽等与亲鸟唾液粘附在岩壁上；巢呈圆杯形；窝卵数 2~3 枚。

Morphological characteristics A small scansorial bird (17-19 cm). Sexes alike. Similar to Common Swift. Bill black, chin and throat white with blackish-brown shaft-streaks. Upperparts, wings and tail mostly greyish-brown with a conspicuous white rump. Belly and undertail-coverts blackish-brown. Tarsi black.

Distribution and habitat Widespread in China except for the northwest. Found near mountains or coastal cliffs, especially near rivers. Can also be found breeding near human habitation. Uncommon migrant in Beijing, also a few pairs breed at Donglingshan in Mentougou District.

Behaviour and ecology Feeds on flying insects. Nests in cliff crevices. Nest is round cup shaped, made of dry grass, leaves, fibrous roots, feathers and other material found on the wing, attached using saliva. Clutch size: 2-3.

北京市一级保护野生动物		IUCN 红色名录　无危 (LC) 2016.10.1

十七、鹃形目 CUCULIFORMES

小鸦鹃	*Centropus bengalensis*	Lesser Coucal	杜鹃科 Cuculidae

未成年 Immature

成鸟 Adult

国家二级保护野生动物

IUCN 红色名录　无危 (LC) 2016.10.1

形态特征　中型攀禽，体长 30~42 厘米。雌雄相似。虹膜深红色。头、颈、上背及下体黑色。两翼栗色，下背和尾上覆羽淡黑色，具蓝色光泽。尾长，黑色。跗跖黑色。

分布与生境　主要分布于中国长江以南地区。栖息于低山丘陵和山麓平原地带的灌丛、草丛、果园中。北京甚罕见于水域附近的灌丛和林地中，仅见于夏季，具体居留型尚不明确。

习性与生态　常单独或成对活动。较为隐蔽，喜茂密的灌木丛或草丛。主要以昆虫和其他小型动物为食，也吃少量植物果实与种子。一般营巢于茂密的灌丛或竹林中；窝卵数 3~5 枚。

Morphological characteristics A medium-sized scansorial bird (30-42 cm). Sexes alike. Iris is dark red. Head, neck, upperparts and underparts are black. Wings chestnut; lower back and rump are light black, glossed with blue. Tail is long and black; tarsi also black.

Distribution and habitat Southern China. Found in shrubby grassland and orchards in low mountains and hills or plains. In Beijing, very rare in shrubby areas and plantations near water. Residential status: recorded in summer but no evidence of breeding.

Behaviour and ecology Feeds alone or in pairs. Prefers to skulk in dense undergrowth, where it feeds on insects and other small creatures, fruit and seeds. Nests in dense shrubs and bamboo. Clutch size: 3-5.

红翅凤头鹃 *Clamator coromandus* Chestnut-winged Cuckoo 杜鹃科 Cuculidae

形态特征 中型攀禽，体长 35~45 厘米。雌雄相似。黑色冠羽显著。喙黑色。喉及胸部橙褐色，颈圈白色与背部的黑色对比明显，翼栗色，腹部近白。尾长，黑色并具蓝色金属光泽。跗跖黑色。

分布与生境 广泛分布于中国南方。在北京于夏季不常见于山区（如小龙门林场、西山、密云水库等地），栖息于低山丘陵和山麓平原的疏林灌丛中，但尚无繁殖证据。

习性与生态 于低矮植被丛中捕食昆虫。振翅与飞行时冠羽收拢。鸣声为响亮的两声或三声一度“ku ku ku”声。一般将卵产于噪鹛和鹊鸲的巢中。

Morphological characteristics A medium-sized scansorial bird (35-45 cm). Sexes alike. Conspicuous black crest, black bill, orange-brown throat and breast. White neck ring contrasts with the black back, chestnut wings and whitish belly. Long, black tail glossed with metallic blue. Tarsi black.

Distribution and habitat Southern China. In Beijing, found in dense undergrowth of low mountains and hills or on the plain (for example at Xiaolongmen, Xishan and Miyun Reservoir). Residential status: uncommon in summer, but no evidence of breeding.

Behaviour and ecology Forages for insects in undergrowth. Gives distinctive double or three notes loud calls "ku ku ku". Brood parasitises laughingthrushes and Oriental Magpie-robins.

北京市一级保护野生动物 | IUCN 红色名录 无危 (LC) 2016.10.1

噪鹃	*Eudynamys scolopaceus*	Common Koel	杜鹃科 Cuculidae

雄鸟 Male

雌鸟 Female

北京市二级保护野生动物		IUCN 红色名录　无危 (LC) 2016.10.1

形态特征　大型攀禽，体长 37~46 厘米。雄鸟喙象牙白色或浅黄绿色略下弯，虹膜为红色，全身黑褐色具蓝绿色金属光泽。雌鸟全身大致为深褐色并密布灰白色斑点，背、翼上覆羽、飞羽及尾羽具横斑。跗跖蓝灰色。

分布与生境　中国除西北部以外，均有分布。在北京为不常见夏候鸟，栖息于山地、丘陵、山脚平原地带林木茂盛之处。

习性与生态　多单独活动。隐藏于茂密的树冠处，连续发出"Ah-"的叫声，声调逐渐提高，可达 12 次。一般将卵产于鸦科鸟类的巢中。

Morphological characteristics A large scansorial bird (37-46 cm). Male has ivory or pale yellowish-green downcurved bill and red eyes; overall plumage black-brown with metallic blue-green gloss. Female is dark brown mottled with white; back, wing-coverts, remiges and tail all barred; tarsi bluish-grey.

Distribution and habitat Widely distributed except northwest region in China. In Beijing, uncommon in wooded areas in the mountains, hills and in lowlands. Residential status: summer breeder.

Behaviour and ecology Solitary, hiding in the canopy of lush woodland. Call is a repeated "Ah-" with increasing emphasis, delivered consecutively up to 12 times. Brood parasitises magpies and crows.

大鹰鹃 | *Hierococcyx sparverioides* | Large Hawk Cuckoo | 杜鹃科 Cuculidae

形态特征 中型攀禽，体长 38~40 厘米。雌雄相似。喙黑灰色，下喙基部淡绿色。头至后颈灰色，上体褐色。尾较长，灰色，具 4~5 条宽阔的黑色横斑和较窄的灰白色端斑。颏黑色，喉、胸部具黑褐色短纵纹，胸具栗棕色宽带，腹部具黑褐色横斑，下体余部大致为白色。跗跖黄色。

分布与生境 主要分布于中国华北、华中及南方地区。在北京为区域性常见夏候鸟和旅鸟。栖于山地阔叶林中。

习性与生态 以昆虫为食，最为喜欢鳞翅目幼虫、直翅目昆虫、鞘翅目昆虫，亦兼吃果类。鸣声为“贵、贵、油”三声一度，愈叫愈响且音调升高，突然停止。一般将卵产于其他鸟类的巢中。

Morphological characteristics A medium-sized scansorial bird (38-40 cm). Sexes alike. Bill blackish-grey with light green at base of lower mandible; plain grey from head to hindneck; upperparts brown; tail long, greyish, with 4-5 broad blackish bars and narrow greyish-white terminal band; chin black; underparts largely white, with blackish-brown short vertical streaks on throat and breast and a broad rufous band on breast; belly barred dark; tarsi yellow.

Distribution and habitat North, central and southern China. Locally common summer breeder and passage migrant in Beijing, primarily found in broadleaved woodland in the hills and mountains in the west and north.

Behaviour and ecology Feeds on insects, especially caterpillars, grasshoppers, and beetles. Sometimes feeds on fruit. Repeated three notes calls "gui gui yo" rising in volume and pitch, stopping abruptly. A brood parasite.

北京市二级保护野生动物 | IUCN 红色名录 无危 (LC) 2016.10.1

北棕腹鹰鹃 *Hierococcyx hyperythrus* Northern Hawk Cuckoo

杜鹃科
Cuculidae

北京市二级保护野生动物

IUCN 红色名录　无危 (LC) 2016.10.1

形态特征　中型攀禽，体长 25~33 厘米。雌雄相似。喙尖黑色，基部黄色。虹膜褐色，金黄色眼圈显著。头侧灰色，枕部具白色条带，颏黑色，喉偏白，上体青灰色，胸棕色。尾灰色，具多条较窄的黑色横斑、宽阔的黑色次端斑和较窄的棕色端斑。跗跖黄色。

分布与生境　在中国主要繁殖于东北，越冬于华南等地。栖于常绿阔叶林、针叶林或山地灌木林中。在北京为罕见夏候鸟和旅鸟。

习性与生态　多单独活动，性隐蔽，不易发现。喜在高树上鸣叫，以昆虫尤其是鳞翅目幼虫为主要食物。其声尖锐而轻，反复鸣叫多次后方停歇。一般将卵产于鹟、鸫等鸟类的巢中。

Morphological characteristics A medium-sized scansorial bird (25-33 cm). Sexes alike. Bill pointed and black with yellow base; iris brown with prominent golden eye-ring; grey on head-sides with white streaks on nape; black chin and white throat; bluish-grey upperparts with a brown breast; tail blackish-brown barred with narrow brown edge on tail feathers; tarsi yellow.

Distribution and habitat Breeds in northeast China, wintering in the south. Found in broadleaved evergreen forest, coniferous woodlands and shrubby hills. Rare summer breeder and passage migrant in Beijing.

Behaviour and ecology Usually solitary, secretive and hard to detect. Often calls from tall trees; preys on insects, especially caterpillars. Call is light and sharp, often repeated more than ten times. Brood parasitises thrushes and flycatchers.

小杜鹃	*Cuculus poliocephalus*	Lesser Cuckoo	杜鹃科 Cuculidae

形态特征 中型攀禽，体长 24~28 厘米。喙黑色，基部黄色。虹膜褐色。头、颈及上体灰色，飞羽黑褐色，翼缘灰色。尾羽深灰色，具白色尖端。上胸浅灰沾棕，下体余部白色，杂以较宽的黑色横斑。跗跖黄色。棕色型雌鸟头及上体棕褐色。

分布与生境 中国除西北部外，均有分布。在北京为不常见夏候鸟。多栖于山中各种林地，有时也到村庄附近的林中。

习性与生态 单独活动，常躲在茂密的枝叶间鸣叫。鸣叫声响亮而急促，音调起伏较大。主要以鳞翅目幼虫为食。巢寄生，将卵产在白腹蓝鹟、莺类和噪鹛等鸟类的巢中。

Morphological characteristics A medium-sized scansorial bird (24-28 cm). Bill black with yellow base; iris brown; head, neck and upperparts grey; remiges blackish-brown with grey fringes; tail dark grey, tipped white; upper breast light grey, washed with brown; remaining of underparts white with broad black bars; tarsi yellow. Hepatic female is brown from head to upperparts.

Distribution and habitat Widespread in China except for the northwest. Uncommon summer breeder in Beijing. Found in woodlands in the hills and mountains, sometimes near villages.

Behaviour and ecology Solitary. Often calls while hiding in thick woods; call is loud and rapid with rising and falling tones. Mainly feeds on caterpillars. Brood parasitises Blue-and-white Flycatchers, warblers and laughingthrushes.

北京市二级保护野生动物 | IUCN 红色名录　无危 (LC) 2016.10.1

棕色型 Hepatic

四声杜鹃	*Cuculus micropterus*	Indian Cuckoo	杜鹃科 Cuculidae

北京市二级保护野生动物

IUCN 红色名录　无危 (LC) 2018.8.6

形态特征　中型攀禽，体长 30~38 厘米。喙黑灰色，基部黄绿色。虹膜红褐色，眼圈黄色。头、颈、上体及上胸灰色（雄鸟）或灰褐色（雌鸟）。尾近末端具宽阔的黑斑，端部近白色。下体白色，具深色横斑。跗跖黄色。

分布与生境　分布于中国东北至西南及东南。在北京为甚常见夏候鸟。栖息于山地或平原地区的密林或城市公园中。

习性与生态　常单独活动，嗜食毛虫。繁殖期内鸣叫频繁，鸣叫声为响亮的四声一度，似“光棍好苦”或“割麦割谷”声，第三声稍高，第四声最低。常巢寄生于多种雀形目鸟类的巢中，在北京的主要寄主为灰喜鹊和黑卷尾。

Morphological characteristics　A medium-sized scansorial bird (30-38 cm). Black upper mandible and blackish green lower mandible; Reddish-brown iris, yellow eye-ring; plain grey (male) or brownish grey (female) on head, neck, upper breast and upperparts; white underparts with dark bars; tail has off-white tips and broad black band near tip; tarsi yellow.

Distribution and habitat　Widely distributed from the northeast to southern China. Abundant summer breeder in Beijing. Usually found in the canopy of dense woodland in the hills and lowlands, including in city parks.

Behaviour and ecology　Usually encountered alone and more often heard than seen. Its loud, four-note call is often transcribed as "one more bottle". Feeds mostly on caterpillars. Brood parasitises a variety of passerine species, especially Azure-winged Magpies and Black Drongos in Beijing.

中杜鹃	*Cuculus saturatus*	Himalayan Cuckoo	杜鹃科 Cuculidae

形态特征 中型攀禽，体长25~34厘米。喙端黑色，喙基黄色。虹膜棕黄色，眼圈黄色。成年雄鸟上体灰色，腹部有黑白色横纹，尾羽深灰色。雌鸟外观通常似雄鸟，但棕色型个体的上体为肉桂褐色，且通体密布黑色横纹，极似东方中杜鹃，无法通过外表与之区分。

分布与生境 主要见于中国南部地区，零星繁殖于华北。在北京为区域性常见夏候鸟。栖息于山区林地。

习性与生态 常单独活动。喜食鳞翅目幼虫和鞘翅目昆虫。鸣声包括三或四个音节，第一音节比后接的四个音节更加尖锐清晰，似戴胜，与东方中杜鹃的鸣声区别明显。巢寄生，将卵产在树莺、柳莺等雀形目鸟类的巢中。

Morphological characteristics A medium-sized scansorial bird (25-34 cm). Bill black at tip, yellow at base; iris brownish yellow, eye-ring yellow. Adult male has grey upperparts, black-and-white barred belly and dark grey tail. Female of the normal morph is similar to male, with a rufous wash on the neck/upper breast; female also occurs in a morph with cinnamon upperparts finely barred black. Very similar to Oriental Cuckoo and unlikely to distinguish by appearance.

Distribution and habitat Mainly southern part of China, scarce in the northern. Locally common summer breeder in Beijing, found in forest in montane areas.

Behaviour and ecology Feeds on insects, especially caterpillars and beetles. Song is a high note followed by two or three lower flat notes, "hoop, hoop-hoop", considered similar to voice of Common Hoopoe (*Upupa epops*). Brood parasitises bush warblers and leaf warblers.

北京市二级保护野生动物 | | IUCN 红色名录 无危 (LC) 2016.10.1

棕色型 Hepatic

东方中杜鹃	*Cuculus optatus*	Oriental Cuckoo	杜鹃科 Cuculidae

北京市二级保护野生动物 *		IUCN 红色名录　无危 (LC) 2016.10.1

形态特征　中型攀禽，体长 25~34 厘米。喙黑色，基部黄色。虹膜棕黄色，眼圈黄色。颏、喉、上胸灰色，上体石板灰色，下胸及腹部白色，满布较宽的黑褐色横斑。尾黑褐色具白斑。跗跖橘黄色。

分布与生境　分布于中国东北、华北等地。在北京为不常见旅鸟。栖息于山区林地。

习性与生态　常单独活动。喜食鳞翅目幼虫和鞘翅目昆虫。鸣叫声为低而浑厚的“呼、呼—”，二声一度。巢寄生，将卵产于莺科鸟类的巢中。

＊注：IUCN 红色名录及《北京市一级保护野生动物名录》中未收录（更新/接受）中杜鹃和东方中杜鹃的物种划分，故此处沿用中杜鹃的数据。

Morphological characteristics　A medium-sized scansorial bird (25-34 cm). Bill blackish with yellow base; brownish-yellow iris, yellow eye-ring; head, upperparts, chin, throat and upper breast slaty grey; lower breast and belly white with broad blackish-brown bars; tail brownish-black with white markings; tarsi orange-yellow.

Distribution and habitat　Northern China. Uncommon passage migrant in Beijing. Found in all types of forests in montane areas.

Behaviour and ecology　Feeds on insects, especially caterpillars and beetles. Call is similar to Himalayan Cuckoo but with fewer, usually two, lower notes "hoop-hoop". Brood parasitises warblers.

大杜鹃	*Cuculus canorus*	Common Cuckoo	杜鹃科 Cuculidae

形态特征 中型攀禽，体长 28~37 厘米。喙近黑色，下喙基部黄色。虹膜及眼圈黄色。雄鸟头、颈、颏、喉、上胸及上体均呈灰色。飞羽黑褐色。腹部白色，具细的黑褐色斑纹。中央尾羽沿羽干两侧缀白色斑点。跗跖黄色。雌鸟似雄鸟，但胸沾棕色。棕色型雌鸟上体棕红色，密布黑色横斑。

分布与生境 广泛分布于中国大部分地区。在北京为甚常见夏候鸟。栖息于开阔的树林中，尤喜近水树林间。

习性与生态 常单独活动，繁殖期频繁鸣叫。鸣声为洪亮的“布谷”二声一度。以昆虫为食，嗜吃毛虫。多见巢寄生于苇莺巢中；在北京主要寄主为东方大苇莺。

Morphological characteristics A medium-sized scansorial bird (28-37 cm). Bill blackish with yellow base at lower mandible; yellow iris and eye-ring. Male has plain grey on head, neck, chin, throat, upper breast and upperparts; remiges blackish-brown; white belly with fine brownish-black bars, white speckles along the side of central tail feathers; tarsi yellow. Female resembles male but has varied rufous wash on breast. Hepatic female has rufous-brown upperparts with black bars.

Distribution and habitat Widely distributed in China. Abundant summer breeder in Beijing. Found in open woodlands and wetlands.

Behaviour and ecology Highly vocal in breeding season. Call is a familiar two-note "cuck-oo". Feeds mostly on soft-bodied insects, especially caterpillars. Brood parasitises reed warblers; in Beijing, the most common host is the Oriental Weed Warbler.

北京市二级保护野生动物 | IUCN 红色名录 无危 (LC) 2016.10.1

棕色型 Hepatic

十八、犀鸟目 BUCEROTIFORMES

戴胜	*Upupa epops*	Common Hoopoe	戴胜科 Upupidae

北京市二级保护野生动物		IUCN 红色名录　无危 (LC) 2020.8.14

形态特征　中型攀禽，体长 24~31 厘米。雌雄相似。喙黑色，细长而弯曲，喙基部淡黄色。头、颈、胸为黄褐色，额至枕部有甚长的扇形冠羽，冠羽栗棕色，羽端缀黑。下背黑色，腰白色，尾上覆羽基部白、端黑，两翼黑褐色具白斑。尾黑色，中央尾羽之半有一道宽阔的白色横斑，下体棕色。跗跖黑色。

分布与生境　中国广泛分布。栖息于城市、平原、山地的林缘中。在北京为常见留鸟、夏候鸟、旅鸟。

习性与生态　常在地面行走觅食。以地表或土层以下昆虫为食，如金针虫、蝼蛄、步甲等。兴奋时冠羽打开呈扇状。叫声为低沉、有明显停顿“hoo-poo-poo”。营巢于树洞中；窝卵数 6~8 枚。

Morphological characteristics A medium-sized scansorial bird (24-31cm). Sexes alike. Unmistakeable. Long, thin and down-curved dark bill, usually with a pale base. Sandy-buff head, neck and breast, fan shaped black-tipped chestnut-brown crest from forehead to nape. Black lower back; white rump; uppertail-coverts are white at the base, black at the tip; blackish-brown wings with white markings. Black tail with wide white band in the middle of the central tail feathers; pale buff underparts. Brown iris and black tarsi.

Distribution and habitat Widely distributed in China. Common resident, summer breeder and migrant in Beijing. Found in open areas with trees in the city, plains and hills alike.

Behaviour and ecology Often found walking on the ground, using its bill to probe for insects underground, such as wireworm, mayfly and beetles. Crest fanned out when excited. Call is a low "hoo-poo-poo". Nests in tree holes. Clutch size: 6-8.

十九、佛法僧目 CORACIIFORMES

蓝胸佛法僧	*Coracias garrulus*	European Roller	佛法僧科 Coraciidae

IUCN 红色名录 无危 (LC) 2019.8.14

形态特征 中型攀禽，体长 30~42 厘米。雌雄相似。喙黑色。虹膜深褐色。成鸟通体淡蓝绿色。头、颈和下体淡蓝色，喉和前颈羽轴白色或暗蓝绿色。肩、背和内侧飞羽棕色，腰部蓝紫色。尾上覆羽、中央尾羽蓝绿色。跗跖暗黄色。

分布与生境 分布于中国西北地区。栖息于海拔 1500 米以下的低山和山脚平原等开阔地方的各种生境中。在北京为甚罕见迷鸟，仅于 2007 年 6 月在首都机场有过一次记载。

习性与生态 常单独或成对活动。多在空中飞翔捕食，上下翻飞，也到地上追捕食物。以甲虫、蟋蟀、蝗虫等无脊椎动物为食，也食小型蛙类、鼠类、蜥蜴等。

Morphological characteristics A medium-sized scansorial bird (30-42 cm). Sexes alike. Dark brown iris, black bill. Adult is overall light blue green; head, neck and underparts light blue with white or dark bluish-green on throat and foreneck; mantle, back and inner side of remiges brown; rump bluish purple; uppertail-coverts and central tail feathers are bluish green; tarsi dark yellow.

Distribution and habitat Northwest China. Found in open areas of low mountains and plains below 1,500m elevation. Very rare vagrant in Beijing with only one record at Beijing Capital International Airport in June 2007.

Behaviour and ecology Usually encountered alone or in pairs. Flycatches for insects in the air and may chase prey on the ground. Feeds on invertebrates such as beetles, crickets and grasshoppers and also on small frogs, rodents and lizards.

三宝鸟	*Eurystomus orientalis*	Dollarbird	佛法僧科 Coraciidae

形态特征 中型攀禽，体长 24~29 厘米。雌雄相似。色彩绚丽，全身羽色大部分为蓝绿色，光线弱时显黑。头大而扁平。喙短而宽，呈朱红色。头、颈、颏黑色，喉中部蓝黑色。翼和尾蓝黑色，翼上有一道显著的天蓝色翼斑。跗跖朱红色。

分布与生境 主要分布于中国东北、华北、华中以及喜马拉雅山脉等地。栖息于山地或平原林中，也喜欢在林区边缘空旷处或林区里的开垦地上活动。在北京为不常见夏候鸟和旅鸟。

习性与生态 夏季单独或成对在林区的边缘或林中空地处活动，捕食飞行昆虫。鸣叫声粗厉，似“ge! ge! ya!”。常营巢于树洞中；窝卵数 3~4 枚。

Morphological characteristics A medium-sized scansorial bird (24-29 cm). Sexes alike. Colourful plumage mostly blue-green; can look blackish in poor light. Big head with short and wide bright red bill. Head, neck and chin black with bluish-black throat, wings and tail. Shows a conspicuous sky-blue wing bar. Tarsi bright red.

Distribution and habitat Northern and central China and the Himalayans. Found in wooded areas in the mountains and on the plain, and in open fields or farmland next to forest. Uncommon summer breeder and passage migrant in Beijing.

Behaviour and ecology Usually found alone or in pairs in open areas at the edge or in forest. Preys on flying insects. Call is a harsh "ge! ge! ya!". Usually nests in tree holes. Clutch size: 3-4.

北京市一级保护野生动物		IUCN 红色名录 无危 (LC) 2016.10.1

白胸翡翠 | *Halcyon smyrnensis* | White-throated Kingfisher | 翠鸟科 Alcedinidae

国家二级保护野生动物 | IUCN 红色名录　无危 (LC) 2016.10.1

形态特征　中型攀禽，体长 26~30 厘米。雌雄相似。喙呈赤红色，粗长基部较宽。虹膜暗褐。颏、喉、胸部中央白色。头、后颈、颈侧以及下体余部均深赤栗色。上背、肩及三级飞羽蓝绿色，下背、腰及尾上覆羽均翠绿色。跗跖珊瑚红色。

分布与生境　主要分布于中国华南各省。栖息于山脚平原河流、湖泊岸边，或池塘、水库等水域岸边，有时亦远离水域活动。在北京为甚罕见迷鸟。

习性与生态　常单独活动，多站在水边树木枯枝上或石头上，有时亦站在电线上。主要以鱼、蟹、软体动物为食，亦食小型陆栖脊椎动物，如蜥蜴等。营巢于河岸土洞中；窝卵数 5~7 枚。

Morphological characteristics A medium-sized scansorial bird (26-30 cm). Sexes similar. Bright red bill thick and long with wide base. Iris dark brown. White on chin, throat and central breast; head, hindneck, neck-sides and remaining underparts dark chestnut; upper back, mantle and tertials bluish green; lower back, rump and uppertail-coverts gleaming jade green. Tarsi coral red.

Distribution and habitat Mainly in southern China. Found along rivers, lakes, ponds and reservoirs; occasionally seen away from water. Very rare vagrant in Beijing.

Behaviour and ecology Often found singly perched on dry branch or rock near water, sometimes on wires. Feeds on fish, crabs, molluscs and small terrestrial creatures, such as lizards. Nests in holes in riverbanks. Clutch size: 5-7.

蓝翡翠	*Halcyon pileata*	Black-capped Kingfisher	翠鸟科 Alcedinidae

形态特征 中型攀禽，体长 28~31 厘米。雌雄相似。喙粗长，壮硕，呈珊瑚红色。虹膜暗褐色。额、头顶、头侧和枕部黑色，颏、颊白色，后颈白色，从两侧向前延伸与喉、上胸部白色相连，形成一宽阔的白色领环。背、腰和尾钴蓝色。初级飞羽黑褐色，内侧基部有一大块白斑。胸以下包括腋羽和翼下覆羽橘黄色。跗跖红色。

分布与生境 主要分布于华北、华东、华中及华南的大部分地区。栖息于林中溪流、山脚与平原地带的河流、水塘和沼泽地带。在北京为不常见夏候鸟。

习性与生态 常在水边的乔木上休息。主要食小鱼、虾和昆虫，亦食小型两栖类及爬行类。筑巢于水域边土洞；窝卵数 3~4 枚。

Morphological characteristics A medium-sized scansorial bird (28-31 cm). Sexes alike. Coral-red bill thick, long and strong. Iris dark brown. Forehead, crown, head-sides and nape black; hind neck white, extending to throat and breast to form a wide collar. Back and rump cobalt blue. Primaries blackish-brown with a large white patch at inner base. Chin, throat, neck-sides, cheek and upper breast white; orange-yellow below breast including axillaries and underwing-coverts. Tarsi red.

Distribution and habitat Most of northern, eastern, central and southern China. Found near wooded streams, ponds and marshy areas. Uncommon summer breeder in Beijing.

Behaviour and ecology Often found perching on tree branches near water. Nests in holes of riverbanks; feeds mainly on fish, shrimps and insects but sometimes takes small reptiles and amphibians. Clutch size: 3-4.

北京市一级保护野生动物

IUCN 红色名录　无危 (LC) 2016.10.1

普通翠鸟	*Alcedo atthis*	Common Kingfisher	翠鸟科 Alcedinidae

雌鸟 Female

雄鸟 Male

IUCN 红色名录　无危 (LC) 2016.10.1

形态特征　小型攀禽，体长 15~17 厘米。雄鸟喙直而细长呈黑色；喉和耳后颈侧为白色，眼下和耳羽栗棕色；额至后颈暗蓝绿色，具翠蓝色细斑，背至尾覆羽为翠蓝色，两翼为蓝绿色；胸部以下为鲜橘黄色；跗跖红色。雌鸟似雄鸟，但下喙橙红色。

分布与生境　主要分布于中国东北、华东、华中、华南及西南地区。栖息于较低海拔城市、郊野的大部分水域。在北京为常见夏候鸟和留鸟。

习性与生态　常单独停栖在灌木、秃枝和露在水面的石块上。主食小鱼、虾和水生昆虫。筑巢于陡直的河岸土洞中；窝卵数 6~7 枚，卵白色。

Morphological characteristics　A small scansorial bird (15-17 cm). Male has long and straight black bill, white on neck sides behind ears and chestnut brown ear-coverts; forehead to hindneck dark blue-green with fine speckles; back to tail-coverts gleaming azure-blue; black-brown remiges with blue green on surface; underparts bright orange-yellow; tarsi red. Female similar to male except lower mandible orange-red with black tip.

Distribution and habitat　Northeast, east, central and south China. Found in most types of wetland habitat with open water, including rivers, lakes and reservoirs. Common summer breeder and resident in Beijing.

Behaviour and ecology　Often found alone perching on a bush, bare branch or rock above water. Usually nests in soft riverbanks, where it excavates a tunnel; feeds on fish, shrimps and aquatic insects. Clutch size: 6-7, white eggs.

攀禽

IV

冠鱼狗	*Megaceryle lugubris*	Crested Kingfisher	翠鸟科 Alcedinidae

IUCN 红色名录　无危 (LC) 2016.10.1

形态特征　中型攀禽，体长 37~42 厘米，具显著的冠羽。喙长、直且粗壮，呈黑褐色。头、冠羽、上体、两翼和尾羽黑色具白色横斑或点斑，颈侧和腹部白色，胸部具一黑白斑的横带。雄鸟翼下覆羽白色，雌鸟翼下覆羽橘黄色。跗跖铅灰色。

分布与生境　广泛分布于中国东部和南部。栖息于低山水边的树枝上。在北京为不常见留鸟。

习性与生态　平时常独栖在近水边的树枝顶上、电线杆顶或岩石上，伺机捕鱼。于水域边挖隧道筑巢；窝卵数 3~5 枚，卵白色。

Morphological characteristics A medium-sized scansorial bird (37-42 cm) with prominent crest. Both sexes have a long, straight, black-brown bill; head, crest, back, wings and tail are black with white bars or speckles; neck-side and belly white with a black and white speckled breast band; tarsi leaden grey. Female separated from male by orange-yellow underwing-coverts (white in males).

Distribution and habitat Widespread in eastern and southern China. Usually found near moving water, such as mountain streams or rivers. Uncommon resident in Beijing.

Behaviour and ecology Often solitary, perching on tree branches, wires or rocks near water, from where it will dive for fish. Nest is usually a tunnel excavated in a soft riverbank. Clutch size: 3-5, white eggs.

斑鱼狗	*Ceryle rudis*	Pied Kingfisher	翠鸟科 Alcedinidae

形态特征 中型攀禽，体长 26~31 厘米。形似冠鱼狗，通体呈杂状黑白斑，头顶冠羽较短。翼上有较宽的白色翅带，飞翔时极明显。尾白色具宽阔的黑色次端斑，下体白色，雄鸟有两条黑色胸带，前面一条较宽，后面一条较窄。雌鸟仅一条胸带。

分布与生境 主要分布于中国长江流域以南。栖息于低山和平原溪流，以及河流、湖泊等开阔水域岸边。罕见于北京，在郊区水域有零散记录，居留型尚不明确。

习性与生态 常成对或结群活动于较大水面，在距水面几米至十几米的低空飞翔觅食，时而贴近水面，时而升起，来回振翅飞翔。食物以小鱼为主，兼吃甲壳类和多种水生昆虫及其幼虫，也啄食小型蛙类和少量水生植物。叫声为尖厉的哨声。

Morphological characteristics A medium-sized scansorial bird (26-31 cm). Superficially similar to crested kingfisher; with black and white plumage and a relatively short crest. Broad white wing bars conspicuous in flight. White tail has broad black subterminal band; underparts white. Male has double breastband with the upper band broad and the lower band thin. Female has only one breastband.

Distribution and habitat Primarily distributed to the south of the Yangtze River in China. Found in open water areas, such as rivers, streams and lakes. Rare in Beijing, scarce in rural wetlands. Residential status: unclear.

Behaviour and ecology Often found in pairs or in groups hovering above water fishing for food. Feeds mainly on small fish and on crustaceans and other aquatic insects and larva; occasionally takes small frogs. Call is a shrill whistle.

IUCN 红色名录　无危 (LC) 2016.10.1

雄鸟 Male

雌鸟 Female

二十、啄木鸟目 PICIFORMES

蚁䴕	*Jynx torquilla*	Eurasian Wryneck	啄木鸟科 Picidae

北京市一级保护野生动物		IUCN 红色名录　无危 (LC) 2016.10.1

形态特征　小型攀禽，体长 16~20 厘米。雌雄相似。喙尖细。后颈、背的中央贯以黑色粗纹，眼后具有明显的暗褐色纹。上体遍布褐灰色斑，似树皮色。两翼暗褐色，具白和褐色斑点。尾有银灰色与褐色不规则斑点，并有数条黑褐色横斑。喉部皮黄有细横纹，至腹以下横纹渐稀。

分布与生境　几乎分布中国全境。栖息于低山、丘陵、平原阔叶林、混交林、灌丛。在北京为不常见旅鸟。

习性与生态　不同于其他啄木鸟，栖于树枝而不攀树，也不錾啄树干取食。受惊扰常做出头部往两侧扭动的动作。通常单独活动，取食地面蚂蚁。于树洞中营巢，但并不自行凿洞；窝卵数 7~12 枚。

Morphological characteristics　A small scansorial bird (16-20 cm). Sexes similar. Short and pointed bill, black central stripe from hindneck to back; distinct dark-brown line runs behind eye; upperparts mottled brownish grey, like tree bark; wings dark brown with rufous-buff spot-bars; silver grey and brown vermiculated tail has several blackish-brown bands; throat buff-yellow with fine bars, sparsely barred below belly.

Distribution and habitat　Widespread in China. Found in broadleaved forest, mixed forest and shrubs of low mountains, hills and plains. Uncommon migrant in Beijing.

Behaviour and ecology　Usually solitary. Unlike other woodpeckers, the Wryneck does not climb trees or drum on tree trunks; instead, perches on branches and often seen foraging on the ground in search of ants. Twists neck in a snake-like pose when disturbed. Nests in pre-existing tree holes. Clutch size: 7-12.

棕腹啄木鸟	*Dendrocopos hyperythrus*	Rufous-bellied Woodpecker	啄木鸟科 Picidae

形态特征 中型攀禽，体长 18~24 厘米。头顶红色（雄鸟）或黑白相杂（雌鸟），眼先至眼周白色。上体和两翼黑色，具白斑；中央尾羽纯黑色；两侧尾羽黑色，具白色横斑，端白色。耳羽、颏、喉至腹部皆为棕黄色，尾下覆羽红色。

分布与生境 广泛分布于中国东部和南部地区。迁徙时在城市、平原的树林中可见。在北京为区域性常见旅鸟。

习性与生态 常单独活动，主要食蚂蚁等昆虫，亦取食树汁。营巢于腐朽或半腐朽的树干洞里；窝卵数 3 枚。

Morphological characteristics A medium-sized scansorial bird (18-24 cm). Sexes similar except for crown, which is dark red in male and white-spotted black in female. Both sexes have white on lores and around eye; upperparts and wings black with white bars; uppertail black with outer two feather pairs barred white; plain brown from ear-coverts, chin and throat to belly; undertail-coverts dark red.

Distribution and habitat Widespread in southern and eastern China. Locally common migrant in Beijing, found in city parks and woodlands on migration.

Behaviour and ecology Usually seen singly. Feeds mainly on insects, including ants, and sometimes feeds on sap. Nests in hole in rotten tree trunk. Clutch size: 3.

北京市一级保护野生动物 | | IUCN 红色名录 无危 (LC) 2016.10.1

雄鸟 Male

雌鸟 Female

小星头啄木鸟	*Dendrocopos kizuki*	Pygmy Woodpecker	啄木鸟科 Picidae

IUCN 红色名录 无危 (LC) 2016.10.1

形态特征 小型攀禽，体长 12~17 厘米。额、头顶至枕灰褐色，雄鸟枕两侧各有一深红色小纵斑。颊纹和眉纹白色，颈侧具白斑，其余上体黑色，具较细密白色横斑。翼黑色而具白色斑点。喉白色，其余下体污白色，具黑褐色纵纹。

分布与生境 主要分布于中国东北和华北地区。栖息于山地针叶林、针阔叶混交林和阔叶林。在北京为不常见留鸟。

习性与生态 主要以各类昆虫为食，偶尔也吃植物果实和种子。营巢于杨树、水曲柳等心材腐朽的阔叶树上；窝卵数 4~7 枚，卵白色。

Morphological characteristics A small scansorial bird (12-17 cm). Sexes similar, except male has a small dark red spot on each side of nape. In both sexes, forehead, crown to nape brownish-grey; white moustachial stripe and supercilium, white marking on side of neck; remainder of upperparts black with dense white bars; black wings with white spots; throat white; and dirty white underparts with brownish-black streaks.

Distribution and habitat Northeast and north China. Found in coniferous, mixed and broadleaved forest in hilly regions. Uncommon resident in Beijing.

Behaviour and ecology Feeds on insects, occasionally fruit and seeds. Nests in holes in rotten trunks of poplar and ash. Clutch size: 4-7, white eggs.

星头啄木鸟	*Dendrocopos canicapillus*	Grey-capped Woodpecker	啄木鸟科 Picidae

形态特征 小型攀禽，体长 14~18 厘米。额和头顶暗灰色，雄鸟枕部两侧具一红色斑点。上体基本黑色杂以白斑，下体淡茶色具黑褐色纵纹，胸部纵纹更为明显。

分布与生境 分布于中国东北、华北和南方大部分地区。在城市、平原及山区的林中可见，在北京为常见留鸟。

习性与生态 主要食昆虫及其幼虫，有时也食杂草种子。飞行时有节奏地交替升降。叫声单调，一连数声。营巢于心材腐朽的树干上，巢位较高；窝卵数 4~5 枚，卵白色。

Morphological characteristics A small scansorial bird (14-18 cm). Sexes similar except for small red streak on each side of nape in male. In both sexes, forehead and crown dark grey; black upperparts with white speckles; light tan underparts with blackish-brown streaks and darker streaks on breast.

Distribution and habitat Northeast, north and southern China. Found in most wooded areas, including city parks, lowland and in the mountains. Common resident in Beijing.

Behaviour and ecology Feeding on insects and larva, sometimes grass seeds. Flight undulating. Call is monotonous several notes. Nests in holes in rotten tree trunks. Clutch size: 4-5, white eggs.

北京市一级保护野生动物

IUCN 红色名录 无危 (LC) 2016.10.1

IV 攀禽

小斑啄木鸟	*Dendrocopos minor*	Lesser Spotted Woodpecker	啄木鸟科 Picidae

IUCN 红色名录 无危 (LC) 2018.8.7

形态特征 小型攀禽，体长 14~18 厘米。雄鸟头顶红色，雌鸟黑色。额和颊白色，后颈至上背黑色，下背白色而具黑色横斑。两翼黑色具白色横斑。尾黑色，外侧尾羽具白色横斑。下体灰白色，两侧具黑色纵纹。

分布与生境 主要分布于中国东北、西北、内蒙古等地。栖息于低山丘陵和山脚平原阔叶林和混交林中。甚罕见于北京，居留型尚不明确，可能为留鸟或冬候鸟。

习性与生态 常单独活动，取食于树的粗枝和枝叶间，多在林冠层活动。主要以天牛成虫和幼虫、双翅目、蚂蚁等昆虫为食。营巢于阔叶树树洞中；窝卵数 3~8 枚，卵白色。

Morphological characteristics A small scansorial bird (14-18 cm). Sexes similar except for colour of cap – red in male and black in female. In both sexes, forehead and cheek white; neck to upper back black, lower back white with black bars.; wings black with white bands; tail black with white bands on outer tail feather; underparts ash-white with black streaks on sides.

Distribution and habitat From northeast China to northwest China. Found in broadleaved forest and mixed forest of low mountain hills and lowland. Very rare in Beijing. Residential status: unclear; a probable resident or winter visitor.

Behaviour and ecology Usually seen singly among tree branches and foliage; often found in the canopy and on smaller thinner branches rather than trunk. Preys on long-horned beetle, dipterans, ants, etc. Nests in hole in broadleaved trees. Clutch size: 3-8, white eggs.

雄鸟 Male

白背啄木鸟 *Dendrocopos leucotos* White-backed Woodpecker

啄木鸟科 Picidae

北京市一级保护野生动物

IUCN 红色名录 无危 (LC) 2020.8.14

形态特征 中等体型的攀禽，体长 22~28 厘米。雄鸟头顶至枕部朱红色，眼先、颊和耳覆羽棕白色；后颈至上背黑色，下背和腰白色；前颈、胸、腹和两胁白色而具黑色羽干纹；下腹和尾下覆羽沾粉红色。雌鸟头顶黑色，其余似雄鸟。

分布与生境 中国间断分布于东北至河北北部、新疆北部；福建西北、江西北部及台湾；陕西南部秦岭至四川中部。多见于山区的针阔叶混交林和阔叶林，亦见于林缘次生林和疏林地带。在北京为不常见留鸟。

习性与生态 常单独或成对活动。主要以天牛成虫和幼虫等昆虫为食，秋冬季也食植物果实和种子。营巢于阔叶树；窝卵数 3~6 枚，卵白色。

Morphological characteristics A medium-sized scansorial bird (22-28 cm). Male has a bright red crown, extending to nape, tannish white lores, cheeks and ear-coverts; hindneck to upper back black, lower back and rump white; white foreneck, breast, belly and flanks with black shaft-streaks; underparts and undertail-coverts washed with pink. Female resembles male but has black crown.

Distribution and habitat Northeast China to north Hebei province, northern Xingjiang; northwest Fujian, northern Jiangxi and Taiwan; southern Shanxi to central Sichuan. Found in broadleaved forest and mixed forest of mountainous regions, secondary forest edge and sparse woods. Uncommon resident in Beijing.

Behaviour and ecology Feeds singly or in pairs. Preys on longhorn beetles and other insects, found by probing under bark. Nests in hole in broadleaved trees. Clutch size: 3-6, white eggs.

大斑啄木鸟 | *Dendrocopos major* | Great Spotted Woodpecker | 啄木鸟科 Picidae

形态特征 中等体型的攀禽，体长 21~26 厘米。成年雄鸟枕部有一红块斑，雌鸟枕部无红块斑。头顶、后颈、背部、翼及尾均为亮黑色，眉纹和颈侧呈纯白色或稍沾淡棕色，肩具大白斑，翼有白色斑，下体从颊至腹部淡棕褐色，下腹和尾下覆羽红色。

分布与生境 在中国主要分布于东部和新疆西部。见于山地和平原林地、农田、城市公园、小区中。在北京为甚常见留鸟。

习性与生态 主要食昆虫，也食植物种子。多营巢于阔叶树树洞中，距地面 8~10 米；窝卵数 4 枚，卵白色。

Morphological characteristics A medium-sized scansorial bird (21-26 cm). Sexes similar except for red patch on nape of male. In both sexes, crown, hindneck, back, wing and tail black; supercilium and side of neck is white or washed with light brown; large white patch on mantle; white bars on wings, underparts tan from cheek to belly; lower belly and undertail-coverts red.

Distribution and habitat Eastern China and Xinjiang Autonomous Region. Found in most types of woodland, including mountains, cultivations, city parks and residential areas. Abundant resident in Beijing.

Behaviour and ecology Feeds on insects and seeds. Nests in tree holes in broadleaved tree 8-10 m above ground. Clutch size: 4, white eggs.

北京市一级保护野生动物 | | IUCN 红色名录 无危 (LC) 2016.10.1

黑啄木鸟 | *Dryocopus martius* | Black Woodpecker | 啄木鸟科 Picidae

北京市一级保护野生动物 | 国家二级保护野生动物 | IUCN 红色名录 无危 (LC) 2016.10.1

形态特征 大型攀禽，体长 41~47 厘米。喙为象牙色，喙端黑灰色，甚为强健。通体黑褐色。雄鸟额、头顶至枕后面朱红色；耳羽、上背黑色，微沾绿色金属光泽。雌鸟似雄鸟但羽色稍淡，仅枕部有朱红色。

分布与生境 分布于中国西北、东北、河北、山西以及西南各地。主要栖息于海拔 1800 米以下的原始针叶林和针阔叶混交林中。甚罕见于北京，具体居留型尚不明确。

习性与生态 常单独活动，飞行不如其他啄木鸟起伏大。主要在树干、粗枝和枯立木上取食，也常到地面和腐朽的倒木上觅食蚂蚁和昆虫，亦食浆果，甚至其他小鸟和鸟蛋。

Morphological characteristics A large scansorial bird (41-47 cm). Bill ivory, strong. Male is black overall, with a red patch from crown to nape; ear coverts and upper back black tinged with green. Female is similar to male but duller; only hindcrown is red.

Distribution and habitat Northwest, northeast China, Hebei, Shanxi province and other regions of southwest China. Found in old-growth coniferous and mixed forest below 1,800 m elevation. Very rare in Beijing. Residential status: unclear.

Behaviour and ecology Usually solitary. Flight lightly undulating. Forages for ants and insects on thick branches or dead wood, including on dead fallen trees. Sometimes eats berries, even chicks and bird eggs.

灰头绿啄木鸟	*Picus canus*	Grey-headed Woodpecker	啄木鸟科 Picidae

形态特征 中等体型的攀禽，体长 27~31 厘米。雄鸟额、头顶前部鲜红色，眼先黑色，其余头、颈灰色，后颈、枕部有黑纹，喉白，下体灰绿色；背、腰、尾上覆羽、两翼覆羽灰绿色，有暗褐色，飞羽具白斑点。雌鸟似雄鸟，头顶不红。

分布与生境 几乎分布中国全境。栖息于低山、丘陵、平原阔叶林、混交林至城市绿地。在北京为常见留鸟。

习性与生态 常单独或成对活动，飞行迅速，呈波浪式前进。常在树干的中下部取食，也常在地面取食蚂蚁等昆虫，偶尔也吃植物果实和种子。巢洞多选择在混交林、阔叶林；窝卵数 2~4 枚。

Morphological characteristics A medium-sized scansorial bird (27-31 cm). Male is bright red on forehead and fore-crown; black lores; the rest of head and neck grey with white throat; greyish-green underparts and black markings on hindneck and nape; greyish-green on back, rump, uppertail-coverts and wing-coverts; dark brown wings; and white spots on remiges. Female is similar to male but lacks red on crown.

Distribution and habitat Widely distributed in China. Found in low mountains, hills, broadleaved and mixed forest. Common resident in Beijing.

Behaviour and ecology Usually feeds singly or in pairs. Flight fast and undulating. Forages at mid to low levels of tree trunks and often forages for ants and insects on the ground, occasionally fruit and seeds. Nests in mixed and broadleaved forests, including large city parks. Clutch size: 2-4.

北京市一级保护野生动物 | IUCN 红色名录 无危 (LC) 2016.10.1

雄鸟 Male

雌鸟 Female

V

北京鸟类图谱

猛禽
RAPTORS

BIRDS OF BEIJING

二十一、鹰形目 ACCIPITRIFORMES

鹗	*Pandion haliaetus*	Osprey	鹗科 Pandionidae

形态特征 中型猛禽，体长 51~64 厘米。雌雄相似。头顶和后颈白色，有黑褐色纵纹。黑褐色的贯眼纹向后通过耳羽区延伸至颈侧部。上体暗褐色，背部有白色小斑。下体近白色，胸部缀赤褐色斑纹。飞翔时翅甚狭长，正面似“M”形，翼下覆羽白色，翼角有明显的黑斑。外趾能反转向后，趾底具刺突。

分布与生境 中国各地均有分布。北京区域性常见于各种大型开阔水域，为旅鸟。

习性与生态 以蛙类、鱼类及小型水禽为食，常盘旋于水域上空，发现猎物后折合双翅直扑猎物，并能潜入水中追捕游鱼。在水域附近乔木上搭筑高大的厚皿状窝；窝卵数 2~4 枚。

Morphological characteristics A medium-sized raptor (51-64 cm). Sexes alike. Crown and hindneck white with brownish-black streaks; brownish-black eye-stripe extends beyond ear-coverts to the neck; upperparts dark brown with white spots on the back; underparts whitish with rufous brown breast-streaks; wings narrow and long, angled in flight forming an M-shape; underwing-coverts white with a distinct black patch at the bend of the wing. Outer toe reversable, with spikes on the bottom of toes.

Distribution and habitat Widespread in China. In Beijing, locally common passage migrant at large open water area.

Behaviour and ecology Feeds on frogs, fish and small water birds. Often hovers over open water. On sight of prey, folds its wings and plunges into water with tarsi first to catch fish. Builds large dish-like nests on tall trees near water. Clutch size: 2-4.

国家二级保护野生动物　IUCN 红色名录　无危 (LC) 2016.10.1

黑翅鸢	*Elanus caeruleus*	Black-winged Kite	鹰科 Accipitridae

	国家二级保护野生动物	IUCN 红色名录　无危 (LC) 2019.6.10

形态特征　中型猛禽，体长 31~34 厘米。雌雄相似。成鸟虹膜红色。喙黑色，黄色蜡膜。前额白色，到头顶逐渐变为灰色。后颈、背、肩、腰直到尾上覆羽蓝灰色。翼上小覆羽和中覆羽黑色，初级飞羽暗灰色，最外侧 7 枚末端黑。尾呈浅叉状。跗跖深黄色，爪黑色。

分布与生境　主要分布于中国西南、华南和东部地区。北京罕见于郊区的各类开阔生境，特别是林缘开阔地及临近湿地之处，居留型尚不明确。

习性与生态　常单独或成对活动，白天常停息在大树树梢或电线杆上。主要以田间鼠类、昆虫、小鸟为食。营巢于平原或山地丘陵地区的树上或高的灌木上；窝卵数 3~5 枚，卵白色或淡黄色。

Morphological characteristics A medium-sized raptor (31-34 cm). Sexes alike. Adult has red iris, black bill with yellow cere; white on forehead shaded to grey at crown; hindneck, back mantle, rump and uppertail-coverts bluish-grey; lesser coverts and medium coverts black; primaries dark grey with black at the tip of the outer seven feathers; shallow-forked tail and dark yellow tarsi with black claws.

Distribution and habitat Southwest, south and eastern China. In Beijing, rare in open country in rural areas, especially woodland edges and wetlands. Residential status: unclear.

Behaviour and ecology Usually found singly or in pairs. Often perched on tall trees, poles or wires. Preys on rodents, insects and small birds. Nests in trees or tall shrubs. Clutch size: 3-5, white or pale yellow eggs.

胡兀鹫	*Gypaetus barbatus*	Bearded Vulture	鹰科 Accipitridae

形态特征 大型猛禽，体长 100~140 厘米。雌雄相似。头两侧多为白色，一条宽阔的黑纹经过眼部向下到颏，与颏部黑色羽毛形成的“胡须”融为一体。眼先和喙基亦被有黑色须状羽，上背、短的肩羽和内侧覆羽暗褐色，具皮黄白色羽轴纹。尾长，楔形，暗褐色或灰褐色。下体黄褐色。跗跖被羽几乎到趾。

分布与生境 分布于中国西北、西南、华北等地。多在高山地带，特别是有悬崖峭壁之处活动。北京甚罕见于西部山区，为迷鸟。

习性与生态 常在山顶或山坡上空缓慢地飞行，主要以大型动物尸体为食，喜新鲜尸体和骨头，有时也猎取水禽、受伤的雉鸡、鹑类和野兔等小型脊椎动物。营巢于高山悬崖岩壁上大的缝隙和岩洞中，为盘状。

Morphological characteristics A large raptor (100-140 cm). Sexes alike. In adult plumage, whitish on the sides of head with a broad black band from lores to chin, merging with black bristles; upper-back, scapulars and inner coverts dark brown with buff white shaft-streaks; tail long and wedge-shaped, dark brown or greyish brown; underparts yellowish brown, brighter on breast; tarsi covered with feathers almost to the toes.

Distribution and habitat Northwest, southwest and north China. Favours mountains, especially cliffs. Very rare vagrant in montane areas in the west of Beijing.

Behaviour and ecology Often glides over top or along mountain ridges. Feeds on dead bodies of large animals, favouring fresh carcasses and bones; sometime also preys on small vertebrates such as water birds, wounded pheasants, quails, hares, etc. Nest is a platform built in a large crevice or cave on mountain crags.

国家一级保护野生动物　IUCN 红色名录　近危 (NT) 2016.10.1

成鸟 Adult

亚成鸟 Immature

凤头蜂鹰 | *Pernis ptilorhynchus* | Oriental Honey Buzzard | 鹰科 Accipitridae

雄鸟 Male

雌鸟 Female

深色型幼鸟 Juvenile (dark morph)

浅色型幼鸟 Juvenile (pale morph)

国家二级保护野生动物　IUCN 红色名录　无危 (LC) 2016.10.1

形态特征　中型猛禽，体长 50~60 厘米。喙黑色。头颈部尤为纤细，具有短而硬的鳞片状羽毛以防蜂类叮咬。具从淡棕白色到黑褐色等多种色型。典型的褐色型雄鸟头灰色，雌鸟头褐色；上体褐色，翼较宽大。尾羽较长，雄鸟尾羽暗褐色，中部淡灰褐色，具甚窄的浅色端斑；雌鸟尾羽淡灰褐色，具两条较窄的深褐色横斑、较宽的深褐色次端斑和甚窄的浅色端斑。跗跖黄色。

分布与生境　中国东北、华中、西南有繁殖，迁徙时经过中国大部分地区。北京迁徙季节常见于西部山区，城市上空亦可见，为旅鸟。

习性与生态　飞行时沉稳缓慢，常形成数量较大的迁徙过境集群。喜食昆虫，尤偏好蜂类，偶尔也食蛇类、蜥蜴、蛙、小型哺乳动物。

Morphological characteristics A medium-sized raptor (50-60 cm). Bill black; slender head and neck with short and hard scaly feathers to help protect from wasp and bee stings. Plumage highly variable from pale whitish-brown to dark brown. Typical adult has grey head (male) or brown head (female) and brown upperparts; wings long and broad; tail also long. Male has dark brown tail with pale greyish-brown tail centre and pale tips; female has pale greyish-brown tail with two narrow dark brown bands, a broad subterminal band and a very narrow pale terminal band. Tarsi yellow in both sexes.

Distribution and habitat Breeds in northeast, central, and southwest China. Seen in most areas of China on migration. In Beijing, often seen in the western mountains and also over the city. Residential status: common migrant.

Behaviour and ecology Flight steady with slow wingbeats. Forms large flocks on migration. Feeds on insects, especially bees and honeycomb, occasionally snakes, lizards, frogs and other small mammals.

高山兀鹫	*Gyps himalayensis*	Himalayan Vulture	鹰科 Accipitridae

形态特征 大型猛禽，体长 120~150 厘米。雌雄相似。喙灰色。头和颈裸露，颈部被有污黄色发状羽毛，颈基部长有皮黄色羽簇呈披针形围绕在颈部。背、翼上覆羽和下体淡黄褐色，飞翔时淡色的下体和黑色的飞羽形成鲜明对比。跗跖灰白色。

分布与生境 分布于中国西北和西南的高海拔地区。北京罕见于西部山区，为迷鸟。

习性与生态 主要以腐肉尸体为食，一般不攻击活动物。视觉和嗅觉敏锐，常在高空盘旋寻找地面上的尸体，通常营巢于高原上悬崖岩壁的凹处和边缘。

Morphological characteristics A large raptor (120-150 cm). Sexes alike. Grey bill and 'bald' appearance, neck is covered with dirty-yellow hairy feathers and is circled with pale brown ruff of long and spiky feathers; back, wing-coverts and underparts light brownish-yellow; pale brown underparts contrasting distinctively with black remiges in flight. Tarsi whitish grey.

Distribution and habitat At high elevation regions of northwest and southwest China. In Beijing, rare vagrant in the western mountains.

Behaviour and ecology Forages for carrion and carcass, rarely attacks live animals. Has sharp vision and keen sense of smell; often circles high searching for carcasses. Nest made of twigs on ledge or in recess of crags on rocky cliffs.

国家二级保护野生动物 | IUCN 红色名录 近危 (NT) 2016.10.1

成鸟 Adult

亚成鸟 Immature

秃鹫	*Aegypius monachus*	Cinereous Vulture	鹰科 Accipitridae

国家一级保护野生动物　IUCN 红色名录　近危 (NT) 2018.8.9

形态特征　大型猛禽，体长 108~116 厘米。雌雄相似。周身黑褐色。喙大，铅灰色，尖端钩曲。头颈部皮肤裸露部分呈铅灰色，头顶被稀疏的褐色绒羽，两侧各有一束明显的蓬松的矛状长羽。翼长而宽大。尾羽较短，展开成楔形。跗跖蓝灰色。

分布与生境　中国大部分地区均有分布。北京区域性常见于西部山区，为留鸟、冬候鸟。

习性与生态　常单只或集小群活动。视觉敏锐，高空窥伺地面上动物尸体、腐肉及活动小动物。通常营巢于悬崖峭壁凹陷处；窝卵数 1~2 枚。

Morphological characteristics A large raptor (108-116 cm). Sexes alike. Overall brownish-black with large blue-grey bill with a hooked tip; 'bald' head actually covered in thin brown down; breast covered with thick hairy down; wings long and broad; wedge-shaped tail; tarsi bluish-grey.

Distribution and habitat Widely distributed in China. In Beijing, locally common in the western mountains. Residential status: resident and winter visitor.

Behaviour and ecology Usually seen singly or in small groups. Has sharp vision; feeds on carrion, carcasses and small animals. Nests in recess of cliffs. Clutch size: 1-2.

蛇雕	*Spilornis cheela*	Crested Serpent Eagle	鹰科 Accipitridae

形态特征 中型猛禽，体长 59~64 厘米。雌雄相似。喙蓝灰色，蜡膜铅黄色。虹膜黄色。头顶具黑色杂白的圆形冠羽，覆盖枕部。上体暗褐色，下体土黄色，颏、喉具暗褐色细横纹，腹部有黑白两色眼斑。飞羽暗褐色，羽端具白色羽缘。尾黑色，中间具一条宽的淡褐色带斑，尾下覆羽白色。跗跖黄色，爪黑色。

分布与生境 分布于中国西南、东南、华南等地。北京迁徙季甚罕见于西部山区，为迷鸟。

习性与生态 主要以蛇类为食，也吃蜥蜴、蛙、鼠类、鸟类和甲壳等动物。营巢于林中高树顶端枝杈上；巢由枯枝构成，呈盘状；窝卵数 1 枚，卵白色、微具淡红色斑点。

Morphological characteristics A medium-sized raptor (59-64 cm). Sexes alike. Bill blue-grey with yellow cere; iris yellow; black and white crest on the back of head reaches nape; upperparts dark brown, chin and throat finely-barred brown; remiges dark brown with white fringing at the tip; tail black with a broad light brown subterminal band; undertail-coverts white; tarsi yellow, claws black.

Distribution and habitat Southern China. In Beijing, very rare in the western mountains on migration. Residential status: vagrant.

Behaviour and ecology Feeds mainly on snakes, sometimes lizards, frogs, rodents, birds and beetles. Nest made of dry twigs built on branches of tall trees in forest. Clutch size: 1, white egg with light red spots.

	国家二级保护野生动物	IUCN 红色名录 无危 (LC) 2016.10.1

成鸟 Adult

亚成鸟 Immature

短趾雕 | *Circaetus gallicus* | Short-toed Snake Eagle | 鹰科 Accipitridae

国家二级保护野生动物　IUCN 红色名录　无危 (LC) 2016.10.1

形态特征　大型猛禽，体长 61~72 厘米。雌雄相似。喙淡蓝灰色，蜡膜白或浅灰色。虹膜黄色。头部宽阔，颈部羽毛浓密而厚实，头、颈和胸部棕褐色，上体余部暗褐色。下体白色，具棕褐色横斑或点斑纹。跗跖上部被羽，趾较短小，呈灰色。

分布与生境　繁殖于中国西北地区，迁徙季过境北方至西南地区。北京迁徙季见于西部山区、北部山区开阔水面附近，为不常见旅鸟。

习性与生态　食物主要为蛇类、蜥蜴类、蛙类及小鸟，亦捕食小型啮齿动物。多营巢于林缘地区，置巢于树顶部枝杈上，窝卵数 2 枚。

Morphological characteristics A large raptor (61-72 cm). Sexes alike. Bill light blue-grey with white or pale cere; iris yellow; wide head with thick feathers around neck; head, neck and breast reddish brown, the remaining upperparts dark brown; underparts white with brown bars or spots; upper tarsi covered with feathers; toes short and grey.

Distribution and habitat Breeds in the northwest, passes through northern to southwest China on migration. Uncommon migrant in Beijing, found near open water in the west and north mountains.

Behaviour and ecology Feeds mainly on snakes, lizards, frogs, small birds and small rodents. Nests on tall trees at forest edge. Clutch size: 2.

乌雕	*Clanga clanga*	Greater Spotted Eagle	鹰科 Accipitridae

形态特征 大型猛禽，体长 61~73 厘米。雌雄相似。喙黑色，基部较浅淡，蜡膜黄色。虹膜为褐色。成鸟全身深褐色近黑，翼下初级飞羽基部有浅色窄月牙状区域，尾上覆羽白色。未成年鸟个体身上有不规则小块淡斑，背部和翼上有浅色斑点组成的条带。跗跖黄色，爪黑色。

分布与生境 中国广泛分布。北京迁徙季见于西部山区和郊区开阔水域上空，为不常见旅鸟。

习性与生态 主要以野兔、鼠类、野鸭、蛙、蜥蜴、鱼和鸟类等小型动物为食。营巢于林中高大的乔木树上；巢的结构较为庞大，主要由枯树枝构成，为平盘状；窝卵数 1~3 枚。

Morphological characteristics A large raptor (61-73 cm). Sexes alike. Bill black with a paler base and yellow cere; iris brown. Adult is dark brown all over with a crescent marking under the wings at the base of the primaries; uppertail-coverts white. Juvenile has irregular light spotting all over the body, with the light spots forming bars on the back and wings. Tarsi yellow with black claws.

Distribution and habitat Widespread in China. In Beijing, uncommon migrant over the western mountains and near large open water areas in the suburbs.

Behaviour and ecology Often found on trees near lowland wet swamps. Feeds mainly on small animals such as hares, rodents, ducks, frogs, lizards, fish and birds. Nest, a huge platform consists of dry twigs, built on branches of tall trees in forest. Clutch size: 1-3.

国家一级保护野生动物 | IUCN 红色名录 易危 (VU) 2016.10.1

成鸟 Adult

亚成鸟 Immature

靴隼雕	*Hieraaetus pennatus*	Booted Eagle	鹰科 Accipitridae

浅色型 Pale morph

深色型 Dark morph

国家二级保护野生动物　IUCN 红色名录　无危 (LC) 2016.10.1

形态特征　中型猛禽，体长 45~54 厘米。雌雄相似。喙苍灰色，蜡膜淡黄色。上体褐色具黑色和皮黄色杂斑。其肩羽前方具两块显著的白色斑块，迎面飞来时容易观察到。翼下最内侧的 3 枚初级飞羽色浅。深色型胸部棕色；浅色型胸部淡皮黄色，翼下覆羽的浅色与飞羽的深色对比明显。尾下覆羽近白色。跗跖淡黄色，爪黑色。

分布与生境　繁殖于中国新疆、东北，迁徙季见于北方至西南等地。北京迁徙季见于山区及附近的低海拔区域，为不常见旅鸟。

习性与生态　主要以小型哺乳类、鸟类及爬行动物为食。通常营巢于森林中高大的乔木上，巢多置于树上部枝杈上，结构较为庞大而简陋；巢呈盘状；窝卵数 1~3 枚。

Morphological characteristics A medium-sized raptor (45-54 cm). Sexes alike. Bill pale grey, cere light yellow. Brown upperparts with black and buff scattered spots. The white patches at the base of the leading edge of the wing are prominent in flight. Underwing plumage of the three most inner primaries are lighter. Dark morph has brown breast while the pale morph has buff breast; the latter has a distinctive underwing pattern with pale underwing-coverts contrasting with the dark remiges. Undertail-coverts off-white; tarsi light yellow with black claws.

Distribution and habitat Breeds in northwest and northeast China and seen on migration from northern to southwest China. In Beijing, uncommon passage migrant in montane areas and lowlands nearby.

Behaviour and ecology Feeds mainly on small mammals, birds and reptiles. Nest is a huge shabby platform consists of dry twigs built on the branches of tall trees in the forest. Clutch size: 1-3.

草原雕 | *Aquila nipalensis* | Steppe Eagle | 鹰科 Accipitridae

形态特征 大型猛禽，体长 71~82 厘米。雌雄相似。喙黑褐色，蜡膜暗黄色，嘴裂长度可达眼中后部，明显较其他雕更深。头显得小而突出。体羽以褐色为主，上体土褐色，飞羽黑褐色，杂以较暗的横斑，下体暗土褐色。跗跖黄色，爪黑色。

分布与生境 在中国北部、西北部繁殖，东南部越冬。北京迁徙季见于西部山区，为不常见旅鸟。

习性与生态 或长时间地栖息于电线杆上，或盘旋于草原和荒地上空。主要以草原或荒漠地带的啮齿类、野兔、沙蜥、鸟类等小型脊椎动物为食。营巢于悬崖上或山顶岩石堆中，也营巢于地面干草堆；巢浅盘状；窝卵数 1~3 枚，卵白色。

Morphological characteristics A large raptor (71-82 cm). Male and female are alike. Bill brownish-black and cere dark yellow; gape extends to the back of eye, further than other eagles. Head appears small and prominent. Primarily brown plumage with upperparts dust-brown. Primaries brownish-black with dark bars; underparts pale brown. Tarsi yellow with black claws.

Distribution and habitat Breeds in northern and northwest China, wintering in southeast China. Uncommon migrant in Beijing, found in the western mountains on migration.

Behaviour and ecology Perches on electricity poles or hangs over grassland, looking for prey. Feeds mainly on small vertebrates such as hares, rodents, lizards and birds. Nests is a shallow platform on rock piles on top of mountains or on piles of dry grass on the ground. Clutch size: 1-3, white eggs.

国家一级保护野生动物 | IUCN 红色名录 濒危 (EN) 2020.7.30

亚成鸟 Immature

成鸟 Adult

亚成鸟 Immature

白肩雕	*Aquila heliaca*	Imperial Eagle	鹰科 Accipitridae

成鸟 Adult

亚成鸟 Immature

国家一级保护野生动物　IUCN 红色名录　易危 (VU) 2016.10.1

形态特征　大型猛禽，体长 73~84 厘米。雌雄相似。喙灰色，蜡膜黄色。虹膜浅褐色。头后颈部有大片的浅色羽，泛黄。上体黑褐色，肩部长形肩羽纯白色，形成显著的白色块斑。尾上覆羽灰褐色，具不规则的黑褐色横斑，近尾端有宽阔的黑色横斑。翼下暗褐色。跗跖黄色。

分布与生境　繁殖于中国新疆，迁徙时经过北方，在长江中下游、华南、西南等地越冬。北京迁徙季见于西部山区，为罕见旅鸟。

习性与生态　主要以啮齿类、雉类和中型哺乳动物为食，多在河谷、沼泽、草地和林间空地等开阔地方觅食。营巢于林中高大乔木之上，偶尔也营巢于悬崖岩石上，巢通常会使用多年；窝卵数 2~3 枚。

Morphological characteristics A large raptor (73-84 cm). Sexes alike. Grey bill, yellow cere and light brown iris. Light-coloured yellowish feathers on the back of neck; brownish-black upperparts; long, white scapulars form a prominent white patch on mantle; greyish brown uppertail-coverts have irregular brownish-black bars; dark brown under wings; tarsi yellow.

Distribution and habitat Breeds in Xinjiang. Seen on migration in northern China; wintering in southern China. In Beijing, rare migrant found in the western mountains.

Behaviour and ecology Feeds mainly on medium-sized mammals, rodents and birds, such as pheasants. Hunts in river valleys, marshes, meadows and open woods. Nest, which is often reused, is built in tall trees in the forest, occasionally on rocky cliff. Clutch size: 2-3.

金雕	*Aquila chrysaetos*	Golden Eagle	鹰科 Accipitridae

形态特征 大型猛禽，体长 79~102 厘米。雌雄相似。喙灰色，蜡膜黄色。体羽深褐色，头后枕部和后颈披针形羽毛且呈金黄色。翼型较白肩雕而言翅基渐窄。下体黑褐色，翼下飞羽较浅。未成年个体飞行时可见初级飞羽基部和尾羽基部为白色。跗跖黄色。

分布与生境 中国广泛分布。北京多见于西部、北部山区岩石地带，为不常见留鸟、旅鸟。

习性与生态 通常单独或成对活动，喜欢在高空盘旋。以雁鸭类、雉类等鸟类以及狍子、獾、野兔等中、小型兽类为食。通常在针叶林或悬崖上筑巢；窝卵数 1~3 枚。

Morphological characteristics A large raptor (79-102 cm). Sexes alike. Grey bill and yellow cere; dark brown plumage with lighter golden-brown on the nape; wings narrower at the wing base compared with Imperial Eagle; underparts brownish-black with lighter underwing. Juvenile has large white patch on the underwing and tail, visible in flight. Tarsi yellow.

Distribution and habitat Widely distributed in China. In Beijing, uncommon migrant and resident in the western and northern mountains.

Behaviour and ecology Often found singly or in pairs. Feeds on birds (e.g., Anseriformes birds and pheasants) and small to medium-sized mammals such as roe deer, badgers and hares. Often nests in coniferous woodland or on cliffs. Clutch size: 1-3.

国家一级保护野生动物 | IUCN 红色名录 无危 (LC) 2016.10.1

亚成鸟 Immature

亚成鸟 Immature

成鸟 Adult

白腹隼雕 | *Aquila fasciata* | Bonelli's Eagle | 鹰科 Accipitridae

成鸟 Adult
亚成鸟 Immature

国家二级保护野生动物

IUCN 红色名录　无危 (LC) 2016.10.1

形态特征　中型偏大的猛禽，体长 68~73 厘米。雌雄相似。喙蓝灰色，蜡膜黄色。上体暗褐色，头顶及后颈部呈棕褐色，颈侧和肩部羽缘灰白色，飞羽灰褐色，灰色的尾羽较长，下体白色，具黑色细纵纹。飞翔时翼显狭长，翼覆羽深色。跗跖柠檬黄色，爪黑色。

分布与生境　主要分布于中国长江以南地区。北京迁徙季见于西部山区，为罕见旅鸟。

习性与生态　性胆大而凶猛，飞行速度很快，偶尔发出尖锐叫声。常单独活动，多在低空鼓翼飞行，很少在高空盘旋。主要以鼠类及中小型鸟类为食，也吃野兔等。营巢于河谷岸边的悬崖上或树上；窝卵数 1~3 枚。

Morphological characteristics　A medium-large raptor (68-73 cm). Sexes alike. Bluish-grey bill and yellow cere. Dark brown upperparts, crown and hindneck brown; greyish white fringing on neck-sides and mantle feathers; remiges brownish grey; tail feathers long and grey; underparts whitish with black fine streaks. Wings look narrow and long in flight, with dark wing-coverts; tarsi lemon yellow with black claws.

Distribution and habitat　In China, distributed south of the Yangtze River. In Beijing, rarely found in the mountains on migration.

Behaviour and ecology　Bold and ferocious, flies at great speed, accompanied by occasional screams. Often found singly flying with wings held up, rarely hovering or gliding high. Hunts for small to medium-sized animals like rodent and bird. Often nests on cliffs or trees in river valley. Clutch size: 1-3.

凤头鹰	*Accipiter trivirgatus*	Crested Goshawk	鹰科 Accipitridae

形态特征 中型猛禽，体长 41~49 厘米。雌雄相似。喙黑褐色，蜡膜黄绿色。虹膜黄色或橘色。头部从前额至后颈灰色，具显著冠羽，其余上体褐色。喉白色，具显著深色喉中线。胸棕褐色。下体白色，具窄的棕褐色横斑。飞行时可见发达的白色尾下覆羽。翼下飞羽具数条宽阔的黑色横带。跗跖部较松雀鹰更为粗壮，跗跖黄色，爪黑色。

分布与生境 主要分布于中国长江以南地区。近年北京迁徙季数次记录于西部山区，罕见，居留型尚不明确。

习性与生态 性隐秘而机警，领域性强，常躲藏在树叶丛中。多单独活动，飞行缓慢，主要在森林中的地面上捕食，以蛙、蜥蜴、小鸟、鼠类、昆虫等为食。营巢于针叶林或阔叶林中高大的树上，位置多离水域不远。

Morphological characteristics A medium-sized raptor (41-49 cm). Sexes alike. Brownish-black bill and greenish-yellow cere. Yellow or orange iris. Grey from forehead to neck with prominent crest, the remaining upperparts brown; throat white with a distinct dark stripe; breast brown, underparts white with narrow brown bars; white undertail-coverts conspicuous in flight; broad black bands shown on remiges under wing. Tari is stronger than that of Besra. Tarsi yellow with black claws.

Distribution and habitat In China, distributed south of the Yangtze River. In Beijing, rare with a few records from the western mountains in recent years on migration. Residential status: unclear.

Behaviour and ecology Secretive, vigilant and territorial. Often hides in the canopy, from where it hunts for frogs, lizards, rodents, birds and insects on the ground. Nest usually in tall trees in coniferous or broadleaved forests not far from water.

	国家二级保护野生动物	IUCN 红色名录 无危 (LC) 2016.10.1

亚成鸟 Immature

成鸟 Adult

赤腹鹰	*Accipiter soloensis*	Chinese Sparrowhawk	鹰科 Accipitridae

雄鸟 Male

雌鸟 Female

国家二级保护野生动物

IUCN 红色名录　无危 (LC) 2016.10.1

形态特征　小型猛禽，体长 25~30 厘米。雄鸟喙黑色，蜡膜橘黄色尤为醒目；虹膜暗褐色；上体淡蓝灰色，胸至两胁橙棕色，下体余部白色；翼下除初级飞羽羽端黑色外，几乎全白；跗跖橘黄色。雌鸟似雄鸟，虹膜黄色。幼鸟上体褐色，下体白，腹部具褐色粗纵纹。

分布与生境　在中国南方广泛分布。北京迁徙季见于山区，在山区林地甚至城市公园中有繁殖，为不常见旅鸟和夏候鸟。

习性与生态　喜开阔林区。性隐秘而机警。主要以蛙、蜥蜴、昆虫等为食，偶尔也吃小型鸟类或鼠类。营巢于林中的树丛上；窝卵数 2~5 枚。

Morphological characteristics A small raptor (25-30 cm). Adult male has black bill, prominent orange-yellow cere, dark brown iris; light blue-grey upperparts, white underparts, breast and flanks washed with pink; wings white underneath except black on primary tips; tarsi orange-yellow. Female is similar to male with yellow iris. Juvenile has brown upperparts and white underparts with wide broad streaks on belly.

Distribution and habitat Widely distributed in southern China. In Beijing, uncommon in mountain areas, breeding in mountain forest and even in city parks. Residential status: migrant and summer breeder.

Behaviour and ecology Favours open forest, usually near water. Secretive and vigilant. Hunts mainly for frogs, lizards and insects, occasionally taking small bird and rodents. Nests on top of trees in woodland. Clutch size: 2-5.

日本松雀鹰	*Accipiter gularis*	Japanese Sparrowhawk	鹰科 Accipitridae

形态特征 小型猛禽，体长 25~34 厘米。成年雄鸟的虹膜橘红色，头、背灰蓝色，胸腹部可见淡红褐色，腹部横纹不甚清晰。雌鸟虹膜黄色，背褐色，胸腹部横纹清晰。未成年鸟喉白色，胸部纵纹，腹部横纹。本种尾较雀鹰短，整体感觉紧凑。

分布与生境 中国各地均有分布。多栖息于针叶林或针阔叶混交林带，也出现在林缘和疏林地带。在北京为区域性常见的旅鸟及罕见夏候鸟。

习性与生态 多单独活动。两翼鼓动甚快，常在快速鼓翼飞翔之后接着又进行一段直线滑翔。主要以小型鸟类为食，也食昆虫、蜥蜴等。常营巢于茂密的山地森林和林缘地带；巢小而坚实，呈圆而厚的皿状；窝卵数 5~6 枚。

Morphological characteristics A small raptor (25-34 cm). Adult male has orange-red iris, bluish-grey head and back, pale rufous breast and belly and fine barring on belly; female has yellow iris, brown back and prominent barring on breast and belly. Juvenile has white throat, streaked breast and barred belly. Shorter-tailed and more compact than Eurasian Sparrowhawk.

Distribution and habitat Widespread in China. In Beijing, locally common passage migrant and rare summer breeder in coniferous and mixed forest, including forest edge and open woodland.

Behaviour and ecology Usually seen singly. Fast wingbeats, interspersed with short glides. Hunts mainly for small birds, and also feeds on lizards and insects. Nest is small and sturdy, built in dense mountain forest or forest edge. Clutch size: 5-6.

国家二级保护野生动物　IUCN 红色名录　无危 (LC) 2016.10.1

雌鸟 Female

雄鸟 Male

亚成鸟 Immature

松雀鹰	*Accipiter virgatus*	Besra	鹰科 Accipitridae

亚成鸟 Immature

成鸟 Adult

国家二级保护野生动物　IUCN 红色名录　无危 (LC) 2016.10.1

形态特征　小型猛禽，体长 24~36 厘米。雄鸟颊、颈侧和其余上体灰褐色，喉白色具显著的深色喉中线；下体白色，具宽而粗的栗色横斑；似凤头鹰而跗跖部较细，跗跖黄色。雌鸟似雄鸟，颊褐色。

分布与生境　广泛分布于中国南方各地。多见于山地和较低海拔的森林。北京迁徙季甚罕见于西部山区，为旅鸟。

习性与生态　性机警，常单独或成对在林缘和丛林边等较为空旷处活动和觅食。以各种小鸟为食，也吃蜥蜴、蝗虫、甲虫等昆虫以及小型鼠类。营巢于茂密森林中枝叶茂盛的高大树木上部；窝卵数 3~4 枚。

Morphological characteristics A small raptor (24-36 cm). Male has brownish-grey face, neck sides and upperparts; throat white with broad throat-stripe; underparts white with broad and dark chestnut band. Overall, similar to Crested Goshawk, but with thinner and yellow tarsi. Female is similar to male but with brown face.

Distribution and habitat Mainly southern China. Seen in montane and lowland forests. In Beijing, very rare in the western mountains on migration.

Behaviour and ecology Vigilant. Feeds singly or in pairs in open areas of forest edge or adjacent to forest. Hunts mainly for small birds and also feeds on lizards, grasshoppers, beetles, other insects and rodents. Nests on top of tall trees with exuberant foliage in the forest. Clutch size: 3-4.

雀鹰	*Accipiter nisus*	Eurasian Sparrowhawk	鹰科 Accipitridae

形态特征 中型猛禽，体长 32~43 厘米。成年雄鸟具清晰的白色眉纹，头部、上体及两翼青灰色，尾较长，具深色横斑；颊橙色，下体白色，具较密的橙色斑纹。雌鸟体型较大，上体灰褐色，下体白色，具细密的深灰褐色横纹。未成年个体上体更偏褐色。

分布与生境 中国广泛分布。北京常见于平原开阔林地、农田和城市公园，主要为旅鸟和冬候鸟，夏季偶见于山区，近年在西部高海拔山区有繁殖记录。

习性与生态 常单独生活。捕食小型啮齿类和小型鸟类。营巢于山区松树顶端；巢呈厚皿状；窝卵数 3~4 枚。

Morphological characteristics A medium-sized raptor (32-43 cm). Adult male has distinctive white supercilium; slate-grey from head to upperparts and wings; tail long with dark bars; cheeks orange; underparts white, with extensive orange bars. Female is larger than male and has greyish brown upperparts; underparts barred dark greyish brown. Juvenile has brownish upperparts.

Distribution and habitat Widely distributed in China. In Beijing, common in open woodland, cultivations and city parks. Residential status: mainly migrant and winter visitor; rare in mountains in summer with few breeding record at western mountains at high elevation.

Behaviour and ecology Feeding solitary. Hunts mainly for small birds and rodents, Nests on top of pine trees of mountain forest. Clutch size: 3-4.

	国家二级保护野生动物	IUCN 红色名录　无危 (LC) 2016.10.1

亚成鸟 Immature

雄鸟 Male

雌鸟 Female

苍鹰	*Accipiter gentilis*	Northern Goshawk	鹰科 Accipitridae

成鸟 Adult

幼鸟 Juvenile

国家二级保护野生动物　IUCN 红色名录　无危 (LC) 2016.10.1

形态特征　中型猛禽，体长 47~57 厘米。成鸟白色眉纹显著；头、上体及两翼深灰色；下体污白色，具细密的深色横纹；跗跖黄色。未成年鸟个体上体褐色，腹部淡皮黄，具显著的深褐色纵纹。本种与雀鹰相似，但体型较大，跗跖更为粗壮，且翼较长而尾较短。

分布与生境　中国大部分地区均有分布。北京多见于丘陵山麓的针叶林、阔叶林和混交林中，亦见于林缘开阔地，为区域性常见旅鸟和冬候鸟，近年来西部山区有繁殖记录。

习性与生态　常单独活动。异常凶猛，为少数不惧鸦科鸟类的猛禽。主要食物为野兔、鼠类及各种鸟类等。营巢于高大乔木上；巢用枯草筑成，呈厚皿状；窝卵数 3~4 枚。

Morphological characteristics A medium-sized raptor (47-57 cm). Adult has prominent white supercilium; head, upperparts and wings dark grey; underparts off-white, with extensive fine dark bars; tarsi yellow. Juvenile has brown upperparts and distinctive heavily-streaked buffish belly. Resembles Eurasian Sparrowhawk but has larger body, longer wings, shorter tail and stronger tarsi.

Distribution and habitat Widely distributed in China. Usually found in coniferous, broadleaved and mixed forest, also in open country near forests. Locally common passage migrant and winter visitor in Beijing with few breeding record at western mountains.

Behaviour and ecology Usually seen singly. Extremely ferocious; one of the few birds of prey that is fearless of corvids. Hunts mainly for hares, rodents and birds. Builds a thick disc-like nest with dry grass in tall trees.. Clutch size: 3-4.

白头鹞	*Circus aeruginosus*	Western Marsh Harrier	鹰科 Accipitridae

形态特征 中型猛禽，体长 49~60 厘米。雄鸟虹膜黄色；头部至胸部为淡黄白色，具褐色纵纹；背部及翼上覆羽主要为深褐色，翼前缘淡黄白色；外侧初级飞羽黑色；内侧飞羽和初级飞羽灰色；尾羽淡灰色；下体大致为红褐色。雌鸟全身大致呈深褐色，仅头顶、颏、喉、翼前缘和胸淡黄白色。幼鸟似雌鸟但翼下覆羽为淡褐色。

分布与生境 繁殖于中国新疆西北部，冬季南迁。栖息于近湿地的各种开阔区域。在北京为甚罕见迷鸟，但因本种幼鸟与白腹鹞幼鸟在野外极难辨别，故北京地区的白头鹞记录有待进一步考证确认。

习性与生态 常单独或成对活动，多于水边草地或沼泽地低空飞行搜索食物。主要以小型鸟类、雏鸟、鸟卵、小型啮齿类、蛙、蜥蜴、蛇等为食。

Morphological characteristics A medium-sized raptor (49-60 cm). Adult male has yellow iris; cream from head and breast, with brownish streaks; mantle and upperwing-coverts dark brown; forewing cream or off-white; outer primaries black; remaining remiges and primary coverts grey; tail pale grey; underparts largely rufous-brown. Female mostly dark brown but has cream crown, chin, throat, forewing and breast. Juvenile like female but has pale brown underwing-covers.

Distribution and habitat Breeds in northwest China, winters to the south. Found in open county near water. Very rare vagrant in Beijing. Existing records in Beijing are encouraged to be reviewed as it is very difficult to distinguish juveniles of Eastern and Western Marsh Harriers in the field.

Behaviour and ecology Feeds singly or in pairs. Often flies low above grassland near water or marsh hunting for prey. Hunts mainly for small birds, chicks, bird eggs, small rodents, frogs, lizards and snakes.

	国家二级保护野生动物	IUCN 红色名录 无危 (LC) 2016.10.1

雄鸟 Male

雌鸟 Female

亚成鸟 Immature

白腹鹞	*Circus spilonotus*	Eastern Marsh Harrier	鹰科 Accipitridae

雄鸟 Male

雌鸟 Female

雄鸟 Male

亚成鸟 Immature

国家二级保护野生动物 IUCN 红色名录 无危 (LC) 2016.10.1

形态特征 中型猛禽，体长 43~54 厘米。成年雄鸟头、颈、胸及上体大致为黑白相间，部分个体头、颈全为黑色；翼上覆羽大致为黑色，具白色斑点，外侧初级飞羽黑色，其余飞羽皆为灰白色；尾上覆羽白色，尾羽银灰色；下体大致为白色，下胸具黑色纵纹。少数雄鸟个体头、胸淡棕色，上体褐色，下体棕红色。雌鸟及幼鸟头及上体深褐色，具疏密不等的白色纵纹；头顶、颏、喉和胸羽色较淡；尾羽褐色，具不清晰的黑色横斑；下体红棕色。

分布与生境 繁殖于中国东北，越冬于东南。北京活动于湿地苇丛区域，为区域性常见旅鸟。

习性与生态 常单独或成对活动。多见在沼泽和芦苇上空低空缓慢飞行。营巢于浓密的苇丛中；窝卵数 4~5 枚。

Morphological characteristics A medium-sized raptor (43-54 cm). Adult male has black-and-white head, neck, breast and upperparts. Some individuals have all-black head and neck; upperwing-coverts largely black, with white spots; outer primaries black; rest of remiges greyish white; uppertail-coverts white; tail silver-grey; underparts mostly white, with black streaks on lower breast. A few males are paler from head to breast; brown above and rufous-brown below. Female and juveniles have dark brown head and upperparts with variable white streaks; crown, chin, throat and breast paler; tail brown with indistinct black bars; underparts rufous-brown.

Distribution and habitat Breeds in northeast China, wintering to the south. Locally common passage migrant in Beijing, often found in reedbeds in open wetlands.

Behaviour and ecology Feeds singly or in pairs; diurnal and vigilant. Often found flying slowly and low over marshes and reedbeds. Nests in extensive reeds. Clutch size: 4-5.

白尾鹞	*Circus cyaneus*	Hen Harrier	鹰科 Accipitridae

形态特征 中型猛禽，体长 45~53 厘米。成鸟喙黑灰色；虹膜黄色。雄鸟整体灰色，尾上覆羽白色；初级飞羽黑色，腹白。雌鸟整体褐色，腹部白有褐色纵纹，翼下飞羽横斑显著。未成年鸟似雌鸟，但虹膜褐色，腹部皮黄色。

分布与生境 繁殖于中国东北、西北。迁徙时过境中国大部分地区，在华北及以南地区越冬。北京区域性常见于西部山区以及较低海拔的开阔荒地和湿地，为旅鸟和冬候鸟。

习性与生态 常沿地面低空飞行，频频鼓动两翼，缓慢地移动。主要以小型鸟类、鼠类、蛙、蜥蜴和昆虫等为食。

Morphological characteristics A medium-sized raptor (45-53 cm). Adult has blackish bill and yellow iris. Male is overall grey with white uppertail-coverts; primaries black and belly white. Female is overall brown; belly white with brown streaks; prominent bars on underside of remiges. Juvenile resembles female but has brown iris and buff belly.

Distribution and habitat Breeds in northeast and northwest China. Found in most regions of China on migration. In Beijing, locally common in the west mountains, and in lowland open country and wetlands. Residential status: passage migrant and winter visitor.

Behaviour and ecology Often seen flying low and slow with wings flapping frequently between glides. Hunts for small birds, rodents, frogs, lizards, and insects.

	国家二级保护野生动物	IUCN 红色名录　无危 (LC) 2016.10.1

雄鸟 Male

雌鸟 Female

亚成鸟 Immature

雌鸟 Female

草原鹞	*Circus macrourus*	Pallid Harrier	鹰科 Accipitridae

雄鸟 Male

雌鸟 Female

	国家二级保护野生动物	IUCN 红色名录 近危 (NT) 2018.8.7

形态特征 中型猛禽，体长 43~52 厘米。雄鸟虹膜黄色；整体似白尾鹞，但灰色更浅；喉部近白色，前几枚初级飞羽黑色。雌鸟虹膜黄色；身体遍布褐色纵纹，尾上覆羽白色；似白尾鹞雌鸟，但具更为明显的浅色领环，体型与翅型也更为纤细和狭窄。未成年鸟似雌鸟，虹膜褐色。

分布与生境 繁殖于中国新疆北部。主要栖息于草原、半荒漠、干旱开阔的平原。在北京为甚罕见旅鸟。

习性与生态 与大部分喜湿地的鹞不同，常在开阔的草原接近地面处缓慢低空飞行。主要以田鼠、黄鼠、跳鼠等啮齿动物以及野兔、蜥蜴、昆虫、鸟类等为食。

Morphological characteristics A medium-sized raptor (43-52 cm). Superficially resembles Hen Harrier but male has lighter grey plumage, white throat and only first few primaries black. Female has brown streaks all over body and white rump, similar to female Hen Harrier but with paler collar ring and slimmer body and narrower wings. Both sexes have yellow iris. Juvenile is similar to female with brown iris.

Distribution and habitat Breeds in northern Xinjiang in China. Mainly found in steppe, semi-desert and dry open country. Very rare migrant in Beijing.

Behaviour and ecology Different from other harriers that are typically found near wetlands, the Pallid Harrier prefers open steppe and grassland. Feeds on rodents such as voles, ground squirrels and jerboas; also hares, lizards, grasshoppers, birds, chicks and bird eggs.

鹊鹞	*Circus melanoleucos*	Pied Harrier	鹰科 Accipitridae

形态特征 中型猛禽，体长 42~48 厘米。雄鸟体羽黑白色甚为分明；头、颈、背和颏、喉部、上胸部羽均为黑色，尾羽灰白色；两翼银灰色为主，外侧 6 枚初级飞羽黑色，具一显著的黑色带自背部延伸至翼上覆羽；腹、两胁、翼下覆羽、尾下覆羽均为白色。雌鸟似白尾鹞雌鸟，但腹部羽色近白色。未成年鸟体色纯褐色，有白腰，翼下有可见的细横纹。

分布与生境 繁殖于中国东北，迁徙时经过东部和南部地区，在长江中下游以南地区越冬。北京多见于各种大型开阔水域，为不常见旅鸟；夏季在野鸭湖、沙河等湿地有记录但无繁殖证据。

习性与生态 常单独活动，多在林边草地和灌丛上空低空飞行，两翼不动，慢慢地飘浮在空中，时高时低地进行上下和向前的活动。主要以小鸟、鼠类、林蛙、蜥蜴、蛇、昆虫等小型动物为食。营巢于湿地草丛或苇丛中；窝卵数 3~6 枚。

Morphological characteristics A medium-sized raptor (42-48 cm). Male has prominent black-and-white plumage; head, neck, back, chin, throat and upper breast all-black; wings largely silver-grey; six outermost primaries black; a prominent black band extends from back to upperwing-coverts; tail greyish white; belly, flanks, underwing-coverts and undertail-coverts all white. Female is similar to female Hen Harrier, but with white belly. Juvenile has chocolate-brown plumage, white rump and visible fine barring on the underwing.

Distribution and habitat Breeds in northeast China, migrating through eastern and southern China to south of the Yangtze River. In Beijing, uncommon migrant near large areas of open water; also rarely found at Yeyahu and Shahe Reservoir in summer, but no evidence of breeding.

Behaviour and ecology Usually seen singly, often flying low over grassland and marshes. Feeds on small birds, rats, frogs, lizards, snakes and insects. Nests in reedbeds or grass at wetlands. Clutch size: 3-6.

国家二级保护野生动物 | IUCN 红色名录　无危 (LC) 2016.10.1

雌鸟 Female

雄鸟 Male

亚成鸟 Immature

黑鸢	*Milvus migrans*	Black Kite	鹰科 Accipitridae

成鸟 Adult

亚成鸟 Immature

国家二级保护野生动物　IUCN 红色名录　无危 (LC) 2020.8.19

形态特征　中型猛禽，体长 54~69 厘米。成鸟整体棕褐色，眼后有明显黑色耳斑，上体暗褐色，尾部呈梯形略内凹，为本种标志性特征。未成年鸟背、覆羽有不规则浅色斑，腹部有棕黄色纵纹，初级飞羽基部发白。

分布与生境　中国各地均有分布。北京迁徙季常见于河湖水边、郊区附近及村庄居住区，为旅鸟。曾为夏候鸟，但近年来已无繁殖记载。

习性与生态　白天活动，常单独在高空盘旋，飞行快而有力，性机警。喜食鱼，亦食腐，有时也捕捉大型昆虫或小型脊椎动物。通常营巢于高大树上；窝卵数 2~3 枚。

Morphological characteristics A medium-sized raptor (54-69 cm). Adult is brown overall with distinct black patch behind eye, dark brown upperparts, slightly forked tail is the iconic character of black kite. Juvenile has irregular pale markings on back and wing coverts, brownish-yellow streaks on belly, white at the base of primaries.

Distribution and habitat Widely distributed in China. In Beijing, common near rivers, lakes and in suburbs on migration. Residential status: migrant. Used to be summer breeder, but no breeding records in recent decades.

Behaviour and ecology Usually seen lazily quartering open areas. Vigilant and diurnal. Mainly feeds on fish, offal, refuse, sometimes large insects and small vertebrates. Nests on tall trees. Clutch size: 2-3.

玉带海雕 *Haliaeetus leucoryphus* Pallas's Fish Eagle 鹰科 Accipitridae

形态特征 大型猛禽，体长 76~88 厘米。雌雄相似。头颈较中国有分布的其他海雕更细长，喙稍小。成鸟头部和颈部有浅褐色披针状羽毛，体羽及翼下均为黑褐色；尾羽具一宽阔醒目的白色横带，因此而得名。幼鸟飞行时可见最内侧几枚浅色初级飞羽，与翼下覆羽的浅色条带组成独特图案。

分布与生境 主要分布于中国中部和西部地区。一般活动于内陆草原和大型开阔水域。北京为甚罕见旅鸟。近年来仅于 2016 年 6 月在延庆野鸭湖有一次记载。

习性与生态 觅食于湖泊岸边，以淡水鱼类和雁鸭等水禽为主。通常营巢于湖泊、河流或沼泽岸边高大乔木树上，也在渔村附近或离水域较远的树上筑巢。

Morphological characteristics A large raptor (76-88 cm). Sexes alike. Head and neck are more slender than other fish eagles in China; smaller bill. Adult has light brown long and needle-like feathers around head and neck, brownish-black plumage on body and underwing; a prominent broad white band on tail is the basis of its scientific and Chinese names. On juvenile, several pale inner primaries contrast with a pale band on the underwing-coverts, forming a unique pattern.

Distribution and habitat Distributed in the central and western region of China. Found in steppes or large wetlands. Very rare migrant in Beijing, with only one recent record at Yeyahu, Yanqing District (June 2016).

Behaviour and ecology Forage along the banks of lakes for freshwater fish and water birds. Nests in tall trees on banks of lakes, rivers or marshes, occasionally on trees near a village or further away from water.

国家一级保护野生动物 | IUCN 红色名录 濒危 (EN) 2018.8.9

亚成鸟 Immature

成鸟 Adult

白尾海雕	*Haliaeetus albicilla*	White-tailed Sea Eagle	鹰科 Accipitridae

亚成鸟 Immature

成鸟 Adult

	国家一级保护野生动物	IUCN 红色名录　无危 (LC) 2020.8.13

形态特征　大型猛禽，体长 74~92 厘米。喙黄色，厚而粗壮。虹膜亦为黄色。上体褐色，头、后颈多为淡黄褐色，胸和腹部褐色。尾短而中央尾羽外凸成楔形，为纯白色。跗跖黄色。幼鸟尾羽杂以褐色和白色斑，喙色较深。

分布与生境　主要在中国东北繁殖，越冬于华北、中部和南方地区。北京不常见于郊区大型开阔水域，为冬候鸟和旅鸟。

习性与生态　常单独或成小群活动，在空中盘旋搜寻猎物。主要以鱼为食，也捕食鸟类和中小型哺乳动物。营巢于海岸高峭岩壁凹处或高大树木顶端；每年将旧巢加以修复，逐年加大，内径可达 1 米；窝卵数 2 枚。

Morphological characteristics　A large raptor (74-92 cm). Yellow bill thick and strong with yellow cere. Upperparts dark brown, head and hindneck tawny, breast and belly brown; white tail short and wedge-shaped; tarsi yellow. Juvenile has brown and white markings on tail and a darker bill.

Distribution and habitat　Breeds in northeast China; wintering in north, central and south China. In Beijing, uncommon in large open areas of water in rural areas. Residential status: winter visitor and migrant.

Behaviour and ecology　Usually encountered singly or in small groups. Preys mostly on fish, also birds and small to medium-sized mammals. Nests on coastal cliffs or on top of a large tree. Nest is re-used and expanded each year and the inner diameter may reach 1 m. Clutch size: 2.

灰脸鵟鹰 | *Butastur indicus* | Grey-faced Buzzard | 鹰科 Accipitridae

形态特征 中型猛禽，体长 39~48 厘米。雌雄相似。喙灰色。虹膜黄色。成鸟颊部灰色明显，有粗白眉纹，具深色喉中线。最外侧几枚初级飞羽端部发黑，胸腹部密布红褐色横纹。尾羽上具黑褐色横斑。跗跖黄色。

分布与生境 分布于中国东部、中部和南部广大地区。北京迁徙季过境于山区，北部山区亦有少量繁殖，为区域性常见旅鸟、夏候鸟。

习性与生态 常单独活动，迁徙期间集群。性情较为胆大，有时也飞至城镇和农村捕食。主要以小型蛇类、蛙、蜥蜴、鼠类和小鸟等食。营巢于阔叶林或混交林中靠河岸的疏林地带或林缘地带的树上；窝卵数 3~4 枚。

Morphological characteristics A medium-sized raptor (39-48 cm). Sexes alike. Bill grey; iris yellow. Adult has distinct grey face with a broad white supercilium; throat white with narrow dark mesial; outer primaries tipped black; breast and belly barred reddish-brown; tail brownish-black barred; tarsi yellow.

Distribution and habitat Distributed in eastern, central and southern China. In Beijing, locally common on migration in mountains. A few breed in the northern mountains. Residential status: migrant, summer breeder.

Behaviour and ecology Feeds singly but often encountered in groups on migration. Fearless; sometimes forages in towns and villages. Feeds on small snakes, frogs, lizards, rats and small birds. Nests on trees at forest edge or in sparse woodlands close to river bank. Clutch size: 3-4.

国家二级保护野生动物 | IUCN 红色名录 无危 (LC) 2016.10.1

成鸟 Adult

亚成鸟 Immature

毛脚鵟	*Buteo lagopus*	Rough-legged Buzzrard	鹰科 Accipitridae

	国家二级保护野生动物	IUCN 红色名录　无危 (LC) 2016.10.1

形态特征　中型猛禽，体长 51~56 厘米。雌雄基本相似。喙深灰，蜡膜黄色。虹膜黄褐色。头颈部淡灰白色，上体多为褐色，羽缘白色或褐色，下体棕褐色。部分成年个体体羽几近纯白色。尾基部白色，翼后缘与尾羽末端具宽阔的黑色条带（雄鸟常 2~3 条尾羽端斑，雌鸟 1 条）。跗跖被羽。

分布与生境　分布于中国北方地区。北京不常见于低山丘陵、林缘地带、荒地、农田，为冬候鸟和旅鸟。

习性与生态　常单独活动，捕食小型啮齿类，偶亦捕食野兔，以及石鸡等鸟类。营巢于苔原河流或森林河流两岸悬崖峭壁上；巢呈盘状；窝卵数 3~4 枚。

Morphological characteristics　A medium-sized raptor (51-56 cm). Sexes similar. Dark grey bill, yellow cere and brownish yellow iris. Head and neck pale greyish-white, upperparts generally brown with white or brown wing-fringing, underparts greyish brown. Some individuals are mostly whitish. Tail base white; black terminal bands on wings and tail (two to three tail bands in males, one in females); tarsi feathered to the toe.

Distribution and habitat　Northern China. In Beijing, uncommon in the hills, forest edge, grassland and cultivations. Residential status: migrant and winter visitor.

Behaviour and ecology　Usually solitary. Forages for small rodents, along with hares and partridges. Often actively hovers. Nests on cliffs or bluffs along the banks of tundra or forest rivers. Clutch size: 3-4.

大鵟	*Buteo hemilasius*	Upland Buzzard	鹰科 Accipitridae

形态特征 中型猛禽，体长 56~71 厘米。雌雄相似。头小而两翼宽大，体型似雕。本种有几种色型，上体及下体自浅褐色到黑褐色皆有。北京常见者为上体暗褐色，胸部发白，下腹部及两胁褐色。飞行时可见初级飞羽基部色浅，在翼双面形成大块白斑。尾羽暗褐色，通常具细横斑。跗跖覆羽棕褐色。

分布与生境 中国除东南部以外，均有分布。北京区域性常见于开阔平原、低山丘陵和农田荒地，为冬候鸟和旅鸟。

习性与生态 常见其在高空盘旋，以食啮齿类动物和雉类为主，亦吃蛙类和较大型昆虫。营巢于高原山区悬崖峭壁顶上或乔木上；窝卵数 2~5 枚。

Morphological characteristics A medium-sized raptor (56-71 cm). Sexes alike. A eagle-like buzzard, with small head and long, broad wings. There are several morphs from pale brown to blackish-brown. Those commonly found in Beijing usually have dark brown upperparts, whitish breast, and brown lower belly and flanks; a large white patch at the base of primaries on the both surface of wing in flight; tail dark brown with fine bands; tarsi covered with brown feathers.

Distribution and habitat Widely distributed in China except for the southeast. In Beijing, locally common in open plains, low mountain hills and farmland. Residential status: winter visitor and migrant.

Behaviour and ecology Often seen hovering and circling high. Feeds on small rodents and pheasants along with frogs and large insects. Nests on top of crags or trees on mountain plateaus. Clutch size: 2-5.

国家二级保护野生动物

IUCN 红色名录　无危 (LC) 2018.8.9

普通鵟　*Buteo japonicus*　Eastern Buzzard　鹰科 Accipitridae

形态特征　中型猛禽，体长 48~56 厘米。虹膜黄色至褐色。蜡膜黄色。头部浑圆，颈短。体色变化多样，常见个体为棕褐色。飞翔时翼下有明显腕状斑。成年个体腹部有棕色横带，尾部较长且宽，飞翔时常呈扇形。跗跖裸露，不被羽。幼鸟下体具显著的深色纵纹。

分布与生境　中国除西北部外，各地均有分布。北京迁徙季大量过境西部山区，冬季常见于郊区开阔地附近稀疏的林地，以及农田、草地和低山地区上空，为旅鸟和冬候鸟。

习性与生态　常见在空中盘旋。主食啮齿类动物，有时亦捕食昆虫、蛙、蜥蜴、蛇、鸟类和小型哺乳动物。营巢于林缘附近树顶或峭壁悬崖上；窝卵数 1~3 枚。

Morphological characteristics　A medium raptor (48-56 cm). Iris yellow to brown, cere yellow. Head and neck short. Plumage variable, the most common morph has brown upperparts; obvious carpal patch on underwing in flight. Adult has brown band on belly; tail relatively large, fanned out in flight; lacks feathers on tarsi. Juvenile has prominent dark streaks on underparts.

Distribution and habitat　Widely distributed in China except for the northwest. In Beijing, common on migration, particularly in the western mountains. Often found in lightly wooded areas near open fields; in autumn and winter also found in farmland, grassland and over hills. Residential status: migrant, winter visitor.

Behaviour and ecology　Often hangs in the wind when hunting or circling high. Feeds on rodents, sometimes on insects, frogs, lizards, snakes, birds and small mammals. Nests on top of crags or trees at forest edge. Clutch size: 1-3.

国家二级保护野生动物　IUCN 红色名录　无危 (LC)　2016.10.1

北领角鸮	*Otus semitorques*	Japanese Scops Owl	鸱鸮科 Strigidae

	国家二级保护野生动物	IUCN 红色名录　无危 (LC) 2016.10.1

形态特征　小型猛禽，体长 23~25 厘米。雌雄相似。虹膜红色。有显著耳羽簇。上体灰褐色，具暗色虫卵状斑和黑色羽干纹。下体灰白色或皮黄色，缀有淡褐色波状横斑和黑色羽干纹。跗跖被羽。

分布与生境　分布于中国东北至华中，与领角鸮的分布界限尚不明晰。北京见于西部山地阔叶林和混交林、山麓林缘附近，为不常见夏候鸟及留鸟。

习性与生态　常单独活动，夜行性。主要以鼠类和昆虫为食。鸣叫声为低沉的“whool”声，间隔时间较长。营巢于天然树洞内，或利用啄木鸟废弃的旧树洞，偶尔也利用喜鹊的旧巢；窝卵数 2~6 枚，卵白色。

Morphological characteristics　A small raptor (23-25 cm). Sexes alike. Red iris and distinct ear-tufts. Upperparts greyish-brown with dark vermiculations and black shaft-streaks; underparts pale or buff with light brown bars and black shaft-streaks; feathered tarsi.

Distribution and habitat　Northeast to central China. The distribution boundary with Collared Scops Owl is unclear. In Beijing, uncommon in broadleaved and mixed forest in the western mountains. Residential status: summer breeder and resident.

Behaviour and ecology　Feeding solitary, nocturnal. Feeds on rats and insects. Call is subdued "whool" with long intervals. Nests in natural tree hole or abandoned holes of woodpeckers, occasionally uses old nest of magpies. Clutch size: 2-6 white eggs.

红角鸮	*Otus sunia*	Oriental Scops Owl	鸱鸮科 Strigidae

灰色型 Grey morph

棕色型 Brown morph

	国家二级保护野生动物	IUCN 红色名录 无危 (LC) 2016.10.1

形态特征 小型猛禽，体长 17~21 厘米。雌雄相似。有灰色和栗棕色两种色型。虹膜黄色。面盘灰褐色，周围棕褐色，具明显的耳羽簇。体羽灰褐色（灰色型）或栗棕色（栗棕色型），肩羽具一由连续的棕白色斑点形成的纵线。飞羽大部为黑褐色。跗跖被羽。

分布与生境 广泛分布于中国东部和南部地区。北京区域性常见于山地和平原的阔叶林、混交林，有时也可见于大学校园、城市公园中，为夏候鸟和旅鸟。

习性与生态 通常单独活动，夜行性。以昆虫、小型鼠类、两栖爬行类等动物为食。叫声为“tuner-tuntun”声，持续时间长。营巢于天然树洞中；窝卵数 3~6 枚，卵白色。

Morphological characteristics A small raptor (17-21 cm). Sexes alike. Two morphs - grey and rufous. Iris yellow; greyish brown facial disc bordered with rufous brown; conspicuous ear-tufts; body largely greyish brown (grey morph) or rufous-brown (brown morph) with a whitish scapular stripe; remiges mostly blackish-brown; tarsi feathered.

Distribution and habitat Widely distributed in eastern and southern China. In Beijing, locally common in broadleaved and mixed forest of mountains and plains, sometimes on university campuses and in city parks. Residential status: summer breeder and migrant.

Behaviour and ecology Usually solitary and nocturnal. Feeds on rodents, insects, amphibians and reptiles. Call is a long lasting "tuner-tuntun". Nests in natural tree holes. Clutch size: 3-6, white eggs.

雕鸮	*Bubo bubo*	Eurasian Eagle Owl	鸱鸮科 Strigidae

形态特征 大型猛禽，体长 55~89 厘米。雌雄相似。双眼较大，虹膜橙黄色，眼上方有一大块斑驳黑斑。耳羽特别发达，显著突出于头顶两侧，通体羽毛黄褐色，具黑色斑点和纵纹。胸和两胁具浅黑色纵纹。跗跖被羽。

分布与生境 中国广泛分布。北京见于山区林地、荒野、平原及裸岩峭壁，为不常见留鸟。

习性与生态 通常在远离人群的偏僻之地。常单独活动，夜行性。主要以鼠类为食，也食野兔等小型兽类、鸟类甚至其他猛禽。鸣声为沉重的“bo-bo”声，带有颤音。营巢于树洞、悬崖峭壁下的凹处或直接产卵于地上；窝卵数 2~5 枚，卵白色。

Morphological characteristics A large raptor (55-89 cm). Sexes alike. Eyes big with yellowish-orange iris, large mottled black patch above eyes; prominent ear-tufts protrude from the sides of the crown. Body is tawny with black speckles and streaks; pale black streaks on breast and flanks. Feathered tarsi.

Distribution and habitat Widely distributed in China. In Beijing, breeds in the mountains, descending to grassland, plain or crags in winter. Residential status: uncommon resident.

Behaviour and ecology Often found singly in remote areas far away from human activity; nocturnal. Feeds mainly on rodents and other small animals like hares, birds and even other raptors. Call is a deep trilling "bo-bo". Nests in tree holes, crevices of cliffs and even on the ground. Clutch size: 2-5, white eggs.

	国家二级保护野生动物	IUCN 红色名录 无危 (LC) 2016.10.1

灰林鸮	*Strix aluco*	Tawny Owl	鸱鸮科 Strigidae

国家二级保护野生动物　IUCN 红色名录　无危 (LC) 2016.10.1

形态特征　中型猛禽，体长 35~40 厘米。雌雄相似。虹膜深褐色；喙淡黄色。无耳羽簇，面盘明显，两眼间有“X”形图案。面盘为灰色（灰色型个体）或棕红色（棕色型个体）；上体暗灰褐色（灰色型个体）或棕红色（棕色型个体），具近黑色纵纹；下体皮黄色；胸腹具浓重的褐色纵纹和细小蠹状纹。跗跖黄褐色。

分布与生境　主要分布于中国东北、华北至南部地区。北京见于西、北部山地阔叶林和混交林中，为不常见留鸟。

习性与生态　多单独活动，夜行性。主要以啮齿类为食，也吃小鸟、蛙和昆虫。鸣叫声为响亮拖长的单声“hu……”或多音节“hu, huhuhuhu……”。主要营巢于树洞中，有时也在岩石下面的地上营巢或利用鸦类巢；窝卵数 2~4 枚。

Morphological characteristics　A medium-sized raptor (35-40 cm). Sexes alike. Dark brown iris and ivory-yellow bill. No ear-tufts but a distinctive facial disc with an "X" between the eyes. Facial disc grey (grey morph) or reddish-brown (brown morph); upperparts dark greyish-brown (grey morph) or reddish-brown (brown morph) with blackish streaks; underparts largely buff; breast and belly heavily streaked and with fine vermiculations; tarsi yellow-brown.

Distribution and habitat　From northeast and north China to southern China. In Beijing, uncommon resident in broadleaved or mixed forest in the western and northern mountains.

Behaviour and ecology　Solitary and nocturnal. Feeds on rodents, also small birds, frogs, insects. Calls include a loud and drawn-out single note "hu..." and multiple notes "hu, huhuhuhu...". Nests in tree holes, sometimes on ground; also utilises nests of crows. Clutch size: 2-4.

长尾林鸮	*Strix uralensis*	Ural Owl	鸱鸮科 Strigidae

形态特征 中型猛禽，体长 50~60 厘米。雌雄相似。虹膜褐色。喙橘黄色。无耳羽簇，面盘显著，呈灰色。整体以浅色调为主。上体褐色具黑色纵纹，杂以棕红色、白色斑点。胸腹部灰白色，具深色稀疏纵纹。两翼和尾羽具横斑。跗跖被羽。

分布与生境 主要分布于中国东北。北京见于山区阔叶林和混交林中，为甚罕见留鸟（1990 年 6 月于小龙门林场有环志个体；1992 年 11 月于十三陵有过记录）。

习性与生态 通常单独活动。白天大多栖息在密林深处，直立地站在靠近树干的水平粗枝上，有时白天也活动和捕食。主要以鼠类为食。叫声为深沉的“huwo-huwo-huwo”声。通常营巢于树洞中，也在树根下的地面上或林中河岸的石崖上营巢；窝卵数 2~6 枚，卵白色。

Morphological characteristics A medium-sized raptor (50-60 cm). Sexes alike. Brown iris and orange-yellow bill. No ear-tufts but a prominent grey facial disc. Overall pale plumage; upperparts brown with black streaks and speckled with reddish brown and white; breast and belly white grey with dark sparse streaks; wings and tail barred; feathered tarsi.

Distribution and habitat Mainly Northeast China. In Beijing, very rare resident in broadleaved or mixed forests in in the mountains (banded at Xiaolongmen in June 1990; the Ming Tombs in November 1992).

Behaviour and ecology Usually solitary. Always perches upright on branches close to a tree trunk in the deep forest when roosting. Feeds on small rodents. Call is a deep "huwo-huwo-huwo". Nests in tree holes, sometimes on the ground at the foot of tree or rock cliff. Clutch size: 2-6, white eggs.

国家二级保护野生动物 | IUCN 红色名录 无危 (LC) 2016.10.1

斑头鸺鹠	*Glaucidium cuculoides*	Asian Barred Owlet	鸱鸮科 Strigidae

	国家二级保护野生动物	IUCN 红色名录　无危 (LC) 2016.10.1

形态特征　小型猛禽，体长 22~26 厘米。雌雄相似。虹膜黄色。无耳羽簇，头、颈和整个上体包括两翼表面暗褐色，具白色横纹。头顶横纹细小而密，并非点状斑。下体白色，胸、上腹及两胁具浓密的褐色横斑。跗跖黄绿色。

分布与生境　分布于中国华中、华南、东南、西南等地。栖息于平原、低山丘陵地带的阔叶林、混交林和林缘灌丛。在北京为甚罕见迷鸟。

习性与生态　大多单独或成对活动。多在白天活动和觅食，也在晚上活动。主要以各种昆虫为食，也吃鼠类、小鸟、蚯蚓、蛙和蜥蜴等。叫声响亮，晨昏时发出快速的颤音，调降而音量增。通常营巢于树洞或天然洞穴中；窝卵数 3~5 枚。

Morphological characteristics　A small raptor (22-26 cm). Sexes alike. Iris yellow; no ear-tufts; head, neck and whole upperparts, including wings, dark brown with white bars. Crown is finely-barred. White underparts with extensive brown bars on breast, upper belly and flanks. Tarsi greenish-yellow.

Distribution and habitat　Central and southern China. Found in broadleaved and mixed forest, or shrubby areas in the plains and at low elevations in the mountains. Very rare vagrant in Beijing.

Behaviour and ecology　Usually solitary or in pairs. Mostly active in daytime, sometimes at night. Feeds mostly on insects and also small rodents, small birds, earth worms, frogs and lizards. Call is a loud and clear quivering trill at dawn and dusk. Nests in tree holes or natural caves. Clutch size: 3-5.

纵纹腹小鸮 | *Athene noctua* | Little Owl | 鸱鸮科 Strigidae

形态特征 小型猛禽，体长 21~26 厘米。雌雄相似。虹膜亮黄色。喙黄绿色。面盘不明显，无耳簇。上体褐色偏灰，头顶具小白斑点，两眼间及眼上下为灰白色。下体棕白色而有褐色纵纹，下腹部至臀部白色。跗跖被白色羽。

分布与生境 广布于中国北方及西部的大多数地区。北京见于低山、林缘，也到村庄附近及农田乃至居民区活动，为区域性常见留鸟。

习性与生态 主要以鼠类、昆虫等为食。白天、夜间皆有活动。常发出"goooek"声，拖长而上扬。巢置于树洞、建筑物屋檐孔洞等处，有时也在自己挖掘的洞穴中营巢；窝卵数 2~8 枚。

Morphological characteristics A small raptor (21-26 cm). Sexes alike. Bright yellow iris and greenish-yellow bill. Indistinctive facial disc and no ear tufts. Upperparts greyish brown with small white spots on crown and greyish white between and around eyes; underparts whitish with brown streaks; lower belly, vent all white; tarsi feathered white.

Distribution and habitat Northern and western China. In Beijing, locally common in low mountain areas, forest edges, farmlands and residential areas both rural and suburban. Residential status: resident.

Behaviour and ecology Feeds on insects and small rodents. Active both during the day and at night. Call is an extended and upward "gooooek". Nests in tree cavity, holes in buildings and sometimes self-excavated holes. Clutch size: 2-8.

国家二级保护野生动物 | IUCN 红色名录 无危 (LC) 2018.8.9

日本鹰鸮	*Ninox japonica*	Northern Boobook	鸱鸮科 Strigidae

	国家二级保护野生动物	IUCN 红色名录　无危 (LC) 2016.10.1

形态特征　小型猛禽，体长 25~32 厘米。雌雄相似。虹膜亮黄色。喙蓝灰色，蜡膜绿色。圆型头，无耳簇，身体修长似鹰样。上体深褐，下体具宽阔卵圆形褐色斑点连成纵纹。跗跖黄色。

分布与生境　繁殖于中国东北至华北，冬季南迁。北京见于林缘地带柏树，偶至城市公园、大学校园中，为不常见夏候鸟、旅鸟。

习性与生态　多于黄昏和晚上单独活动，有时白天也活动。主要以鼠类、小鸟和昆虫等为食。叫声为 2 或 3 声“whoop”音。通常营巢于树木的天然洞穴中，也利用鸳鸯和啄木鸟的旧洞；窝卵数 3 枚。

Morphological characteristics　A small raptor (25-32 cm). Sexes alike. Bright yellow iris and slaty-grey bill with greenish cere. Head round, no ear-tufts and body slender like a hawk. Upperparts dark brown; underparts streaked with broad brown oval spots; tarsi yellow.

Distribution and habitat　Breeds in northern China, wintering in the south. In Beijing, uncommon at forest edge, often with cypress trees; rarely found in city parks and university campuses. Residential status: summer breeder and migrant.

Behaviour and ecology　Mostly solitary and active at dusk and at night, occasionally in the daytime. Feeds on insects, rodents and small birds. Song consists of two or three "whoop" calls. Nests in natural tree cavity and also utilises old holes of woodpeckers and mandarin ducks. Clutch size: 3.

长耳鸮	*Asio otus*	Long-eared Owl	鸱鸮科 Strigidae

形态特征 中型猛禽，体长 29~39 厘米。雌雄相似。虹膜橙红色。头顶两侧具一对长形耳羽簇，飞行时耳羽簇不可见。棕黄色圆形面盘显著，胸腹部皮黄有黑褐色不连贯纵纹。翼下初级飞羽有多道横纹有别于短耳鸮。跗跖被羽。

分布与生境 中国除西南高海拔地区，几乎皆有分布。北京见于针叶林、针阔混交林或岸边林地，为不常见冬候鸟和旅鸟；近年亦有繁殖鸟记录于通州。天坛公园在 20 世纪以来曾有持续稳定的长耳鸮过冬记录（10~100 只），近些年数量锐减，甚至整个冬季未有记录。

习性与生态 集小群越冬。夜行性。多见其立于柏树或柳树之上。主要以鼠类为食，食物匮乏时也食麻雀或蝙蝠。通常利用乌鸦、喜鹊或其他猛禽的旧巢，有时也在树洞中营巢；窝卵数 3~8 枚。

Morphological characteristics A medium-sized raptor (29-39 cm). Sexes alike. Orange-red iris, a pair of long ear-tufts on crown which is invisible in flight; a prominent tawny round facial disc. Buffish breast and belly with brownish-black streaks. More heavily barred on underwing than Short-eared Owl. Feathered tarsi.

Distribution and habitat Widespread in China except for the Tibetan Plateau. In Beijing, found in coniferous and mixed woodland and wooded river banks. A stable wintering population (10-100 individuals) had been recorded in the Temple of Heaven in the 20th century, but the numbers have sharply declined in recent decade, with none recorded last several winters. Breeding pairs recorded in Tongzhou District recent years. Residential status: uncommon winter visitor and migrant, rare summer breeder.

Behaviour and ecology Winters in small groups, nocturnal. Often perched in cypress or willow trees. Feeds mainly on rodents, also tree sparrows and bats. Often utilises used nest of corvids or raptors. Clutch size: 3-8.

	国家二级保护野生动物	IUCN 红色名录　无危 (LC) 2018.8.9

短耳鸮	*Asio flammeus*	Short-eared Owl	鸱鸮科 Strigidae

	国家二级保护野生动物	IUCN 红色名录 无危 (LC) 2016.10.1

形态特征 中型猛禽，体长 34~40 厘米。雌雄相似。虹膜亮黄色，眼周黑色。喙黑色。面盘明显。耳羽簇短，野外常不可见。身体黄褐色。下体棕黄色，具黑色羽干纵纹。飞行时翼下初级飞羽仅有两道较粗横纹。跗跖被棕色羽。

分布与生境 中国广泛分布。北京冬季不常见于水边开阔荒地、农田中，为旅鸟和冬候鸟。

习性与生态 多在黄昏和晚上活动和猎食，但也在白天活动。平时多栖息于干草丛中，很少栖于树上。主要以鼠类为食，也吃昆虫、蜥蜴和小鸟，偶尔吃植物果实和种子。通常营巢于沼泽附近地上草丛中；窝卵数 3~8 枚。

Morphological characteristics A medium-sized raptor (34-40 cm). Sexes alike. Bright yellow iris, overall tawny-brown plumage, black bill, black around eyes with a distinct facial disc. Ear-tufts short, often invisible in the field. Underparts tawny with black shaft-streaks and two broad bands on underwing. Tarsi covered with brown feathers.

Distribution and habitat Widespread in China. In Beijing, uncommon in open country and farmland near water. Residential status: winter visitor and migrant.

Behaviour and ecology Active at dusk and at night, sometimes in the daytime. Often roosts in dry grass, rarely on trees. Feeds on rodents, along with insects, lizards, and small birds, occasionally fruit and seeds. Nests in grasses near marsh. Clutch size: 3-8.

二十三、隼形目 FALCONIFORMES

黄爪隼	*Falco naumanni*	Lesser Kestrel	隼科 Falconidae

形态特征 中型猛禽，体长 28~34 厘米。雄鸟头、颈灰色；背及内侧翼上覆羽棕红色，无斑；大覆羽灰色；飞羽和初级覆羽黑褐色；腰、尾上覆羽和尾羽灰色，尾羽具宽阔黑色次端斑和近白色端斑；颏、喉淡黄白色，下体余部淡棕黄色，胸及两胁具黑色圆形斑点；跗跖淡黄色，爪近白色。雌鸟头淡褐色，具深色细纵纹；上体棕红色，具黑色横斑；下体具黑褐色纵纹。雌鸟甚似红隼雌鸟，但髭纹更细弱。

分布与生境 繁殖于中国北方，迁徙时见于华北至西南。北京迁徙季见于山区及开阔原野，为不常见旅鸟。

习性与生态 习性似红隼。营巢于山区河谷悬崖峭壁上，或占据喜鹊、乌鸦等鸟类的巢；窝卵数 4~5 枚。

Morphological characteristics A medium-sized raptor (28-34 cm). Male has blue-grey head and neck; back and inner upperwing-coverts rufous-brown without dark bars; greater coverts grey; remiges and primary coverts blackish-brown; rump, uppertail-coverts and tail greyish; tail has prominent black subterminal band and whitish tips; chin and throat pale yellowish white; remaining of underparts pale buff, with black spots on breast and flanks. Female has pale brownish head with fine dark streaks; upperparts rufous-brown with black bars; underparts have blackish-brown streaks. Female closely resembles female Common Kestrel but has more indistinct moustachial stripe.

Distribution and habitat Breeds in northern China, migrants to north and southwest China. In Beijing, uncommon migrant in montane areas and in lowland open country.

Behaviour and ecology Habits similar to Common Kestrels. Nests in on cliff-ledged river valley in montane areas or occupies nests of magpies and crows. Clutch size: 4-5.

	国家二级保护野生动物	IUCN 红色名录　无危 (LC) 2018.8.9

雄鸟 Male

雌鸟 Female

红隼 *Falco tinnunculus* Common Kestrel

隼科 Falconidae

雄鸟 Male

雌鸟 Female

国家二级保护野生动物

IUCN 红色名录　无危 (LC) 2016.10.1

形态特征　中型猛禽，体长 33~38 厘米。雄鸟头部灰色，眼下有显著髭纹；背部淡砖红色缀以黑色点斑；下体淡棕黄色有不长的黑色细纵纹及点斑，尾羽灰色具宽阔的黑色次端斑；跗跖深黄色，爪黑色。雌鸟上体羽红褐色密布深色横斑，头部同背色，尾具黑色次端斑和多道较窄的深色横斑。

分布与生境　广布于中国除沙漠腹地以外的几乎所有地域。北京常见于山地、平原的各种生境及城市中。北京种群包括夏候鸟、冬候鸟和旅鸟多种居留型。

习性与生态　飞行迅速而敏捷。常在空中捕食昆虫，也吃啮齿类、蜥蜴、蛙、小型鸟类等脊椎动物。营巢于悬崖上或建筑物上，亦常占据喜鹊和乌鸦的巢；窝卵数 4~5 枚。

Morphological characteristics　A medium-sized raptor (33-38 cm). Male has finely streaked grey head; upperparts deep chestnut with black spots; pale rufous-buff underparts with short black fine stripes and spots; tail grey with wide dark terminal band; tarsi tawny and claws black. Female has warm brown upperparts barred dark; head and back brownish, finely barred dark; tail rufous with a broad dark subterminal band and several narrow dark bars.

Distribution and habitat　Widespread in China except hinterland desert. In Beijing, common in montane and lowland areas, also found in urban areas. Residential status: summer breeder, winter visitor and migrant.

Behaviour and ecology　Flight fast and agile. Often catches insects on the wing. Also feeds on rodents, lizards, frogs and small birds. Nests in on cliffs or buildings, also occupies nests of magpies and crows. Clutch size: 4-5.

红脚隼	*Falco amurensis*	Amur Falcon	隼科 Falconidae

形态特征 小型猛禽，体长 26~30 厘米。雄鸟全身大致为深灰色，眼圈、跗跖及尾下覆羽橘红；具白色的翼下覆羽，与黑灰色的飞羽形成鲜明的对比。雌鸟头顶及上体黑灰色，部分个体上体具淡褐色横纹；下体及翼下近白色，具深色斑点和纵纹。雌鸟似燕隼，但头部羽色略淡，且下体深色斑纹不及燕隼浓密。

分布与生境 中国除西部地区，几乎均有分布。北京为常见旅鸟和夏候鸟，活动于低山和平原的各种开阔生境。

习性与生态 飞翔时两翅快速扇动，间或进行一阵滑翔。主要以昆虫为食，有时也捕食小型鸟类、蜥蜴、蛙、鼠类等小型脊椎动物。常营巢于疏林中高大乔木树的顶枝上，也占喜鹊巢；窝卵数 4~5 枚。

Morphological characteristics A small raptor (26-30 cm). Male has dark blue-grey body, with deep orange eye-ring, tarsi and undertail-coverts; white underwing-coverts contrasts with blackish-grey remiges. Female has blackish grey head and upperparts; upperparts sometimes barred pale brown; white underwing and underparts with dark spots and streaks. Female resembles Eurasian Hobby but has paler head and less heavily streaked underparts.

Distribution and habitat Widespread in China except for the west. In Beijing, common migrant and summer breeder in open areas in lowlands and low hills.

Behaviour and ecology Active flight with rapid wingbeats and often hovers. Feeds mostly on insects, sometimes small birds, lizards, frogs and rodents. Nests in trees canopy in open woodlands, sometimes occupies magpie nests. Clutch size: 4-5.

国家二级保护野生动物　IUCN 红色名录　无危 (LC) 2016.10.1

雄鸟 Male

雄鸟 Male

雌鸟 Female

灰背隼 | *Falco columbarius* | Merlin | 隼科 Falconidae

雄鸟 Male

雌鸟 Female

国家二级保护野生动物 | IUCN 红色名录 无危 (LC) 2016.10.1

形态特征 小型猛禽，体长 24~32 厘米。雄鸟头及上体青灰色，具不甚清晰的淡棕白色眉纹和灰褐色髭纹；飞羽黑色；尾羽灰色，具宽阔的黑色次端斑和近白色端斑；下体淡黄白色，具褐色纵纹；跗跖黄色。雌鸟上体褐色，尾羽淡褐色，具数道深褐色横斑及黑褐色次端斑。

分布与生境 繁殖于中国东北和西北地区，越冬于中部及南部地区。北京见于平原旷野、荒漠草原灌丛地带等开阔地，为不常见冬候鸟、旅鸟。

习性与生态 飞行敏捷而灵活。常在空中飞行捕食小型鸟类，亦食昆虫、啮齿类、蜥蜴、蛙类等。多占用鸦、鹊的旧巢或至草原地面上营巢；窝卵数 3~4 枚。

Morphological characteristics A small raptor (24-32 cm). Male has blue-grey head and upperparts; supercilium pale brownish white; indistinct moustachial stripe greyish brown; remiges blackish; tail grey with broad black subterminal band and whitish tips; underparts pale buff, streaked brown; tarsi yellow. Female has brownish upperparts; tail pale brown with dark brown bars and blackish-brown subterminal band.

Distribution and habitat Breeds in northeast and northwest China; winters in central and southern China. In Beijing, uncommon in open areas such as semi-deserts, grassland and scrub. Residential status: winter visitor and migrant.

Behaviour and ecology Active flight with fast wingbeats. Feeds on small birds caught on the wing, also insects, rodents, lizards and frogs. Nests in grassland on ground, often in abandoned nest of crows and magpies. Clutch size: 3-4.

燕隼	*Falco subbuteo*	Eurasian Hobby	隼科 Falconidae

形态特征 中型猛禽，体长 29~33 厘米。雌雄相似。眼圈黄色。头顶黑色，从眼下和耳部伸出两道较粗重的近黑色髭纹，酷似头盔。两翼狭长，停落时翼尖超过尾端。胸腹部具粗壮的黑色纵纹，尾下覆羽、下腹红棕色。

分布与生境 中国几乎全境分布。在北京为区域性常见旅鸟、夏候鸟。栖息于山地森林及较低海拔的开阔区域。

习性与生态 常单只或成对活动，飞行快速而敏捷似雨燕。捕食对象主要为空中的昆虫，有时亦捕食小鸟和蝙蝠。营巢于疏林或林缘和田间的高大乔木上，通常自己很少营巢，而是侵占乌鸦和喜鹊的巢；窝卵数 2~4 枚。

Morphological characteristics A medium-sized raptor (29-33 cm). Sexes alike. Eye-ring yellow; crown black, with two relatively bold blackish moustachial stripes like a helmet; wings long, narrow and pointed; primary projection extends beyond tail when perched; black-streaked breast and belly; vent and undertail-coverts rufous.

Distribution and habitat Widespread in China. In Beijing, locally common in western montane forests and lowland open areas. Residential status: migrant and summer breeder.

Behaviour and ecology Often solitary or in pairs. Dashing flight like a swift. Feeds on insects on the wing, sometimes small birds and bats. Nests on in canopy of trees in open woodland or forest edges, usually in abandoned nests of other birds, such as a crow or magpie. Clutch size: 2-4.

国家二级保护野生动物

IUCN 红色名录 无危 (LC) 2016.10.1

亚成鸟 Immature

成鸟 Adult

猎隼	*Falco cherrug*	Saker Falcon	隼科 Falconidae

形态特征 中型猛禽，体长 46~56 厘米。雌雄相似。颈背偏白，头顶浅褐。眼下可见清晰髭纹。上体多褐色，翼下大覆羽具黑色细纹，翼尖黑色。下体偏白，胸腹部具黑褐色点状斑。跗跖黄色。未成鸟上体深褐色；下体满布黑色纵纹；跗跖浅蓝灰色。

分布与生境 分布于中国西部、中部、东北及华北。栖息于山区开阔地带、水库周边荒地。在北京为不常见旅鸟、冬候鸟。

习性与生态 性凶猛。主要以野兔为食，有“兔虎”之称。大多在悬崖峭壁上的缝隙中营巢，或者营巢于树上，有时也利用其他鸟类的旧巢；窝卵数 3~5 枚。

Morphological characteristics A medium-sized raptor (46-56 cm). Sexes alike. Crown usually rather pale brown with whitish nape, distinct moustachial stripe; upperparts mostly brown, greater underwing-coverts finely striped black; wing-tip black; underparts whitish; breast and belly with dark brown spots; tarsi yellow. Juvenile has tawny earth-brown upperparts; underparts boldly streaked black; tarsi pale bluish-grey.

Distribution and habitat West, central, north and northeast China. Inhabits open habitats in montane areas and around reservoirs. Residential status: uncommon migrant and winter visitor.

Behaviour and ecology Aggressive. Mainly feeds on hares, hence the nicknamed "Hare Tiger". Nests in crevice on cliffs or in old stick nests in trees, sometimes in abandoned nest of other birds. Clutch size: 3-5.

国家一级保护野生动物 | IUCN 红色名录 濒危 (EN) 2016.10.1

成鸟 Adult

亚成鸟 Immature

游隼	*Falco peregrinus*	Peregrine Falcon	隼科 Falconidae

成鸟 Adult
亚成鸟 Immature

国家二级保护野生动物　IUCN 红色名录　无危 (LC) 2016.10.1

形态特征　中型猛禽，体长 42~50 厘米。头顶及脸颊黑色，有粗重而显著的黑色髭纹，整体如头盔。上体深青黑色。腹部、翅下覆羽及尾下覆羽近白色，具黑色横纹。眼圈、跗跖黄色。未成年个体上体深灰褐色。胸腹皮黄，密布深色纵纹。

分布与生境　广布于中国东部至西南。北京迁徙季见于西部山区，为不常见旅鸟；亦有少量繁殖记录于城区附近及山区。

习性与生态　多单独或成对活动。俯冲飞行迅速。捕猎偏好野鸭，或在城市里袭击鸽群。繁殖期营巢于高山悬崖峭壁顶端或直接置于地面隐蔽处；窝卵数 2~6 枚。

Morphological characteristics A medium-sized raptor (42-50 cm). Crown and cheeks black, with distinct bold black moustachial stripes like a helmet; upperparts slate-black; belly, underwing- and undertail-coverts whitish barred black; eye-ring and tarsi yellow. Juvenile has dark greyish-brown upperparts; pale buff-brown breast and belly darkly streaked.

Distribution and habitat Widespread form eastern to southwest China. In Beijing, uncommon migrant in montane and lowland areas, especially open wetlands such as reservoirs or lakes; also a rare summer breeder in urban area and northern mountains.

Behaviour and ecology Solitary or in pairs. Feeds on water birds and sometimes attacks feral or domestic pigeon flocks in urban areas. Nests on ledge on high cliffs or concealed places on ground. Clutch size: 2-6.

VI

鸣禽
PASSERINES

BIRDS OF BEIJING

二十四、雀形目 PASSERIFORMES

黑枕黄鹂 *Oriolus chinensis* Black-naped Oriole 黄鹂科 Oriolidae

形态特征 中型鸣禽，体长 22~27 厘米。喙粉红色。虹膜褐红色。雄鸟羽色金黄，宽阔的黑色贯眼纹延伸至头枕部。翼和尾羽大都黑色缀以金黄色羽缘和斑块；跗跖铅灰色。雌鸟似雄鸟而体色偏黄绿色。未成年个体背部黄绿色，下体具黑色纵纹。

分布与生境 广泛分布于中国东部地区。北京区域性常见于低山丘陵和山脚平原地带的茂密乔木上，为夏候鸟、旅鸟。

习性与生态 常单独或成对活动，有时也见 3~5 只的松散小群。主要在高大乔木的树冠层活动，很少下到地面。捕食大量昆虫，亦食浆果。营巢于树上；窝卵数 4 枚，卵呈粉红色。

Morphological characteristics A medium-sized passerine (22-27 cm). Male has pink bill pink and brownish-red iris. Body largely golden-yellow with a contrasting broad black band from lores to hindcrown and nape. Wings and tail patterned black and yellow; tarsi grey. Female like male but mostly olive-yellow. Juvenile has yellowish-olive crown and upperparts; underparts streaked black.

Distribution and habitat Widespread in eastern China. In Beijing, locally common in dense woodland in the hills and foothills. Residential status: summer breeder and migrant.

Behaviour and ecology Often solitary or in pairs, sometimes in loose flocks of 3-5. Mainly active in canopy, rarely on ground. Mainly feeds on insects, also berries. Nests in trees. Clutch size: 4 pinkish-red egg.

北京市二级保护野生动物 | IUCN 红色名录 无危 (LC) 2018.8.7

成鸟 Adult

幼鸟 Juvenile

暗灰鹃鵙	*Lalage melaschistos*	Black-winged Cuckoo-shrike	山椒鸟科 Campephagidae

IUCN 红色名录 无危 (LC) 2018.8.6

形态特征 中型鸣禽，体长 20~25 厘米。雄鸟喙黑褐色，全身基本灰色调；两翼灰黑，尾下覆羽白色，尾羽黑色，三对外侧尾羽末端有明显白色圆斑；跗跖黑色。雌鸟似雄鸟，但色浅，下体及耳羽具较淡的白色横斑。

分布与生境 主要分布于中国长江流域和东南沿海。北京罕见于平原及低山林地，为夏候鸟及旅鸟。

习性与生态 常单独活动，多在高大的树冠层活动，很少到地上活动和觅食。主要以昆虫为食，也吃少量植物果实和种子。营巢于高大乔木树冠层的水平枝上，较隐蔽；巢呈浅杯状；窝卵数 2~4 枚。

Morphological characteristics A medium-sized passerine (20-25 cm). Male has dark brown bill; greyish overall but wings uniformly blackish; undertail-coverts white; black rectrices with white dots on three outer feather-tips; tarsi black. Female similar to male but slightly paler overall with paler bars/scales on underparts and ear-coverts.

Distribution and habitat Mainly along Yangtze River valley and southeast coast in China. In Beijing, rare on wooded hills. Residential status: summer breeder and migrant.

Behaviour and ecology Often solitary, mostly in canopy of tall trees, rarely on ground. Feeds mostly on insects, with small amounts of fruit and seeds. Shallow, cup-shaped nest hidden on horizontal branch of canopy. Clutch size: 2-4.

小灰山椒鸟	*Pericrocotus cantonensis*	Swinhoe's Minivet	山椒鸟科 Campephagidae

形态特征 中型鸣禽，体长 16~20 厘米。雄鸟额和头顶前部白色，并延伸至眼后，形成一短的眉纹，眼先黑色；头顶后部、枕暗灰或灰黑色，背灰黑色，腰至尾上覆羽沙褐色。两翼黑褐色；尾长，呈黑色，外侧尾羽白色；下体颏、喉、腹为白色。雌鸟似雄鸟，但头顶至背部呈较淡的灰色。

分布与生境 分布于中国长江流域和东南地区。北京见于低山丘陵和平原地带的茂密林地，为罕见旅鸟及夏候鸟。

习性与生态 常成小群活动在高大的乔木上，有时亦见在树丛间飞翔，边飞边叫，鸣声清脆。主要以昆虫为食。营巢于松树或其他高大乔木上；窝卵数 3~4 枚。

Morphological characteristics A medium-sized passerine (16-20 cm). Male has white forehead-patch which extends in marked short eyebrow to just behind eye, as a short supercilium; lores black; dark grey or greyish black hindcrown and nape. Back greyish-black; rump and uppertail-coverts pale drab brown; wings tinged dark brown. Tail long, black; outer rectrices white. Chin, throat and belly whitish. Female similar to male but has paler greyish crown and mantle.

Distribution and habitat Mainly along Yangtze River valley and southeast China. In Beijing, rare in dense woodlands in low hills and lowlands. Residential status: migrant and summer breeder.

Behaviour and ecology Often in small flocks in tall trees; sometimes flies amongst trees giving a clear whistle in flight. Mainly feeds on insects. Nest is bowl-shaped built in pine or other tall trees. Clutch size: 3-4.

IUCN 红色名录　无危 (LC) 2016.10.1

灰山椒鸟	*Pericrocotus divaricatus*	Ashy Minivet	山椒鸟科 Campephagidae

IUCN 红色名录　无危 (LC) 2018.8.7

形态特征　中型鸣禽，体长 18~21 厘米。喙黑色。雄鸟前额至头顶前部白色，头顶后部及贯眼纹黑色；上体深灰色；飞羽黑褐色，近基部具白斑，形成斜形翼斑；尾羽较长，呈黑色，外侧尾羽白色；下体近白色；跗跖黑色。雌鸟似雄鸟但头为灰色。与小灰山椒鸟的区别在于白色眉纹不延伸至眼后。

分布与生境　分布于中国东部、中部、西南等地。北京见于低山丘陵至平原地带，偶至城市公园，为不常见旅鸟。

习性与生态　常成群在林冠层上空飞翔，边飞边叫，鸣声清脆。主要以鞘翅目、鳞翅目等昆虫为食。营巢于落叶阔叶林和红松阔叶混交林中，巢多置于高大树木侧枝上；巢呈碗状；窝卵数 4~5 枚。

Morphological characteristics A medium-sized passerine (18-21 cm). Bill black. Male has white forehead and forecrown. Hindcrown and eye-stripe black. Upperparts dark grey. Flight-feathers blackish-brown with white patches, forming distinctive wingbars. It has long, black tail with white outer rectrices. Underparts whitish. Tarsi black. Female similar to male but with grey crown. Resembles Swinhoe's Minivet but the white forehead patch is smaller and hardly extends behind eye.

Distribution and habitat Eastern, central and southwest China. In Beijing, uncommon migrant in the hills and foothills, occasionally in parks.

Behaviour and ecology Often flies in flocks over canopy, delivering a clear whistling call while flying. Feeds mostly on insects. Bowl-shaped nest built on side branches of tall trees in deciduous broadleaved and mixed forests. Clutch size: 4-5.

长尾山椒鸟 | *Pericrocotus ethologus* | Long-tailed Minivet | 山椒鸟科 Campephagidae

形态特征 中型鸣禽，体长 17~20 厘米。体型纤细修长。雄鸟头顶和背部辉黑色，颊和喉黑色，腰和尾上覆羽、腹部均为猩红色；两翼黑色，具显著的红色翼斑；跗跖黑褐色。雌鸟前额黄色，头至背灰黄色，腰及外侧尾羽黄色；耳羽、颈侧灰白色，喉黄白色；胸、腹黄色；两翼黑色，具黄色翼斑。

分布与生境 分布于中国西南、中部至华北地区。北京区域性常见于海拔 1000 米以上的山地森林中，为夏候鸟。

习性与生态 常集小群活动，叫声尖锐单调，边飞边鸣。觅食于树上，很少下到地上或低矮的灌丛中。主要以昆虫为食。通常营巢于林间树上；巢呈杯状；窝卵数 2~4 枚。

Morphological characteristics A medium-sized passerine (17-20 cm) with slim and slender body. Male has glossy black crown and upperparts. Rump, uppertail-coverts and underparts cinnamon. Wings black with distinctive red wing-patch. Cheek and throat black. Tarsi blackish-brown. Female has greyish-yellow upperparts. Forehead, rump and outer rectrices yellow; ear-coverts and neck-sides greyish-white, throat yellowish-white, breast and belly yellow, black wings with yellow wing-patch.

Distribution and habitat Southwest, central and north China. In Beijing, locally common in montane forests (above 1,000 m). Residential status: summer breeder.

Behaviour and ecology Often in flocks; delivers a sharp and monotonous call, whistling while flying. Forages on trees, rarely on ground or in low shrubs. Feeds mainly on insects. Builds cup-shaped nests in trees in forests. Clutch size: 2-4.

北京市二级保护野生动物 | IUCN 红色名录 无危 (LC) 2018.8.7

雄鸟 Male

雌鸟 Female

黑卷尾	*Dicrurus macrocercus*	Black Drongo	卷尾科 Dicruridae

北京市一级保护野生动物 | | IUCN 红色名录 无危 (LC) 2016.10.1

形态特征 中型鸣禽，体长 24~30 厘米。雌雄相似。喙黑色。虹膜棕红色，全身体羽呈黑色。尾较长，明显呈叉形，尾下覆羽隐约可见浅色的斑纹。跗跖黑色。

分布与生境 分布于中国东部和南部。北京见于低山林地、郊区村庄附近，为常见夏候鸟。

习性与生态 不甚惧人，繁殖期间性凶猛好斗，如红脚隼、乌鸦、喜鹊等鸟类侵入或临近其巢附近时，则奋起冲击入侵者，直至驱出巢区为止。食物以昆虫为主。鸣声响亮粗厉。巢置于榆、柳等树上；巢呈碗状；窝卵数 3~4 枚。

Morphological characteristics A medium-sized passerine (24-30 cm). Sexes alike. Bill small, black; iris brownish-red; overall blackish plumage and long, deeply forked tail distinctive; some show vague paler scales on undertail-coverts; tarsi black.

Distribution and habitat Eastern and southern China. In Beijing, common in woodlands, near villages in suburbs. Residential status: summer breeder.

Behaviour and ecology Not afraid of human beings; can be aggressive during breeding, driving away potential predators such as Amur Falcon and corvids. Feeds mainly on insects. Bowl-shaped nest often built in elm and willow tree. Clutch size: 3-4.

灰卷尾	*Dicrurus leucophaeus*	Ashy Drongo	卷尾科 Dicruridae

形态特征 中型鸣禽，体长 24~32 厘米。雌雄相似。喙黑色。虹膜橙红色。眼先、眼周、颊和耳羽白色。其余头部、上体及下体皆为灰色，但下体灰色略淡。尾羽较长，呈深叉形，向外卷曲。跗跖黑色。

分布与生境 分布于中国东部和南部地区。北京甚罕见于平原丘陵地带的乔木树冠顶端和林间旷野，居留型尚不明确。

习性与生态 捕捉飞虫或俯冲捕捉飞行中的猎物。食物以昆虫为主。巢置于阔叶高大乔木树冠岔枝间，呈浅杯状；窝卵数 3~4 枚。

Morphological characteristics A medium sized passerine (24-32 cm). Sexes alike. Bill black; iris orange-red; lores, eye-ring, cheeks and ear-coverts white. Rest of head, upperparts and underparts mostly grey, but underparts paler. It has long, deeply forked, outward-curved tail. Tarsi black.

Distribution and habitat Eastern and southern China. In Beijing, very rare in tree canopy in plains and hills, open areas and clearings in woodland. Residential status: unclear.

Behaviour and ecology Feeds mainly on insects; often "fly-catches" from perch. Shallow, cup-shaped nest built between branches of broadleaved tree. Clutch size: 3-4.

北京市一级保护野生动物 | IUCN 红色名录 无危 (LC) 2016.10.1

发冠卷尾	*Dicrurus hottentottus*	Hair-crested Drongo	卷尾科 Dicruridae

北京市一级保护野生动物

IUCN 红色名录　无危 (LC) 2016.10.1

形态特征　大型鸣禽，体长 27~35 厘米。雌雄相似。喙黑色。虹膜红色。额顶部有多条较细的毛发状饰羽。全身羽毛大多黑色，上体及胸部具蓝色金属光泽。外侧尾羽末端向外曲，并向内上方卷曲。跗跖呈黑色。

分布与生境　主要分布于中国南方至华北。北京区域性常见于西北部山区近水（如溪流）的森林中，为夏候鸟；迁徙过境时偶至城市公园。

习性与生态　单独或成对活动；非繁殖期有时集松散的小群。主要在树冠层活动和觅食，以各种昆虫为食，偶尔也吃少量植物果实、种子等。营巢于较高的乔木上；窝卵数 3~4 枚。

Morphological characteristics A large passerine (27-35 cm). Sexes alike. Bill black; iris red. Hair-like plumes extend from forehead. Body mostly all-black with bluish gloss on upperparts and breast. Outer rectrices have upward- and inward-curled tips. Tarsi black.

Distribution and habitat Southern and north China. In Beijing, locally common in northern and western montane forests near water (e.g. streams), occasionally in parks on migration. Residential status: summer breeder.

Behaviour and ecology Solitary or in pairs; sometimes forms small and loose flocks in non-breeding season. Usually forages in the tree canopy, where it feeds on various insects, occasionally on fruit and seeds. Deep plate-shape nest built in trees. Clutch size: 3-4.

寿带	*Terpsiphone incei*	Amur Paradise Flycatcher	王鹟科 Monarchidae

形态特征 中型鸣禽，体长 18~39 厘米。喙和眼圈辉蓝色。头、冠羽和喉部蓝黑色。雄鸟两根中央尾羽极度延长。雄鸟有两种色型，栗色型：上体、尾羽栗红色，下体羽灰白色为主；白色型：除头部蓝黑色外全身纯白。雌鸟上体、尾羽为栗色，无延长的中央尾羽。

分布与生境 广泛分布于中国东部及南部地区。北京不常见于低山地带树林中，为夏候鸟、旅鸟。

习性与生态 常隐匿在林冠下层，单独或成对活动，飞行飘逸，很少落地。主要以昆虫为食。营巢于树杈间；巢呈圆锥形，距地面较高；窝卵数 2~4 枚。

Morphological characteristics A medium-sized passerine (18-39 cm). Bill and eye-ring blue. Head, crest and throat bluish-black. Male has extremely long central rectrices. Male of most common (rufous) morph has rufous-chestnut upperparts and tail; remainder of underparts whitish. Male of white morph has glossy black head contrasting with overall all-white plumage. Female has rufous upperparts and tail and lacks tail-streamers.

Distribution and habitat Eastern and southern China. In Beijing, uncommon summer breeder and passage migrant in low hill and lowland woodlands.

Behaviour and ecology Often stays hidden in the canopy; solitary or in pairs. Feeds mostly on insects. Cone-shaped nest built between branches, often high above ground. Clutch size: 2-4.

北京市一级保护野生动物		IUCN 红色名录 无危 (LC) 2016.10.1

雄鸟 Male/ 白色型 White morph

雄鸟 Male/ 栗色型 Rufous morph

雌鸟 Female

虎纹伯劳	*Lanius tigrinus*	Tiger Shrike	伯劳科 Laniidae

形态特征 中型鸣禽，体长 18~19 厘米。喙黑色，喙形粗厚，喙端呈显著的钩状。虹膜褐色。贯眼纹粗黑色。雄鸟头顶至上背蓝灰色，下背、肩至尾上覆羽栗红色，并杂以黑色的鳞状斑纹；下体几乎纯白色；跗跖黑褐色。雌鸟似雄鸟，但下体具细横纹。

分布与生境 分布于中国东北、华北、华南。北京见于开阔的平原旷野、山麓丘陵树木繁茂的地区，为不常见夏候鸟、旅鸟。

习性与生态 常栖于树上枯枝顶端，窥伺着猎物活动。性凶猛，不但捕食昆虫，亦常袭击小型鸟类。营巢于树上或灌丛中；窝卵数 5~6 枚。

Morphological characteristics A medium-sized passerine (18-19 cm). Black bill strongly hooked and iris brown. Male has broad, black eye-stripe. Crown and upper back grey-blue; diagnostic blackish bars/scales on deep rufous-brown upperparts; underparts all whitish. Tarsi dark brown. Female similar to male but somewhat duller with more prominently barred/scaled underparts.

Distribution and habitat Northeast, north and south China. In Beijing, uncommon in wild plains, wooded areas in foothills and hills. Residential status: summer breeder and migrant.

Behaviour and ecology Often perches on top of dead branches on trees, watching for prey. Ferocious; hunts not only for insects but also small birds. Nests in trees or shrubs. Clutch size: 5-6.

北京市二级保护野生动物 | IUCN 红色名录 无危 (LC) 2016.10.1

雄鸟 Male

雌鸟 Female

牛头伯劳	*Lanius bucephalus*	Bull-headed Shrike	伯劳科 Laniidae

雌鸟 Female

雄鸟 Male

北京市二级保护野生动物　　IUCN 红色名录　无危 (LC) 2016.10.1

形态特征　中型鸣禽，体长 18~23 厘米。整体与其他伯劳相比头较大而喙较短。喙深灰色而端黑，呈钩状。雄鸟头顶到枕部栗红色；贯眼纹黑色，眉纹白色；上体灰褐色，羽缘棕红色；两翼大致为黑色，初级飞羽基部白色，形成小块白色翼斑；尾羽黑色；胸、胁部棕红色，下体余部白色；跗跖黑褐色。雌鸟似雄鸟，但耳羽棕色，下体具深色鳞状斑。

分布与生境　中国东部广大地区皆有分布。北京不常见于低山、丘陵和平原地带的疏林和林缘灌丛，为夏候鸟、旅鸟。

习性与生态　性活跃，常在林缘或路边灌丛中跳上跳下，有时站在小树或灌木枝头鸣叫，鸣声粗厉洪亮。主要以昆虫为食，如甲虫、蟋蟀等。常建巢于疏林或灌丛；巢呈杯状；窝卵数 4~7 枚。

Morphological characteristics　A medium-sized passerine (18-23 cm). Compared with other shrikes, it has relatively large head and short bill. Bill dark grey with black hooked tip. Male is rufous from crown to nape. Eye-stripe black; supercilium white; upperparts greyish-brown with rufous fringes; wings largely black. There is a white patch at base of primaries. Tail black. Breast and flanks rufous; rest of underparts white; tarsi blackish-brown. Female similar to male but has brown ear-coverts and dark scaled underparts.

Distribution and habitat　Widespread in eastern China. In Beijing, uncommon summer breeder and migrant in open forests, forest edge and scrub on the hills and plains.

Behaviour and ecology　Active, often moving along forest edge or roadside shrubs, sometimes perching on branches of small trees or shrubs; often utters a noisy chatter. Feeds mainly on insects, such as beetles and crickets. Cup-shaped nest built in open forest or scrub. Clutch size: 4-7.

红尾伯劳	*Lanius cristatus*	Brown Shrike	伯劳科 Laniidae

形态特征 中型鸣禽，体长 17~21 厘米。喙黑色，钩状。虹膜暗褐色。雄鸟具显著的白色眉纹和黑色贯眼纹；头顶灰色或棕红色；背肩部灰褐色，腹面棕白色，尾羽棕褐色；跗跖灰色。雌鸟似雄鸟，但上体羽色较暗淡，且胸侧至两胁具深色横斑。

分布与生境 广泛分布于中国东部及南部地区。北京见于山区和平原的林地和灌丛，为常见夏候鸟、旅鸟。

习性与生态 单独或成对活动，性活泼，常在枝头跳跃或飞上飞下。常突然飞去捕猎，然后再飞回原来栖木上栖息。巢多置于幼树和灌木上，呈杯状；窝卵数 5~7 枚。

Morphological characteristics A medium-sized passerine (17-21 cm). Bill black, hooked; iris dark brown. Male has prominent white supercilium, black eye-stripe; crown greyish or rufous-brown; mantle brownish-grey; underparts brownish-white; rectrices greyish-brown. Tarsi grey. Female similar to male but has duller upperparts and barred breast-sides and flanks.

Distribution and habitat Eastern and southern China. In Beijing, common summer breeder and migrant in woodlands and shrubs in mountains and lowlands.

Behaviour and ecology Solitary or in pairs; active, often perches on bare branch from where it hunts for large insects. Cup-shaped nest built in young tree or shrub, such as larch or black locust. Clutch size: 5-7.

北京市二级保护野生动物 | IUCN 红色名录 无危 (LC) 2016.10.1

成鸟 Adult

幼鸟 Juvenile

棕背伯劳	*Lanius schach*	Long-tailed Shrike	伯劳科 Laniidae

IUCN 红色名录　无危 (LC) 2016.10.1

形态特征　中型鸣禽，体长 22~28 厘米。雌雄相似。喙粗壮，呈黑色，先端具显著的利钩和齿突。虹膜暗褐色。前额黑色，具黑色贯眼纹，头顶至上背灰色。下背、肩、腰和尾上覆羽棕色，尾羽黑色。颏、喉和腹中部白色，其余下体淡棕色或棕白色，两胁和尾下覆羽棕红色或浅棕色。跗跖黑色。

分布与生境　主要分布于中国华东及南方广大地区。北京罕见于平原农田、湿地附近的灌丛，居留型尚不明确。近年来（2015 年后）在京记录频率有所增加，各月皆有记录，但仍无确切的繁殖记载。

习性与生态　多单独活动。性凶猛，不仅擅捕食昆虫，也能捕杀小鸟、蛙和鼠类。尾羽常向左右两边不停地摆动。

Morphological characteristics A medium-sized passerine (22-28 cm). Sexes alike. Thick, black bill with sharp hook at tip; iris dark brown; forehead and eye-stripe black, crown and upper back grey; lower back, shoulder, rump and uppertail-coverts brown; black rectrices; white chin, throat and central belly; remainder of underparts pale brown or brownish-white with rufous or pale brown flanks and undertail-coverts; black tarsi.

Distribution and habitat Mainly from east China to southern China. In Beijing, rare in lowland farmlands and shrubs near water. Its residential status is still unclear. Records in Beijing have been increasing in different months since 2015. But there has not any confirmed breeding records.

Behaviour and ecology Mostly solitary. Aggressive. Feeds on insects, birds, frogs and rodents. Often swings tail from side to side.

灰背伯劳	*Lanius tephronotus*	Grey-backed Shrike	伯劳科 Laniidae

形态特征 中型鸣禽，体长 22~25 厘米。雌雄相似。喙黑色，钩状。虹膜暗褐色。具黑色贯眼纹，头顶至下背暗灰色，腰灰色，尾上覆羽转为锈棕色。尾羽黑褐色。下体白色为主，但胸侧、两胁至尾下覆羽为橙褐色。跗跖黑色。

分布与生境 主要分布于中国中部至西南地区。繁殖季多见于海拔 2500~4000 米的林缘和开阔灌丛，非繁殖季则迁移至较低海拔。北京仅 2020 年 7 月见于东灵山，为甚罕见迷鸟。

习性与生态 单独活动。主要以昆虫为食，也食鼠类。营巢于小树或灌木侧枝上；巢为杯状；窝卵数 4~6 枚。

Morphological characteristics A medium-sized passerine (22-25 cm). Sexes alike. Bill black, hooked; iris dark brown; eye-stripe black, mantle and scapulars uniform grey (concolours with crown); grey rump; rufous uppertail-coverts. Underparts largely white; breast-sides, flanks and undertail-coverts orange-brown. Tarsi black.

Distribution and habitat Mainly central and southwest China. Occurs in forest edges and shrubs between 2,500-4,000 m; descends lower in non-breeding season. Very rare vagrant in Beijing; only once recorded at Donlingshan in Mentougou District in July 2020.

Behaviour and ecology Solitary. Mainly feeds on insects, also on mice. Nests on branches of small trees or shrubs, cup-shaped. Clutch size: 4-6.

IUCN 红色名录 无危 (LC) 2016.10.1

灰伯劳	*Lanius excubitor*	Great Grey Shrike	伯劳科 Laniidae

北京市二级保护野生动物　　IUCN 红色名录　无危 (LC) 2016.10.1

形态特征　中型鸣禽，体长 23~26 厘米。雌雄相似。喙黑色，呈钩状。虹膜暗褐色。具黑色贯眼纹。头顶至尾上覆羽烟灰色，腰白色或浅灰色；两翼黑色，仅初级飞羽基部白色。尾羽较长，大部分为黑色，外侧尾羽白色。下体灰白色，颈侧、喉、胸、胁及腹羽均具细密的暗褐色鳞纹。跗跖黑色。

分布与生境　主要分布于中国北方。北京罕见于西北部山区开阔林地和灌丛，为冬候鸟、旅鸟。

习性与生态　单独活动，性凶猛，喜食小型兽类、鸟类、蜥蜴及各种昆虫等。巢置于不高的树上；窝卵数 4~7 枚。

Morphological characteristics A medium-sized passerine (23-26 cm). Sexes alike. Bill black and hooked; iris pale brown; eye-stripe black. Grey from crown to uppertail-coverts, white or pale greyish rump; black wings with prominent white patch at base of primaries. Tail long, black; outer rectrices white. Underparts greyish white with pale brown bars/scales on neck-sides, throat, breast, flanks and belly; black tarsi.

Distribution and habitat Mainly northern China. In Beijing, rare in open woodland and shrubs in northern and western montane areas. Residential status: winter visitor and migrant.

Behaviour and ecology Solitary and aggressive. Feeds on birds, lizards, various insects, etc. Nests on low trees. Clutch size: 4-7.

楔尾伯劳 *Lanius sphenocercus* Chinese Grey Shrike

伯劳科
Laniidae

形态特征 中型鸣禽，体长 25~31 厘米。雌雄相似。喙黑色，短而粗厚，端部呈钩状。具黑色贯眼纹。上体灰色，初级飞羽基部至次级飞羽基部具白色带斑，腰部灰色可区别于灰伯劳的白色腰。尾羽长，羽端呈凸状尾，中央尾羽黑色，其余尾羽大多端部白色，最外侧 3 对尾羽全为白色。下体羽白色。跗跖黑褐色。

分布与生境 主要分布于中国北部和东部。北京一般于秋冬季见于芦苇丛中，也出现于低山、平原和丘陵地带的疏林和林缘灌丛，为区域性常见冬候鸟、旅鸟和偶见夏候鸟。

习性与生态 常单独或成对活动。食物主要为昆虫，亦常捕食小型脊椎动物，如蜥蜴、小鸟及鼠类。在乔木或灌木上筑巢；窝卵数 5~6 枚。

Morphological characteristics A medium-sized passerine (25-31 cm). Sexes alike. Black bill short, thick and hooked; Iris dark brown; eye-stripe black; upperparts grey; white band at base of primaries and secondaries; grey rump (cf. Great Grey Shrike's white rump); long rectrices with convex tip; central rectrices black, inner outer rectrices have white tip; and outermost three pairs of rectrices white; underparts white; dark brown tarsi.

Distribution and habitat Mainly northern and eastern China. In Beijing, locally common in wetlands, as well as open woodland and scrub at the forest edge in lowland and hilly areas, in autumn and winter. Residential status: winter visitor and migrant, also a scarce summer breeder.

Behaviour and ecology Often solitary or in pairs. Feeds mainly on insects but also preys on small vertebrates such as lizards, birds and rodents. Nests on trees or in shrubs. Clutch size: 5-6.

北京市二级保护野生动物

IUCN 红色名录 无危 (LC) 2016.10.1

松鸦	*Garrulus glandarius*	Eurasian Jay	鸦科 Corvidae

IUCN 红色名录　无危 (LC) 2016.10.1

形态特征　大型鸣禽，体长 30~36 厘米。雌雄相似。喙较短，为黑色。整体近浅棕色，头顶粉棕色并具黑色纵纹，具较粗的黑色髭纹。飞羽黑色为主，但次级飞羽基部白色；翼上覆羽具闪亮的蓝色横斑及点斑，极为醒目。飞行时白色的腰和黑色的尾形成鲜明对比。跗跖肉色。

分布与生境　中国广泛分布，但于东部及南部更为常见。北京见于山区针叶、阔叶林中，为不常见留鸟。

习性与生态　除繁殖期多见成对活动外，其他季节多集成 10 只以内的小群四处游荡，性嘈杂吵闹，食性较杂。通常营巢于高大乔木顶端较为隐蔽的枝杈处；巢呈杯状；窝卵数 5~8 枚。

Morphological characteristics　A large passerine (30-36 cm). Sexes alike. It has black, relatively short bill. Head and body largely pinkish-brown; crown streaked black; broad black moustachial stripe at throat-side; flight-feathers mainly black; secondaries based white. Upperwing-coverts have distinctive turquoise-blue bars and spots. In flight, conspicuous white rump contrasts with black rectrices; flesh-coloured tarsi.

Distribution and habitat　Widespread in China; commoner eastern and southern China. In Beijing, uncommon in coniferous and broadleaved forests in montane areas, rarely in lowland. Residential status: resident.

Behaviour and ecology　In pairs in breeding season. Often found in small and noisy groups (less than 10 individuals). Omnivorous. Cup-shaped nest usually hidden in canopy. Clutch size: 5-8.

灰喜鹊	*Cyanopica cyanus*	Azure-winged Magpie	鸦科 Corvidae

形态特征 大型鸣禽，体长 32~42 厘米。雌雄相似。喙黑色。成鸟头顶、头侧、后枕为黑色，肩背部灰色，喉部至腹部为灰白色，翼及尾羽均为淡蓝色。两根中央尾羽甚长且具白色端斑，飞翔时尤为明显。跗跖黑色。幼鸟体色大多较暗，头顶花白较为斑驳。

分布与生境 中国除青藏高原外广泛分布。北京见于山地、田野、村庄附近和城内公园、居民区树木较多地带，为甚常见留鸟。

习性与生态 多成群活动，有时甚至集成多达数十只的大群，食性杂但以动物性食物为主，兼食一些乔灌木的果实及种子。鸣声宏亮且粗厉，无韵律。多营巢于次生林和人工林中，也在村镇附近和路边树上营巢；窝卵数 4~9 枚。

Morphological characteristics A large passerine (32-42 cm). Sexes alike. Bill black. Adult has black crown, head-sides and nape; grey mantle and back; greyish-white from throat to belly, azure-blue on wings and tail; long central rectrices have distinctive white tips, obvious in flight; black tarsi. Juvenile somewhat duller with mottled whitish crown.

Distribution and habitat Widespread in China except for the Tibetan Plateau. Abundant resident in Beijing. Occurs in woodlands in lower mountains and fields, often near human habitation, as well as in parks and residential areas with trees.

Behaviour and ecology Often in flocks. Omnivorous, but feeds mainly on small animals (especially insects), fruits and seeds. Calls loud and harsh. Nests in secondary forest and plantations, as well as in trees in villages and towns. Clutch size: 4-9.

北京市一级保护野生动物 | IUCN 红色名录 无危 (LC) 2016.10.1

红嘴蓝鹊 *Urocissa erythroryncha* Red-billed Blue Magpie

鸦科 Corvidae

北京市一级保护野生动物

IUCN 红色名录　无危 (LC) 2018.8.7

形态特征　大型鸣禽，体长 51~63 厘米。雌雄相似。喙呈朱红色。成鸟头、颈和前胸黑色，胸部以下白色，头顶至枕部灰白色。上体蓝灰色，翼及尾羽蓝紫色，两根中央尾羽甚长且具白色端斑，外侧尾羽依次渐短，末端有黑白相间的带状斑。跗跖红色。幼鸟喙色暗淡，且尾羽较短。

分布与生境　主要分布于中国华北至南方广大地区。北京主要见于山地及丘陵林区，也见于一些较大的城市公园中，为常见留鸟。

习性与生态　喜集小群活动，性活泼而嘈杂，飞行呈大波浪状，尾羽常展开，甚为飘逸。食性杂，主要以昆虫等动物性食物为食，也吃植物果实、种子等。鸣声婉转多样，有时亦较嘈杂。营巢于树木侧枝上；巢呈碗状；窝卵数 3~6 枚。

Morphological characteristics A large passerine (51-63 cm). Sexes alike. Adult has prominent red bill; black head, neck and breast. Greyish-white from crown and nape. Grey-blue upperparts, purplish-blue and green gloss on wings and tail, outer rectrices graduated, with black and white banded tip; tarsi red. Juvenile has duller bill and shorter tail.

Distribution and habitat Mainly from north to southern China. In Beijing, common in montane and low hill forests and also some of the larger urban parks. Residential status: resident.

Behaviour and ecology Often in small flocks, lively and noisy, flying through trees with spread, flowing tail. Omnivorous, mainly feeds on animals such as insects, as well as fruits and seeds. Utters a variety of melodious, sometimes harsh, calls; can be noisy. Bowl-shaped nest built on branch of tree. Clutch size: 3-6.

喜鹊	*Pica pica*	Common Magpie	鸦科 Corvidae

形态特征 大型鸣禽，体长 37~48 厘米。雌雄相似。喙黑色。成鸟头、胸、颈、背至尾均为黑色。翼有蓝绿色金属光泽，尾羽具暗绿色金属光泽。肩和腹部为白色。飞行时可见两翼初级飞羽白色，羽端黑色。尾羽较长，呈凸状。跗跖黑色。幼鸟羽色似成鸟，但黑羽部分染有褐色，金属光泽也不显著。

分布与生境 中国几乎全境分布。北京于山区、平原、城市公园、居民区都有栖息，为甚常见留鸟。

习性与生态 常集成小群或成对活动，多从地面取食，食性杂。鸣声一般为响亮而单调的喳喳声。常筑巢在高大乔木接近树顶处；巢体甚大，主要以树枝筑成，顶部封闭，其内层具一泥制碗状内巢，并垫以草、羽毛、苔藓等；窝卵数 5~8 枚。

Morphological characteristics A large passerine (37-48 cm). Sexes alike. Bill black. Adult has black head, breast, neck, mantle and back; glossy blue and green wings and green-glossed tail; white scapulars and belly; convex rectrices longer than wings; black tarsi. In flight, white primaries show black tips. Juvenile similar to adult but black feathers tinged brown and less glossy.

Distribution and habitat Widespread in China. In Beijing, abundant resident in mountains and lowlands, especially farmland, rural areas, parks and residential areas.

Behaviour and ecology Often in small flocks or in pairs. Omnivorous. Gives loud, repetitive calls "cha cha". Builds large, domed nests near the top of tall trees. The nest is made up of twigs and includes an interior mud and moss cup lined with grasses, feathers and mosses. Clutch size: 5-8.

IUCN 红色名录 无危 (LC) 2016.10.1

星鸦	*Nucifraga caryocatactes*	Spotted Nutcracker	鸦科 Corvidae

IUCN 红色名录　无危 (LC) 2016.10.1

形态特征　大型鸣禽，体长 28~38 厘米。雌雄相似。喙黑色，呈楔形。成鸟体羽咖啡褐色，头侧、胸及上背具较小白色斑点或条纹。头顶为很暗的咖啡褐色，枕部则逐渐变为咖啡色。两翼较宽，近黑色。尾较短，呈黑褐色，最外侧尾羽具宽的白色端斑。臀和尾下覆羽白色。跗跖黑色。

分布与生境　主要分布于中国东北、华北、西北等地。北京见于山地针叶林中，为不常见留鸟。

习性与生态　单独或成对活动，偶成小群。主要以松子、昆虫等为食。飞行起伏而有节律。营巢于针叶树上；窝卵数 3~4 枚。

Morphological characteristics A large passerine (28-38 cm). Sexes alike. Black, wedged bill. Adult is mostly dark brown overall, sprinkled with small white spots or streaks on head-sides, breast and upper back; pale brown crown and brown nape; black wings; Tail short, blackish-brown, outermost rectrices have wide white tips. Vent to undertail-coverts white. Black tarsi.

Distribution and habitat Mainly in northeast, north and northwest China. In Beijing, uncommon in montane coniferous forests. Residential status: resident.

Behaviour and ecology Solitary or in pairs; and occasionally in small flocks. Feeds mainly on pine nuts and insects. Undulating in flight. Nests in coniferous trees. Clutch size: 3-4.

VI 鸣禽

红嘴山鸦	*Pyrrhocorax pyrrhocorax*	Red-billed Chough	鸦科 Corvidae

形态特征 大型鸣禽，体长 36~47 厘米。雌雄相似。喙鲜红色，细长而下弯。虹膜暗褐色。周身黑色略有蓝紫色光泽。跗跖红色。幼鸟似成鸟，但喙色较暗淡。

分布与生境 主要分布于中国淮河以北的大部分地区，以及青藏高原。北京区域性常见于山地裸岩地带，有时散见于近山平原的田地，为留鸟。

习性与生态 常集大群在山头上空和山谷间飞翔。主要以昆虫为食，也食植物性食物。鸣声尖锐、嘈杂。通常营巢于悬崖峭壁上的岩石缝隙、岩洞和岩边往内的凹陷处；巢呈碗状；窝卵数 3~6 枚。

Morphological characteristics A large passerine (36-47 cm). Sexes alike. Red bill is long, pointed and decurved. Iris pale brown. Adult is overall black with glossy metallic bluish-purple sheen; red tarsi. Juvenile similar to adult with duller bill.

Distribution and habitat In most northern areas north of Huai River in China, and the Tibetan Plateau. In Beijing, locally common in bare rocky areas in mountains, and sometimes feeds in lowland fields near mountains. Residential status: resident.

Behaviour and ecology Sociable, often in large flocks around mountains and ravines. Feeds mainly on insects, but also vegetation. Calls sharp and piercing. Bowl-shaped nest built in rocky crevice, cave or niche on cliff-ledge. Clutch size: 3-6.

IUCN 红色名录 无危 (LC) 2016.10.1

寒鸦	*Corvus monedula*	Eurasian Jackdaw	鸦科 Corvidae

IUCN 红色名录 无危 (LC) 2018.8.9

形态特征 大型鸣禽，体长 30~36 厘米。雌雄相似。喙黑色，较短。虹膜白色。前额至头顶黑色而富有光泽，枕、后颈、颈侧灰白色。背暗灰色。肩、腰、翼以及尾上覆羽和尾羽为黑色。喉和上胸纯黑色，其余下体暗灰黑色。跗跖黑色。

分布与生境 主要分布于中国新疆西北部、西藏西部。北京仅 2020 年 4 月见于十三陵，为甚罕见迷鸟。

习性与生态 常成群活动，有时也和其他鸦类组成大群。食物较杂。营巢于树洞、河谷土崖崖壁洞中；窝卵数 4~7 枚。

Morphological characteristics A large passerine (30-36 cm). Sexes alike. Bill short, blackish; iris white. Forehead and crown glossy black; neck-sides and nape lighter grey; back pale grey; black upperparts, including shoulder, rump, wings, uppertail-coverts and rectrices; throat and breast black; remainder of underparts pale greyish-black; black tarsi.

Distribution and habitat Mainly in northwest Xinjiang and western Tibetan Plateau in China. Very rare vagrant in Beijing, with only one record at Shisanling Reservoir, Changping District (April 2020).

Behaviour and ecology Often in flocks, sometimes with other crows. Omnivorous. Usually nests in tree holes or on sandy cliffs in river valleys. Clutch size: 4-7.

达乌里寒鸦 | *Corvus dauuricus* | Daurian Jackdaw | 鸦科 Corvidae

形态特征 大型鸣禽，体长 27~35 厘米。雌雄相似。喙黑色，较短粗。虹膜黑褐色。头侧耳羽灰黑色。从后颈向两侧延伸至胸侧和腹部为白色，其余体色均为黑色。跗跖近黑色。另有通体黑色者，为达乌里寒鸦的幼鸟亦或一种色型，目前尚无研究确认。

分布与生境 主要分布于中国北部、中部、西南和东南。北京常见于山地、丘陵、平原、农田、旷野等各类生境，为冬候鸟、留鸟，亦有少量夏候鸟繁殖于山区。

习性与生态 冬季喜欢集多至数千只的大群活动。杂食性，非繁殖期主食植物种子，但繁殖期主食昆虫。叫声较小嘴乌鸦更显短促而尖锐。通常营巢于悬崖崖壁洞穴中；窝卵数 4~8 枚。

Morphological characteristics A large passerine (27-35 cm). Sexes alike. Adult has short, black bill; iris dark brown. Greyish-black around ear-coverts; white from nape to breast-sides and belly, rest black. Tarsi blackish. Some birds have all black plumage, paler and duller; whether such birds are juvenile or a dark morph, is not yet fully understood.

Distribution and habitat Mainly in northern, central and southern China. In Beijing, common in mountains, hills, plains and farmland. Residential status: winter visitor, resident, and also scarce summer breeder in montane areas.

Behaviour and ecology Often in large flocks in winter, sometimes numbering even thousands. Omnivorous; feeds mainly on seeds and grain outside breeding season, but mostly insects during breeding season. Calls shorter and sharper than Carrion Crow. Usually nests in caves on cliff-ledge. Clutch size: 4-8.

IUCN 红色名录 无危 (LC) 2016.10.1

成鸟 Adult

幼鸟 Juvenile

成鸟 Adult

秃鼻乌鸦 | *Corvus frugilegus* | Rook | 鸦科 Corvidae

幼鸟 Juvenile

成鸟 Adult

IUCN 红色名录 无危 (LC) 2016.10.1

形态特征 大型鸣禽，体长 41~51 厘米。雌雄相似。喙呈圆锥形且显尖细，喙基部周围没有羽毛覆盖，裸露灰白色皮肤，在黑色体羽的映衬下非常显眼。除喙基部外，通体漆黑，并伴有蓝紫色金属光泽。尾羽较宽。飞行时头显突出，两翼较长窄。跗跖为黑色。

分布与生境 中国除青藏高原和西南地区外广泛分布。北京见于平原、丘陵、低山地形的农田、村庄附近，为不常见留鸟。

习性与生态 常成群活动，冬季也和其他鸦群混合成大群，常于地面取食，食性杂。营巢于林缘、河岸、水塘和农田附近小块树林中，巢多置于高大乔木的树杈上；窝卵数 5~6 枚。

Morphological characteristics A large passerine (41-51 cm). Sexes alike. Bill slender and pointed. Bare greyish-white skin around bill-base and lack of nostril feathering, distinctly contrasted with black plumage. Except for bill-base, plumage all-black, with bluish-purple metallic gloss; tail broad. Head protrudes in flight. Tarsi black.

Distribution and habitat Widespread in China, except for southwest China and the Tibetan Plateau. In Beijing, uncommon in farmland and areas near village in lowlands and hills. Residential status: resident.

Behaviour and ecology Often in flocks, in winter mixed with other crows. Often feeds on ground, omnivorous. Nests in small woods near forest edge, riverbanks, ponds and farmland; mostly on branches of tall trees. Clutch size: 5-6.

小嘴乌鸦	*Corvus corone*	Carrion Crow	鸦科 Corvidae

形态特征 大型鸣禽，体长 41~53 厘米。雌雄相似。喙黑色，较粗大。全身羽色漆黑而有光泽。尾较宽。跗跖黑色。似大嘴乌鸦，但本种喙稍细，且前额较平，与喙峰无明显转折。

分布与生境 中国广泛分布。北京为甚常见留鸟和冬候鸟，活动于城市、郊野乡村和低山区域，冬季集群多见于城区。

习性与生态 喜集群活动，常与大嘴乌鸦和达乌里寒鸦等混群。在较开阔的农田、荒地、垃圾场等处觅食，食性杂。冬季常集大群在市中心街道两旁的树木夜宿。鸣声为响亮而略显圆润的“kar kar kar”声。营巢于高大乔木近树顶处；窝卵数 4~6 枚。

Morphological characteristics A large passerine (41-53 cm). Sexes alike. Bill strong, thick, black; body all-black with slight metallic gloss; tail broad; tarsi black. Resembles Large-billed Crow but has relatively slender bill and continuous slope from culmen to forehead.

Distribution and habitat Widespread in China. In Beijing, abundant in urban areas, lowland countryside and low hills. Forms flocks and moves to city centre in winter. Residential status: winter visitor and resident.

Behaviour and ecology In flocks. Sometimes joins mixed flocks with Large-billed Crows and Daurian Jackdaws. Forages in open farmland, wastelands, garbage dumps; omnivorous. Often roosts on trees along streets in city centre overnight in winter. Gives loud and rolling calls "kar kar kar". Nests near the top of tall trees. Clutch size: 4-6.

IUCN 红色名录 无危 (LC) 2016.10.1

冠小嘴乌鸦 *Corvus cornix* Hooded Crow 鸦科 Corvidae

IUCN 红色名录　无危 (LC)* 2016.10.1

形态特征　大型鸣禽，体长 41~53 厘米。雌雄相似。喙黑色。虹膜褐色。全身大部分为黑色，但上背和下体粉灰色。跗跖黑色。

分布与生境　主要分布于中国新疆西部。北京地区甚罕见，为迷鸟，仅于 2016 年 10 月在首都国际机场附近有过一次记载。

习性与生态　喜集群栖息。取食于矮草地及农耕地，以无脊椎动物为主要食物，但喜吃尸体，常在道路上吃被车辆碾死的动物。

● *注：IUCN 红色名录未收录（更新 / 接受）冠小嘴乌鸦和小嘴乌鸦的物种划分，故此处沿用小嘴乌鸦的数据。

Morphological characteristics　A large passerine (41-53 cm). Sexes alike. It has black bill; brown iris; body largely black, with pinkish-grey upper back and underparts. Tarsi black.

Distribution and habitat　Mainly in the west of Xinjiang in China. Very rare in Beijing with only one record near the Beijing Capital International Airport in October 2016. Residential status: vagrant.

Behaviour and ecology　Often in flocks, foraging on open grassland and farmland. Feeds mainly on invertebrates but also scavenges; often seen eating roadkill.

白颈鸦	*Corvus pectoralis*	Collared Crow	鸦科 Corvidae

形态特征 大型鸣禽，体长 45~54 厘米。雌雄相似。成鸟喙黑色。后颈、颈侧至下胸为白色，形成一白色领环，其余体羽全为黑色。上体及翼具紫蓝色金属光泽。跗跖黑色。

分布与生境 主要分布于中国华北、华中、华南及东南。北京罕见于平原和低山开阔地区的湿地、农田和村庄附近，居留型尚不明确。

习性与生态 性机警，多单独行动。在地面取食，食性杂。通常营巢于郊野乡村区域的高大乔木上；窝卵数 3~7 枚。

Morphological characteristics A large passerine (45-54 cm). Sexes alike. Black bill. Adult has all-black plumage except for white collar from hindneck and neck-sides to lower breast. Glossy purplish-blue upperparts and wings; black tarsi.

Distribution and habitat North, central, south and southeast China. Rare in Beijing. Seen at wetlands, farmlands and villages in lowlands and low hills. Residential status: unclear.

Behaviour and ecology Vigilant and often solitary. Feeds on the ground; omnivorous. Nests in trees at countryside. Clutch size: 3-7.

IUCN 红色名录 易危 (VU) 2018.8.7

大嘴乌鸦	*Corvus macrorhynchos*	Large-billed Crow	鸦科 Corvidae

IUCN 红色名录　无危 (LC) 2016.10.1

形态特征　大型鸣禽，体长 44~54 厘米。雌雄相似。喙黑色粗大且厚，喙峰显著。喙上缘与前额交界处几成直角，显得额头尤为突出，此与小嘴乌鸦有明显不同。头颈部略具蓝紫色金属光泽。全身羽色漆黑而有光泽。跗跖为黑色。

分布与生境　主要分布于除西北部以外的中国广大地区。北京甚常见于农田、村庄、城市公园等人类居住地附近，亦见于山地，为留鸟。

习性与生态　少集大群，常单独或集小群活动。食性杂。叫声为单调、响亮而粗犷的“ah ah”声，常边飞边叫。营巢于高大乔木顶部枝杈处；窝卵数 3~5 枚。

Morphological characteristics　A large passerine (44-54 cm). Sexes alike. Bill black, thick and rather long, with downcurved upper mandible; steep forehead (cf. Carrion Crow's continuous slope from culmen to crown), glossy bluish-purple head and neck and peaked crown; overall plumage gloss-black; black tarsi.

Distribution and habitat　Widespread in China except for the northwest. In Beijing, abundant resident in farmland, villages and parks, in mountains as well as lowlands.

Behaviour and ecology　Often solitary or in small flocks. Omnivorous. Gives loud, harsh call "ah ah". Often calls in flight. Nests in tree canopy. Clutch size: 3-5.

方尾鹟	*Culicicapa ceylonensis*	Grey-headed Canary Flycatcher	玉鹟科 Stenostiridae

IUCN 红色名录　无危 (LC) 2016.10.1

形态特征　小型鸣禽，体长 10~13 厘米。雌雄相似。喙扁平，喙基宽阔几成等边三角形，上喙黑色，下喙褐色。头、胸灰色，有灰白色眼圈，略具冠羽，上体和尾羽黄绿色，下体偏黄色。跗跖黄褐色。

分布与生境　主要分布于中国中部至南部。北京甚罕见于平原和山区的林地和公园，为迷鸟。

习性与生态　常单独、成对或与其他雀形目鸟类混群活动，多在林冠层和林缘灌丛中捕食小型无脊椎动物。鸣声婉转而复杂。

Morphological characteristics A small passerine (10-13 cm). Sexes alike. Flat bill with broad base, almost forming an equilateral triangle; black upper mandible and brown lower mandible; grey head and breast with greyish-white eye-ring; crested appearance; greenish upperparts and tail; yellowish underparts and yellowish-brown tarsi.

Distribution and habitat Central and southern China. In Beijing, very rare in woodlands and parks. Residential status: vagrant.

Behaviour and ecology Often Solitary, in pairs or joins mixed flocks of other passerines, foraging in small branches of trees and shrubs at forest edge. Feeds mainly on insects. Song melodious.

煤山雀	*Periparus ater*	Coal Tit	山雀科 Paridae

形态特征 小型鸣禽，体长 10~11 厘米。雌雄相似。头、颈、喉至上胸黑色，具不明显的冠羽。上体深蓝灰色。翼上覆羽深褐色，具蓝灰色羽缘，大覆羽和中覆羽具白色端斑，形成两道较窄的白色翼斑。颊和枕部白色，腹部污白色或淡皮黄色。

分布与生境 主要分布于中国西北、东北、华北、华南、西南。北京不常见于低山针叶林和混交林中，冬季有时下至平原林地，为留鸟。

习性与生态 性较活泼，行动敏捷，常在树枝间穿梭跳跃。主要以昆虫为食。通常营巢于天然树洞中，有时也在土崖和石隙中营巢；巢呈杯状；窝卵数 8~10 枚。

Morphological characteristics A small passerine (10-11 cm), Sexes alike. Head, neck, throat and upper breast black; slightly crested; upperparts dark blue-grey; upperwing-coverts dark brown, fringed blue-grey; greater coverts and median coverts have white tips, forming two distinctive white wingbars; cheek and nape white; belly off-white or pale buff-white.

Distribution and habitat Mainly northeast, northwest, north, south and southwest China. In Beijing, uncommon in coniferous and mixed forests in the hills; sometimes in seen in woodlands in winter. Residential status: resident.

Behaviour and ecology Active and agile, moving quickly between branches. Feeds mainly on insects in summer and seeds in winter. Cup-shaped nest built in natural tree hole, sometimes on earth cliff or rock crevice. Clutch size: 8-10.

北京市二级保护野生动物 | | IUCN 红色名录 无危 (LC) 2016.10.1

PASSERINES

黄腹山雀	*Pardaliparus venustulus*	Yellow-bellied Tit	山雀科 Paridae

雄鸟 Male

雌鸟 Female

北京市一级保护野生动物 | IUCN 红色名录 无危 (LC) 2016.10.1

形态特征 小型鸣禽，体长 9~11 厘米。雄鸟前额、头顶至上背及喉和上胸部亮黑色；颊和耳羽白色，后颈中间有白斑；下背至腰蓝灰色；两翼黑褐色，具两道白色翼斑；飞羽具灰绿色羽缘；尾黑色，较短，外侧尾羽外翈具白斑；下胸和腹部亮黄色，翼上两道白点斑。雌鸟与雄鸟相似，但上体色淡，且颏、喉为淡黄白色。

分布与生境 中国特有种，主要分布于中国东部、南部、中部地区。北京常见于低山和山脚平原地带的林地和平原城市公园中，为西北部山地的夏候鸟，城区、郊区平原和山地的旅鸟，近年来低海拔林地有少量越冬群体。

习性与生态 常集小群出没，穿梭活跃于树冠间。主要以昆虫为食，也吃植物果实和种子等。于树洞中营杯状巢；窝卵数 5~7 枚。

Morphological characteristics A small passerine (9-11 cm). Adult male has shiny black forehead, crown, upper back, throat and upper breast; cheeks and ear-coverts white; white patch on hindneck; cheeks and ear-coverts white; lower back and rump bluish-grey; wings blackish-brown with two white wingbars. Tail short, black, with white patches on outer vanes of outer rectrices. Lower breast and belly bright yellow. Female similar but has paler upperparts and pale yellowish-white chin and throat.

Distribution and habitat Endemic to China, mainly in eastern, central and southern China. In Beijing, common in montane forest, as well as wooded hills and even lowland parks, especially on migration. Residential status: passage migrant in montane areas and lowlands (in both city and countryside); summer breeder in montane areas in the north and west of Beijing. In recent years small wintering population have been recorded in lowland woodlands.

Behaviour and ecology Often in noisy and mobile flocks in the tree canopy. Feeds mainly on insects and also fruits and seeds. Cup-shaped nests built in tree holes. Clutch size: 5-7.

杂色山雀	*Sittiparus varius*	Varied Tit	山雀科 Paridae

形态特征 小型鸣禽，体长 13~14 厘米。雌雄相似。头顶、枕至后颈黑色，前额、眼先、脸颊乳黄色。上背栗红色，下背、腰、两翼和尾羽均为蓝灰色。颏、喉黑色，两胁栗红色，胸、腹中央至尾下覆羽污白色。

分布与生境 主要分布于中国辽宁东部、吉林西南部和台湾，零星记载于东部其他地区。北京见于海拔 1000 米以下的林地，有时亦见于城市公园中，为罕见迷鸟。

习性与生态 性活泼，多在树冠中下层枝叶间活动，偶尔也下到地下觅食。营巢于树洞中；窝卵数约 5 枚。

Morphological characteristics A small passerine (13-14 cm). Sexes alike. Crown, nape and hindneck black; forehead, lores and cheeks milky-yellow; upper back chestnut red; lower back, rump, wings and tail blue-grey; chin and throat black; chestnut flanks; Off-white from central breast to central belly and undertail-coverts.

Distribution and habitat Mainly in eastern Liaoning Province, southwest Jilin Province and Taiwan Province, scarce in other areas in eastern China. In Beijing, rare vagrant in forests below 1,000 m, sometimes in urban parks.

Behaviour and ecology Lively, often in middle and lower branches of tall trees; occasionally forages on ground. Nests in tree holes. Clutch size: 5.

IUCN 红色名录 无危 (LC) 2016.10.1

沼泽山雀	*Poecile palustris*	Marsh Tit	山雀科 Paridae

北京市二级保护野生动物		IUCN 红色名录　无危 (LC) 2016.10.1

形态特征　小型鸣禽，体长 10~13 厘米。雌雄相似。喙黑色，较短而厚实。头顶黑色，略具光泽，头侧近白色。上体和两翼灰褐色。颏、喉部具一小块黑色斑，其余下体污白色，两胁略沾皮黄色。

分布与生境　主要分布于中国东北、华北及黄河、长江流域。北京常见于海拔 800 米以下的低山和平原的林缘疏林灌丛、农田以及城市公园中，为留鸟。

习性与生态　性较活泼，常穿梭于树林或灌丛间，主要以昆虫为食，也食植物果实、种子及嫩芽。鸣叫声有数种，其中比较典型的是甚具节奏感的“zi zi her her”或“zi zi her”；鸣啭为响亮的哨音“biu biu biu”。营巢于树洞中；窝卵数 6~10 枚。

Morphological characteristics A small passerine (10-13 cm). Sexes alike. Black bill short and thick; glossy black crown; whitish head-sides; greyish-brown upperparts and wings; a small black patch on chin and throat. Remainder of underparts off-white; washed pale buff on flanks.

Distribution and habitat Mainly in northeast and north China, also along Yellow River and Yangtze River valley in China. In Beijing, common in most habitats with trees, including lower to mid-level mountains (usually below 800 m). Residential status: resident.

Behaviour and ecology Actively moves between branches in trees or shrubs. Feeds mainly on insects in summer; fruit and seeds in winter. Gives various calls, such as "zi zi her her" or "zi zi her". Songs loud whistling "biu biu biu". Nests in tree holes. Clutch size: 6-10.

褐头山雀	*Poecile montanus*	Willow Tit	山雀科 Paridae

形态特征 小型鸣禽，体长 11~13 厘米。雌雄相似。头顶褐色或黑褐色，缺乏光泽，头两侧白色。上体灰褐色。颏、喉部黑色。下体余部污白色或淡皮黄白色，两胁淡皮黄色。甚似沼泽山雀，但本种头顶偏褐色，且缺乏光泽，喉部的黑色区域面积较大；此外二者鸣声亦有显著差异。

分布与生境 广泛分布于中国北部、中部至西南地区。北京区域性常见于海拔 800 米以上的山地森林中，但非繁殖期常向较低海拔移动，为留鸟。

习性与生态 性活泼，行动敏捷，在枝桠间穿梭寻觅食物，有时倒悬枝头。主要以昆虫为食，亦食少量种子和嫩芽等植物性食物。鸣声为略显粗哑的"zi chur chur chur"或仅为"chur chur chur"。巢多筑于树洞中和树裂隙中；窝卵数 6~10 枚。

Morphological characteristics A small passerine (11-13 cm). Sexes alike. Brown or blackish-brown crown lacks gloss; head-sides white; upperparts greyish-brown; chin and throat blackish; remaining underparts off-white or pale buff-white; flanks pale buff. Closely resembles Marsh Tit but has duller, brownish crown (i.e. lacks gloss), and has a larger blackish patch on throat. It is also easy to differ the two species in vocalisations.

Distribution and habitat Mainly in northern, central and southwest China. In Beijing, locally common in forests above 800 m. Often descends in non-breeding season. Residential status: resident.

Behaviour and ecology Active and agile, moves between branches searching for food, sometimes hanging upside-down. Feeds mainly on insects and also seeds and shoots. Gives harsh calls "zi chur chur chur" or simply "chur chur chur". Nests in tree holes and tree cracks. Clutch size: 6-10.

北京市二级保护野生动物 | IUCN 红色名录 无危 (LC) 2019.7.31

欧亚大山雀 | *Parus major* | Great Tit | 山雀科 Paridae

雌鸟 Female　雄鸟 Male

北京市二级保护野生动物 *　　IUCN 红色名录　无危 (LC) 2016.10.1

形态特征　小型鸣禽，体长 13~14 厘米。成鸟前额、头顶及眼先黑色，颊至耳羽白色。上体及内侧翼上覆羽黄绿色。两翼余部大致为黑色，大覆羽处具一道清晰的白色翼斑。尾羽较长，近黑色，外侧尾羽白色。喉黑色，腹中央贯以黑色宽带与喉部相连，雄鸟的黑色带较雌鸟长且宽。下体余部黄色（雄鸟更为鲜艳）。与大山雀相似，但本种下体为黄色而非白色。

分布与生境　主要分布于中国新疆西北部。北京甚罕见于平原林地，仅于 2019 年 3 月在温榆河有过一次记载。

习性与生态　性较活泼，行动敏捷，不甚惧人。主要以昆虫为食，也食植物果实、种子等。通常营巢于树洞中；窝卵数 6~10 枚。

* 注：《北京市二级保护野生动物名录》未收录（更新 / 接受）欧亚大山雀和大山雀的物种划分，故此处沿用大山雀的数据。

Morphological characteristics　A small passerine (13-14 cm). Adult has black forehead, crown, and lores, white cheek and ear-coverts; upperparts and inner upperwing-coverts yellowish-green. Remainder of wings largely blackish, with a distinctive wingbar on greater coverts. Tail long, blackish; outer rectrices white. Chin and throat black; a black stripe from throat to central belly. The black stripe is longer and broader in males. Rest of underparts yellow (brighter in males, paler in females). Resembles Cinereous Tit but has yellow underparts.

Distribution and habitat　Mainly in northwest Xinjiang in China. In Beijing, very rare vagrant in wooded areas, with only one record at Wenyu River in March 2019.

Behaviour and ecology　Relatively lively, agile and bold. Feeds mainly on insects, also fruits and seeds. Nests in tree holes. Clutch size: 6-10.

大山雀	*Parus cinereus*	Cinereous Tit*	山雀科 Paridae

形态特征 小型鸣禽，体长 13~15 厘米。雌雄相似。前额、眼先、头顶皆为黑色，颊至耳羽白色，后颈具一较小的白色斑块。上背黄绿色，下背、腰及尾上覆羽蓝灰色。两翼黑褐色，具一道白色翼斑。尾羽深蓝灰色，外侧尾羽白色。颏、喉黑色，具一显著的黑色带自喉延伸至腹部中央。下体余部污白色。

分布与生境 中国广泛分布。北京常见于山区至平原各类林地及城市公园中，为留鸟。

习性与生态 性活泼，行动敏捷。鸣声为“zi her”、“zi-zi her”或“zi-zi her-her”等。营巢于树洞中；窝卵数 6~10 枚。

● *注：郑光美（2017）认为大山雀（*Parus major*）复合种应包含两个物种：①欧亚大山雀（*Parus major*，英文名 Great Tit），含欧亚大陆西部及北部诸亚种，其中 *turkestanicus*（北京迷鸟）和 *kapustini* 亚种在中国有分布；②大山雀（*Parus cinereus*，英文名 Cinereous Tit），含欧亚大陆东部及南部诸亚种，计有五个亚种在中国有分布。值得一提的是，Gill 等（2021）认为其中的四个亚种 *tibetanus*、*subtibetanus*、*minor*（北京留鸟）和 *commixtus* 应被视为远东山雀（*P. minor*）的亚种，而 *hainanus* 则应被视为苍背山雀（*P. cinereus*）的亚种。

Morphological characteristics A small passerine (13-15 cm). Sexes similar. Black from forehead and lores to crown; cheek and ear-coverts white; a small white patch on hindneck; upper back yellowish-green; lower back, rump and uppertail-coverts bluish-grey; wings blackish-brown with a white wingbar; tail dark blue-grey with white outer rectrices; chin and throat black. A black stripe extends from throat to central belly. Remainder of underparts off-white.

Distribution and habitat Widespread in China. In Beijing, common resident in montane and lowland forests; sometimes in urban parks.

Behaviour and ecology Active and agile. Calls "zi her", "zi-zi her" or "zi-zi her-her". Nests in tree holes. Clutch size: 6-10.

北京市二级保护野生动物		IUCN 红色名录 无危 (LC) 2016.10.1

中华攀雀	*Remiz consobrinus*	Chinese Penduline Tit	攀雀科 Remizidae

雄鸟 Male

雌鸟 Female

IUCN 红色名录　无危 (LC) 2016.10.1

形态特征　小型鸣禽，体长 10~12 厘米。喙铅灰色。虹膜暗褐色。雄鸟头部灰色，宽阔的黑色贯眼纹延至耳羽；后颈和颈侧栗红色；背、腰和尾下覆羽沙褐色，飞羽、尾羽暗褐色；颏、喉近白色，胸、腹部皮黄色。雌鸟似雄鸟，头淡褐色，具深褐色贯眼纹。

分布与生境　主要分布于中国东北、华北和长江中下游地区。北京不常见于湿地附近的柳树林和芦苇丛中，为夏候鸟、冬候鸟和旅鸟。

习性与生态　常见其成小群，攀于芦苇、灌丛之中。主要以昆虫为食，冬季多食植物种子等。营巢在杨树、榆树、柳树等阔叶树上；巢呈囊袋状，主要由树皮纤维、羊毛、蒲绒、杨絮、柳絮等编织而成，结构精巧，悬于乔木柔软的细枝末梢；窝卵数 7~8 枚。

Morphological characteristics A small passerine (10-12 cm). Lead grey bill; dark brown iris. Male has grey head, broad black eye-stripe reaching through ear-coverts; chestnut red hindneck and neck-sides; buff-brown back, rump and undertail-coverts; dark brown flight-feathers and rectrices; whitish chin and throat; yellowish breast and belly. Female similar to male but with paler brown head and dark brown eye-stripe.

Distribution and habitat Mainly in northeast, north and along the lower Yangtze River valley in China. In Beijing, uncommon in willows and reeds at rural wetlands. Residential status: summer breeder, winter visitor and migrant.

Behaviour and ecology Often in small flocks, climbing among reeds and shrubs. Feeds mainly on insects in summer and seeds in winter. Spectacular hanging woven nest made of bark fibre, wool and catkins, often in willow trees or other broad-leaved trees. Clutch size: 7-8.

蒙古百灵	*Melanocorypha mongolica*	Mongolian Lark	百灵科 Alaudidae

北京市二级保护野生动物	国家二级保护野生动物	IUCN 红色名录　无危 (LC) 2016.10.1

形态特征　中型鸣禽，体长 18~22 厘米。雌雄相似。喙黄色而端黑，较为粗厚。头顶栗棕色，头顶中央淡棕黄色。眉纹淡黄白色，甚长，自眼延伸至枕部。头两侧淡棕色。上体以褐色为主，杂有灰白色斑纹。外侧初级飞羽黑褐色，次级飞羽和内侧初级飞羽皆为白色。中央尾羽棕色，最外侧尾羽白色，其余尾羽黑褐色，尖端羽色较淡。下体污白色为主。胸具断开的黑色横带。

分布与生境　主要分布于中国西部和东北地区。北京区域性常见于大型水域周边荒地、草地、干涸河床，为旅鸟、冬候鸟。

习性与生态　常集大群，特别是迁徙和越冬期间，擅奔跑，亦擅飞翔，常在空中边飞边鸣。主要以草籽为食，也捕食少量昆虫等。营巢地面稍凹处或草丛间；窝卵数 3~5 枚。

Morphological characteristics　A medium-sized passerine (18-22 cm). Sexes alike. Bill thick, yellow, with black tip. Crown rufous-brown; pale yellowish-brown at centre of crown. The pale yellowish-white supercilium extends from eye to nape. Head-sides pale brownish. Upperparts largely brown with greyish-white streaks. Outer primaries blackish-brown; secondaries and inner primaries white. Central rectrices rufous-brown; outermost rectrices mostly white; rest of rectrices blackish-brown with pale tips. Underparts largely off-white, with a broken black band on breast.

Distribution and habitat　Mainly western and northeast China. In Beijing, locally common in open rough ground, grassland and dry riverbeds. Residential status: winter visitor and migrant.

Behaviour and ecology　Often in large flocks, especially during migration and in winter; often gives a whistling call in flight. Feeds mainly on seeds and insects. Nests on the ground. Clutch size: 3-5.

大短趾百灵 | *Calandrella brachydactyla* | Greater Short-toed Lark | 百灵科 Alaudidae

形态特征 小型鸣禽，体长 14~16 厘米。雌雄相似。喙浅褐色，较粗壮。眉纹白色，耳羽褐色。上体棕褐色，具黑色纵纹。第四枚初级飞羽显著比前三枚初级飞羽短，三级飞羽则显著长于邻近的次级飞羽。尾黑褐色，最外侧尾羽白色。下体白色微沾黄，无显著斑纹。跗跖肉色。

分布与生境 主要分布于中国西部和东北地区。北京不常见于近水的荒草地、干燥沙地和农田地带，为旅鸟、冬候鸟。

习性与生态 常集小群活动。主要以杂草种子、昆虫等为食。营巢于地面凹坑；窝卵数 3~4 枚。

Morphological characteristics A small passerine (14-16 cm). Sexes alike. Pale greyish-brown, stout bill; white supercilium; brown ear-coverts; black-streaked tan upperparts; the fourth primary is significantly shorter than three outermost primaries; tertials distinctively longer than secondaries; dark brown tail with white outermost rectrices; white underparts tinged yellow; flesh tarsi.

Distribution and habitat Mainly western and northeast China. In Beijing, uncommon in grassland, dry sandy habits and farmland. Residential status: migrant and winter visitor.

Behaviour and ecology Often in small flocks. Feeds mainly on seeds and insects. Nests on ground. Clutch size: 3-4.

北京市二级保护野生动物 | | IUCN 红色名录 无危 (LC) 2018.8.9

短趾百灵	*Alaudala cheleensis*	Asian Short-toed Lark	百灵科 Alaudidae

北京市二级保护野生动物 | IUCN 红色名录 无危 (LC) 2018.8.9

形态特征 小型鸣禽，体长 13~15 厘米。喙浅黄色，较为短小。眉纹、眼周棕白色，颊、耳羽棕褐色。上体羽浅沙棕色，具黑褐色纵纹。外侧第四枚初级飞羽仅比前三枚略短，三级飞羽与邻近的次级飞羽长度无显著差异。尾黑褐色，最外侧尾羽为白色。下体乳白色，胸部具清晰的黑色细纵纹。跗跖较短。

分布与生境 主要分布于中国秦岭－淮河以北地区。北京不常见于近水的沙质荒草地等环境，为旅鸟，亦有甚零散的繁殖和越冬种群。

习性与生态 常成小群活动，能垂直起飞，边飞边鸣，有时也呈波浪形往前飞。主要以杂草种子为食，有时也吃少量昆虫。营巢于草丛中地面；窝卵数 3~4 枚。

Morphological characteristics A small passerine (13-15 cm). Sexes alike. Bill short and pale yellow; brownish-white supercilium and eye-ring; brown cheeks and ear-coverts; pale buff-brown upperparts with dark brown streaks; the fourth primary is slightly shorter than the three outermost three primaries; tertials and secondaries are of similar length; dark brown tail with clean white outermost rectrices; fine dark streaking across breast; milky white underparts; short tarsi.

Distribution and habitat Mainly north from the Qinling–Huaihe Line in China. In Beijing, uncommon in open, dry sandy habitats, often close to water. Residential status: migrant, also scarce summer breeder and winter visitor.

Behaviour and ecology Often in small flocks; utters a distinctive whistle while flying. Feeds mainly on seeds in winter and insects in summer. Nests on ground. Clutch size: 3-4.

凤头百灵	*Galerida cristata*	Crested Lark	百灵科 Alaudidae

形态特征 中型鸣禽，体长 15~20 厘米。雌雄相似。喙细且长，呈黄粉色，喙峰灰色。头顶中央几枚黑褐色长羽形成冠羽。上体棕褐色，翼后缘褐色，无白色边缘。尾羽黑褐色，最外侧尾羽淡褐色。下体淡皮黄白色，胸部黑褐色纵纹甚显著。

分布与生境 主要分布于中国北方至长江流域。北京不常见于近水的荒草沙地和农田、旷野，为留鸟

习性与生态 常单独或集松散的小群生活。主食植物种子和少量昆虫。营巢于荒漠沙砾地上或岩石草地上；窝卵数 2~3 枚。

Morphological characteristics A medium-sized passerine (15-20 cm). Sexes alike. Bill long and slim, yellowish-pink, culmen grey; overall brown plumage with long, pointed crest, visible at rear crown; buff-brown upperparts, brown wing-edge lacking white trailing edge; dark brown central rectrices and pale brown outer rectrices. Underparts pale buff-white with prominent dark brown streaks on breast.

Distribution and habitat Mainly from northern China to the Yangtze. In Beijing, uncommon in rough grassland and dry, stony habitats, often near water. Residential status: resident.

Behaviour and ecology Usually solitary or in small groups. Feeds mainly on seeds in winter and insects in summer. Nests on ground among stones or rocky grassland. Clutch size: 2-3.

北京市二级保护野生动物

IUCN 红色名录 无危 (LC) 2016.10.1

云雀	*Alauda arvensis*	Eurasian Skylark	百灵科 Alaudidae

北京市二级保护野生动物 | 国家二级保护野生动物 | IUCN 红色名录 无危 (LC) 2018.8.9

形态特征 中型鸣禽，体长 15~19 厘米。雌雄相似。喙较短而尖，上喙灰色，下喙角质色。虹膜暗褐色。头顶具冠羽。上体褐色，具显著的黑色纵纹。飞行时能观察到翼后缘和尾羽两侧有比较明显的白色边缘。下体灰白，胸部密布不连贯黑色纵纹。跗跖肉褐色。

分布与生境 主要分布于中国北方、华东、华南。北京常见于开阔的草地和农田荒地，为旅鸟、冬候鸟。

习性与生态 多集群在地面急走、觅食、追逐活动，极少上树。头顶羽冠兴奋时立起。常骤然自地面垂直地起飞，直冲天空。繁殖季主食昆虫，非繁殖季主食植物种子。于草地上营巢；巢以草编织而成；窝卵数 3~5 枚。

Morphological characteristics A medium-sized passerine (15-19 cm). Sexes alike. Bill short and pointed; upper mandible dark grey, lower mandible horn; dark brown iris; crested crown; brown upperparts with blackish streaks; greyish-white underparts; breast with dark streaks; in flight, shows a thin, diffuse white trailing edge to upperwing and white sides of tail; tarsi flesh-brown.

Distribution and habitat Mainly northern, east and south China. In Beijing, common in open grassland and farmland. Residential status: migrant and winter visitor.

Behaviour and ecology Often in flocks, foraging on open ground. When excited or alarmed, raises crest. Often "catapults" up to sky. Feeds mainly on seeds in non-breeding season and insects in breeding season. Builds grass-lined nest on grasslands. Clutch size: 3-5.

角百灵	*Eremophila alpestris*	Horned Lark	百灵科 Alaudidae

形态特征 中型鸣禽，体长 15~19 厘米。喙短，上喙黑褐色，下喙黄褐色。雄鸟头顶两侧各具一簇突出的黑色角状羽，头侧至喉部为白色，具一较宽黑色条带横贯眼先至颊部；上体沙褐色，胸、腹部白色，胸部具一黑色横带向两侧延伸而渐细；跗跖黑色。雌鸟似雄鸟，头顶角状羽不明显，胸部黑色横带亦较窄小。

分布与生境 主要分布于中国北方和青藏高原。北京不常见于亚高山草甸、荒草地、沙石质干涸河床等地，为冬候鸟和旅鸟。

习性与生态 平时多单独或成对活动，有时亦成小群，主要在地面活动，擅在地面短距离急走。以食草籽等植物性食物为主，亦食昆虫等。营巢于草丛基部的地面上或灌木丛中；窝卵数 2~5 枚。

Morphological characteristics A medium-sized passerine (15-19 cm). Short bill with dark brown upper mandible and buff-brown lower mandible. Male has elongated black feathers on sides of crown, forming narrow "horns"; white head-sides and throat with wide dark patch from lores to cheeks; buff-brown upperparts; white breast and belly; black stripe across breast; black tarsi. Female similar to male but "horns" indistinct and has a narrower black stripe on breast.

Distribution and habitat Northern China and the Tibetan Plateau. In Beijing, uncommon in subalpine meadows, grassland and sandy, stony habitats such as dry riverbeds. Residential status: winter visitor and passage migrant.

Behaviour and ecology Solitary or in pairs, sometimes in small flocks. As with most larks. forages on the ground; often scurries. Feeds on seeds and insects. Nests on ground at grasses base or in shrubs. Clutch size: 2-5.

北京市二级保护野生动物 | IUCN 红色名录 无危 (LC) 2018.11.5

文须雀	*Panurus biarmicus*	Bearded Reedling	文须雀科 Panuridae

雄鸟 Male

雌鸟 Female

IUCN 红色名录　无危 (LC) 2016.10.1

形态特征　中型鸣禽，体长 16~19 厘米。喙黄色，较短。雄鸟前额、头顶、头侧蓝灰色，眼先、眼周黑色并向下延伸形成一簇髭纹，似两撇胡须；背、肩、腰、尾棕色，一对中央尾羽甚长，外侧尾羽依次逐渐缩短。颏、喉和前胸灰白，两胁淡棕黄色；跗跖黑色。雌鸟似雄鸟，但头淡黄棕色，眼下亦无黑色髭纹。

分布与生境　主要分布于中国北方。北京见于较大湿地的苇丛中，为不常见冬候鸟和罕见夏候鸟。

习性与生态　多成对或成小群活动，性活泼。常见其攀在芦苇茎上啄食。食物主要为昆虫和芦苇种子与草籽等。通常营巢于芦苇或灌木下部；窝卵数 5~6 枚。

Morphological characteristics　A medium-sized passerine (16-19 cm). Bill short and yellow. Male has bluish-grey forehead, crown and head-sides; head light blue-grey with long, black drooping "moustache"; brown back, mantle, rump and tail; graduated tail with long pair of central rectrices and shorter outer rectrices; chin, throat and upper breast white; pale buff-brown flanks; black tarsi. Female similar to male but head buff-brown without black "moustache".

Distribution and habitat　Mainly northern China. Uncommon winter visitor and rare summer breeder in Beijing. Occurs in reedbeds at large wetlands.

Behaviour and ecology　Usually in pairs or small groups. Often seen feeding on stems of reeds. Feeds mainly on insects and reed seeds. Nests low in reeds or shrubs. Clutch size: 5-6.

棕扇尾莺 | *Cisticola juncidis* | Zitting Cisticola | 扇尾莺科 Cisticolidae

形态特征 小型鸣禽，体长9~11厘米。雌雄同色。喙粉色，喙峰黑色。成鸟繁殖羽头褐色，头顶具黑色细纹，眉纹棕白色，较短，上体褐色，具显著的黑色纵纹；圆尾，中央略凸出，飞行时显著扇开，尾羽褐色，具淡黄白色端斑和宽阔的黑色次端斑；下体大致为白色，胸两侧及两胁为棕色；跗跖粉色。非繁殖羽似繁殖羽，但头顶黑色纵纹更为显著。

分布与生境 分布于中国华北及其以南区域。一般活动于海拔1000米以下的灌丛、草丛和苇丛中，于湿地附近更为易见。北京为不常见夏候鸟。

习性与生态 繁殖期雄鸟的炫耀行为极具特点：在领域上方进行悬停及急速地上升和下降，上升时发出尖细鸣声似"ji ji ji"，下降时则发出"cha cha cha"的鸣声。多于草丛中营编织巢；窝卵数4~5枚。

Morphological characteristics A small passerine (9-11 cm). Sexes alike. Bill pink with blackish culmen. Breeding adult has brown head with fine black streaks on crown; supercilium short, off-white; upperparts largely mid-brown, distinctively streaked black; brown tail rounded, with whitish tip and broad black subterminal band; tail usually fanned in flight; underparts mostly white; breast-sides and flanks pale brown; tarsi pink. Non-breeding adult similar but has heavier black streaks on crown.

Distribution and habitat Widespread in north and southern China. Uncommon summer breeder in Beijing. Usually seen in scrub, grass and reeds below 1,000 m; commoner at wetland margins.

Behaviour and ecology Breeding male gives undulating courtship display: first rises and hovers over its territory with flight calls "ji ji ji"; then descends and gives "cha cha cha" calls. Nests in reeds or tall grass. Clutch size: 4-5.

IUCN 红色名录 无危 (LC) 2016.10.1

黄腹山鹪莺	*Prinia flaviventris*	Yellow-bellied Prinia	扇尾莺科 Cisticolidae

IUCN 红色名录　无危 (LC) 2016.10.1

形态特征　小型鸣禽，体长 10~13 厘米。雌雄同色。喙黑色。虹膜暗红色。头深灰色，眉纹白色，较短，前额和眼先羽色较深，接近黑色。上体、两翼及尾羽橄榄绿色。尾羽较长，呈显著的凸形。颏、喉、胸白色，腹部至尾下覆羽黄色。跗跖较长，呈黄褐色。

分布与生境　主要分布于中国长江以南。活动于低山及平原的灌丛、苇丛、草地等。在北京为甚罕见迷鸟，仅 2019 年 6 月于宣武艺园内有过一次记载，但存在此为逃逸个体的可能性。

习性与生态　多集 3~10 只的小群活动，性活跃，尾羽常上下摆动，或竖直向上翘起。鸣叫声似猫叫。鸣啭似棕扇尾莺。营巢于灌丛或高草丛中；窝卵数 3~6 枚。

Morphological characteristics A small passerine (10-13 cm). Sexes alike. Bill black; eyes deep red; dark grey head with distinctive short white supercilium; forehead and lores darker; upperparts, wings and tail mostly olive-green; tail long and graduated; chin, throat and breast white; belly and undertail-coverts yellow; arsi relatively long and yellowish-brown.

Distribution and habitat Mainly south of the Yangtze River. Scarce in north China. Occurs in lowland scrub, reeds and grasses. Very rare vagrant in Beijing with only one record at Xuanwu Park in June 2019. Note that the possibility of escape could not be dismissed.

Behaviour and ecology Usually in small groups (less than 10 individuals). Active. Tail often flicking or upright. A common call somewhat like a cat. Song similar to Zitting Cisticola. Nests in scrub or tall grass. Clutch size: 3-6.

纯色山鹪莺 | *Prinia inornata* | Plain Prinia | 扇尾莺科 Cisticolidae

形态特征 小型鸣禽，体长 11~14 厘米。雌雄同色。喙黑色，喙基粉色。头及上体黄褐色，具淡皮黄色眉纹和红色虹膜。尾羽较长，呈明显的凸形，具不清晰的近白色端斑和深褐色次端斑。下体淡黄白色。跗跖粉色。

分布与生境 主要分布于中国秦岭－淮河以南地区。主要活动于海拔 1500 米以下的灌丛、苇丛、草丛和农田中。在北京为甚罕见迷鸟，仅于 2019 年 5 月在温榆河有一次记载。

习性与生态 性活跃。常单独、成对或集小群活动，鸣声轻快，似“di di di”。营巢于草丛中；窝卵数 4~6 枚。

Morphological characteristics A small passerine (11-14 cm). Sexes alike. Bill black with pink base; head and upperparts brown; supercilium pale buff; eyes red; tail long, significantly graduated, with indistinct pale tip and dark brown subterminal band; upperparts pale yellowish-white; tarsi pink.

Distribution and habitat Mainly south of the Qinling–Huaihe Line. Occurs in scrub, reeds, grassland and farmland below 1,500 m. Very rare vagrant in Beijing with only one record at the Wenyu River in May 2019.

Behaviour and ecology Active. Often alone, in pairs or in small flocks. Calls varied, mainly a soft and plaintive "di di di". Usually nests in tall grass. Clutch size: 4-6.

IUCN 红色名录 无危 (LC) 2016.10.1

东方大苇莺	*Acrocephalus orientalis*	Oriental Reed Warbler	苇莺科 Acrocephalidae

北京市二级保护野生动物 | IUCN 红色名录　无危 (LC) 2016.10.1

形态特征　中型鸣禽，体长 16~19 厘米。雌雄同色。喙较长，粉色，喙峰黑色。头顶、上体及两翼橄榄褐色，具较宽阔的淡黄白色眉纹，眼先黑色。飞羽黑褐色，羽缘棕黄色，尾羽暗褐色。颏、喉、颊白色，胸具甚不显著的褐色纵纹。下体余部白色。跗跖灰褐色。

分布与生境　中国除西部地区外广泛分布。见于城区和郊区植被丰富的各种湿地。在北京为甚常见夏候鸟。

习性与生态　擅鸣，喜站在湿地中的苇丛、灌丛和附近的树上高声鸣啭，声似“kra kra kra”。常营巢于苇丛中；窝卵数 3~6 枚。

Morphological characteristics　A medium-sized passerine (16-19 cm). Sexes alike. Bill relatively long, pink with black culmen; head, upperparts and wings largely olive-brown; supercilium broad, pale yellowish-white or off-white; lores black; flight-feathers blackish-brown with brown fringes; tail dark brown; chin, throat and cheek white; breast has indistinct brown streaks; remainder of underparts white; tarsi greyish-brown.

Distribution and habitat　Widespread except for western China. Abundant summer breeder in Beijing. Found at well-vegetated wetlands in both urban and rural areas.

Behaviour and ecology　Highly vocal. Often gives loud, repeated song including chuckles "kra kra kra" from top of reed, shrub or tree. Nests in dense reeds. Clutch size: 3-6.

黑眉苇莺	*Acrocephalus bistrigiceps*	Black-browed Reed Warbler	苇莺科 Acrocephalidae

形态特征 小型鸣禽，体长 12~13 厘米。雌雄同色。喙较短而细，上喙黑灰色，下喙粉色。头顶、上体及两翼黄褐色或橄榄褐色，眉纹白色，于眉纹上方另有一宽阔的黑色条纹自喙基延伸至头后，具较细的黑褐色贯眼纹。尾羽深褐色，具淡褐色羽缘。下体大致为白色，两胁和尾下覆羽淡棕色。跗跖灰褐色。

分布与生境 分布于中国东部地区。在北京见于海拔 900 米以下的水库、河流、湖泊、水塘的苇丛中，在迁徙季节有时远离水域，为区域性常见旅鸟。

习性与生态 常单独或成对活动，隐蔽地在苇丛中穿梭，但雄鸟在鸣啭时会站在芦苇高处或枝头显著处，鸣声为清脆而富有变化的“ji ji jiu jiu”声。多营巢于近水的苇丛和草丛中；窝卵数 4~6 枚。

Morphological characteristics A small passerine (12-13 cm). Sexes alike. Bill short and fine, with blackish upper mandible and pink lower mandible; crown, upperparts and wings brown or olive-brown; supercilium white with long, black brow above; eye-stripe blackish; tail dark brown with pale brown fringes; underparts mostly white; flanks and undertail-coverts pale brown or buff; tarsi greyish-brown.

Distribution and habitat Eastern China. Locally common passage migrant in Beijing, Seen at wetland (e.g. reservoirs, rivers, lakes, ponds) margins with dense reeds (below 900 m). Sometimes far from water on migration.

Behaviour and ecology Alone or in pairs. Skulks in dense reedbeds. Singing male may perch on top of reed or shrub near water. Song is variable churring like "ji ji jiu jiu". Nests in extensive reedbeds or tall grass near water. Clutch size: 4-6.

北京市二级保护野生动物		IUCN 红色名录　无危 (LC) 2016.10.1

细纹苇莺	*Acrocephalus sorghophilus*	Streaked Reed Warbler	苇莺科 Acrocephalidae

北京市二级保护野生动物	国家二级保护野生动物	IUCN 红色名录 濒危 (EN) 2016.10.1

形态特征 小型鸣禽，体长 12~13 厘米。雌雄同色。喙尖而细，喙峰黑色，余部淡黄色。头顶及背部具黑色细纵纹，具较宽的皮黄色眉纹和较细的黑色条纹于眉纹上缘。上体、两翼及尾羽褐色，下体近白色。跗跖褐色。与黑眉苇莺相似，但本种上体具清晰的黑色纵纹。

分布与生境 中国零星见于华北及华东地区。主要活动于各种水域附近的苇丛和草丛中。在北京为甚罕见旅鸟。

习性与生态 常单独活动，在苇丛中快速而隐秘地穿梭，捕食昆虫。

Morphological characteristics A small passerine (12-13 cm). Sexes alike. Bill short, fine, largely pale yellow, with black culmen; crown and mantle finely streaked black; supercilium broad, buff with a narrow black brow bordered above; remainder of head, upperparts and wings brown; off-white below; tarsi brown. Resembles Black-browed Reed Warbler but has streaked mantle.

Distribution and habitat Scarce in east and north China. Very rare passage migrant in Beijing. Occurs in extensive reedbeds and tall grass near water.

Behaviour and ecology Usually alone. Shy. Skulks in dense vegetation and feeds on insects.

钝翅苇莺	*Acrocephalus concinens*	Blunt-winged Warbler	苇莺科 Acrocephalidae

形态特征 小型鸣禽，体长 12~14 厘米。雌雄同色。喙较长，上喙黑灰色，下喙粉色。具较短而细的近白色眉纹，贯眼纹深褐色。头及上体灰褐色或棕褐色，具短而较圆的两翼（初级飞羽突出部较短）。下体大致为近白色，胸侧、两胁和尾下覆羽皮黄色。跗跖淡褐色。

分布与生境 中国见于华北及其以南地区。活动于海拔 1200 米以下水域附近的灌丛、草丛和苇丛中。北京为罕见旅鸟。

习性与生态 单独活动，多数时间隐蔽在浓密的植被中。鸣声为响亮而多变的颤音。多营巢于近水苇丛和灌丛中；窝卵数 3~4 枚。

Morphological characteristics A small passerine (12-14 cm). Sexes alike. Bill relatively long with blackish-grey upper mandible and pink lower mandible; short, narrow off-white supercilium and dark brown eye-stripe; head and upperparts greyish-brown or mid-brown; wings short and blunt (i.e. short primary projection); underparts largely off-white; breast-sides, flanks and undertail-coverts buff; tarsi pale brown.

Distribution and habitat North China and southern China. Rare passage migrant in Beijing. Seen in reeds, grass and scrub at wetland margins below 1,200 m.

Behaviour and ecology Usually solitary in dense vegetation. Song is loud and variable, consisting of churs and whistles. Nests in reeds or shrubs in wetlands. Clutch size: 3-4.

北京市二级保护野生动物		IUCN 红色名录　无危 (LC) 2016.10.1

远东苇莺	*Acrocephalus tangorum*	Manchurian Reed Warbler	苇莺科 Acrocephalidae

北京市二级保护野生动物

IUCN 红色名录　易危 (VU) 2016.10.1

形态特征　小型鸣禽，体长 12~14 厘米。雌雄同色。喙较长，上喙黑灰色，下喙淡黄色。具细长的白色眉纹和黑色贯眼纹，眉纹之上另有一较窄的黑色细纹。头和上体橄榄褐色。尾羽较长，暗褐色，羽缘淡棕色。颏、喉近白色，胸、腹及尾下覆羽淡褐色或皮黄色，胸、腹中央羽色较淡。跗跖粉褐色。

分布与生境　中国见于东北地区至华东沿海。生境似钝翅苇莺，见于植被丰富的湿地边缘。在北京为罕见旅鸟。

习性与生态　习性似钝翅苇莺。鸣声为一连串尖锐的颤音。

Morphological characteristics A small passerine (12-14 cm). Sexes alike. Bill relatively long with blackish-grey upper mandible and pale yellow lower mandible; long, narrow, off-white supercilium with black border above; eye-stripe black; crown and upperparts olive-brown; tail long, rounded, dark brown and fringed pale brown; chin and throat off-white; remainder of underparts largely buff or pale brown; central breast and central belly paler; tarsi pinkish-brown.

Distribution and habitat From northeast to east China (mainly along coast). Rare passage migrant in Beijing. Occurs in extensive vegetation at margins of wetlands.

Behaviour and ecology Similar to Blunt-winged Warbler. Song a series of churs.

布氏苇莺	*Acrocephalus dumetorum*	Blyth's Reed Warbler	苇莺科 Acrocephalidae

形态特征 小型鸣禽，体长 12~15 厘米。雌雄同色。喙长而强健，上喙黑灰色，下喙黄色。具短而细的近白色眉纹和不甚清晰的暗褐色贯眼纹。头、上体至尾羽橄榄褐色，无深色斑纹，两翼较短。下体大致为白色，仅胸至两胁呈淡褐色。跗跖灰褐色。

分布与生境 分布于中国西北地区。多活动于树木、灌丛丰富的湿地，亦有时见于远离水域的林地或灌丛。北京为甚罕见迷鸟，仅于密云水库和昌平区沙河水库有零星记载。

习性与生态 多单独活动，在植被中穿梭、捕食昆虫及其他小型无脊椎动物。鸣声甚为尖锐而响亮。营巢于灌丛或草丛中；窝卵数 4~6 枚。

Morphological characteristics A small passerine (12-15 cm). Sexes alike. Bill long, strong, with blackish-grey upper mandible and yellow lower mandible; off-white supercilium short and narrow; eye-stripe dark brown and indistinct; head and upperparts unmarked olive-brown; wings short; underparts largely off-white; breast and flanks pale brown; tarsi greyish-brown.

Distribution and habitat Mainly northwest China. Usually occurs at well-vegetated wetlands with rich scrub and trees. Sometimes seen in woodlands far from water. Very rare vagrant in Beijing with few records from rural wetlands such as Miyun Reservoir, Miyun District and Shahe Reservoir, Changping District.

Behaviour and ecology Usually alone. Sulks in vegetation and feeds on insects as well as other small invertebrates. Song loud and harsh. Nests in shrubs or tall grass. Clutch size: 4-6.

IUCN 红色名录 无危 (LC) 2016.10.1

厚嘴苇莺	*Arundinax aedon*	Thick-billed Warbler	苇莺科 Acrocephalidae

北京市二级保护野生动物		IUCN 红色名录　无危 (LC) 2016.10.1

形态特征　中型鸣禽，体长 18~20 厘米。雌雄同色。喙较东方大苇莺短，但较厚且钝，喙峰深灰色，其余粉色。头、上体、两翼、尾羽为均一的褐色，无眉纹，眼先白色。两翼较短而圆，尾羽较长，呈显著的凸形。颏、喉白色，胸、两胁及尾下覆羽皆为淡棕色，腹部中央近白色。跗跖灰褐色。

分布与生境　广泛分布于中国东部、南部和西北地区。生境迥异于其他苇莺：常活动于海拔 800 米以下的林地、灌丛和草丛，常远离水域，亦见于植被丰富的城市公园。在北京为不常见夏候鸟和旅鸟。

习性与生态　喜单独活动。营巢于茂密的林间或林缘灌丛中；窝卵数 5~6 枚。

Morphological characteristics A medium-sized passerine (18-20 cm). Sexes alike. Bill shorter than Oriental Reed Warbler, thick and blunt; culmen dark grey; rest of bill pink; head, upperparts, wings and tail plain brown; lacks supercilium; wings short and blunt; tail long and graduated; chin and throat white; breast, flanks and undertail-coverts pale brown; central belly off-white; tarsi greyish-brown.

Distribution and habitat Widespread in eastern, southern and northwest China. Uncommon summer breeder and passage migrant in Beijing. Occurs in woodland, scrub and tall grass below 800 m, often far from water. On migration, also seen also in urban parks with abundant vegetation.

Behaviour and ecology Usually solitary. Nests in dense shrubs in woodlands or woodland edge. Clutch size: 5-6.

北短翅蝗莺	*Locustella davidi*	Baikal Bush Warbler	蝗莺科 Locustellidae

形态特征 小型鸣禽，体长 11~12 厘米。雌雄同色。喙黑色。头顶棕褐色。眉纹较细长，呈淡皮黄色或近白色。颊淡褐色。上体及两翼褐色，两翼较短且圆。尾羽暗褐色。颏、喉白色，胸灰白色具黑色斑点，腹部白色，胸侧至两胁褐色。尾下覆羽暗褐色，端部白色。跗跖淡褐色。

分布与生境 广布于中国东部至南部地区。繁殖于海拔 700 米以上的山区林地和灌丛，非繁殖期见于较低海拔的灌丛和草地，亦见于城市中植被丰富的区域。在北京为不常见旅鸟。

习性与生态 喜单独活动，常在植被中紧贴地面处穿梭。一般营巢于灌丛中；窝卵数 3~6 枚。

Morphological characteristics A small passerine (11-12 cm). Sexes alike. Bill black; mid brown crown and long, buff or off-white supercilium; cheek pale brown; upperparts and wings brown; wings short and blunt; tail dark brown; chin and throat white; breast greyish-white, spotted black; belly white; plain brown from breast-sides to flanks; undertail-coverts dark brown with white tips and tarsi pale brown.

Distribution and habitat Widespread in eastern and southern China. Uncommon passage migrant in Beijing. Breeds in woodlands and scrub above 700 m. In non-breeding season, usually occurs in lowland scrub and tall grass. Seen also in well-vegetated urban areas.

Behaviour and ecology Usually alone. Skulks in dense low vegetation. Nests in bushes. Clutch size: 3-6.

IUCN 红色名录　无危 (LC) 2016.10.1

中华短翅蝗莺	*Locustella tacsanowskia*	Chinese Bush Warbler	蝗莺科 Locustellidae

北京市二级保护野生动物 | | IUCN 红色名录 无危 (LC) 2016.10.1

形态特征 小型鸣禽，体长12厘米。雌雄同色。上喙黑灰色，下喙粉色。眉纹较短，甚不显著，呈淡皮黄色。颊皮黄色，具白色细纵纹。头顶、上体、两翼橄榄褐色，尾羽深褐色。颏、喉、胸至腹中央棕白色，胸侧及两胁淡褐色。尾下覆羽褐色，端部白色。跗跖粉色。

分布与生境 中国见于东北、华北、中部至西南地区。活动于山地至平原的草地和苇丛中。在北京为罕见旅鸟。

习性与生态 常单独活动，性极羞怯，一般在浓密的植被下层隐秘地穿行。鸣声为单调似虫鸣的“zi zi zi”声。一般营巢于灌丛和草地近地面处；窝卵数约5枚。

Morphological characteristics A small passerine (12 cm). Sexes alike. Upper mandible blackish-grey; lower mandible pink; pale buff supercilium short and indistinct; cheek buff, finely streaked white; crown, upperparts and wings olive-brown; tail dark brown; chin, throat, central breast and belly off-white; breast-sides and flanks pale brown; undertail-coverts brown, with white tips and tarsi pink.

Distribution and habitat From northeast and north China to central and southwest China. Rare passage migrant in Beijing. Occurs in dense scrub and reedbeds in lowland and montane areas.

Behaviour and ecology Solitary and shy. Prefers to skulk in extensive low grass. Gives insect-like song "zi zi zi". Nests close to the ground in dense scrub and grass. Clutch size: 5.

矛斑蝗莺	*Locustella lanceolata*	Lanceolated Warbler	蝗莺科 Locustellidae

形态特征 小型鸣禽，体长 11~14 厘米。雌雄同色。喙黑灰色，下喙基部粉色。眉纹淡黄色，不甚清晰。头顶、上体及两翼褐色，具显著的黑色纵纹。尾羽黑褐色。颏、喉白色，下体余部亦大致为白色，胸至两胁具黑色纵纹。尾下覆羽褐色，亦具黑色纵纹。跗跖粉色。

分布与生境 分布于中国东部至南部地区。多栖息于低山至平原的灌丛、草丛以及湿地的苇丛中。在北京为不常见旅鸟。

习性与生态 多单独活动，喜在浓密的植被中以及地面上快速穿梭。鸣声为重复的尖细"chi chi"声。营巢于草丛中；窝卵数 3~5 枚。

Morphological characteristics A small passerine (11-14 cm). Sexes alike. Bill blackish-brey, with pink-based lower mandible; indistinct supercilium pale buff; crown, upperparts and wings brown, with distinctive black streaks; tail blackish-brown; chin and throat white; remainder of underparts largely white with black streaks from breast to flanks; undertail-coverts brown, also streaked black; tarsi pink.

Distribution and habitat Eastern and southern China. Uncommon passage migrant in Beijing. Usually occurs in tall grass, scrub and sometimes reedbeds in lowlands.

Behaviour and ecology Solitary. Usually moves fast in dense vegetation, often running on the ground. Gives thin, insect-like song, a repeated "chi chi". Nests in grass. Clutch size: 3-5.

北京市二级保护野生动物		IUCN 红色名录 无危 (LC) 2016.10.1

小蝗莺	*Locustella certhiola*	Pallas's Grasshopper Warbler	蝗莺科 Locustellidae

北京市二级保护野生动物		IUCN 红色名录 无危 (LC) 2016.10.1

形态特征 小型鸣禽，体长 14~16 厘米。雌雄同色。喙深灰色，下喙基部粉色。眉纹近白色，较宽。贯眼纹暗褐色。但不甚清晰。头顶、上体及两翼褐色，头顶后部、后颈至背部具显著的黑色纵纹，腰棕褐色，尾羽褐色，具近白色端斑和黑色次端斑。下体通常近白色，仅胸侧、两肋和尾下覆羽淡褐色。跗跖粉色。似矛斑蝗莺但本种具清晰的眉纹，且下体无深色纵纹。

分布与生境 中国广泛分布，繁殖于北方，迁徙时旅经东部及南部地区。主要活动于湿地及其附近的灌丛、草丛和苇丛中。在北京为区域性常见旅鸟。

习性与生态 迁徙时一般单独活动，行为隐秘，多数时间于浓密的苇丛和草丛中穿行。营巢于浓密的草丛中；窝卵数 4~6 枚。

Morphological characteristics A small passerine (14-16 cm). Sexes alike. Bill dark grey with pink-based lower mandible; supercilium broad off-white; indistinct eye-stripe dark brown; crown, upperparts and wings brown; hindcrown, hindneck and mantle heavily streaked black; rump rufous; tail brown, with off-white tips and black subterminal band; underparts largely off-white with pale brown breast-sides, flanks and undertail-coverts; tarsi pink. Resembles Lanceolated Warbler but has distinctive supercilium and lacks dark streaks on underparts.

Distribution and habitat Widespread in China. Breeds in northern China; migrates across eastern and southern China. Locally common passage migrant in Beijing. Usually seen in scrub, grass and reeds in wetland habitats.

Behaviour and ecology Usually solitary on migration. Shy. Mostly skulks in extensive reedbeds and lush grass in non-breeding season. Nests in dense grass. Clutch size: 4-6.

斑背大尾莺 *Locustella pryeri* Marsh Grassbird

蝗莺科 Locustellidae

形态特征 小型鸣禽，体长 13~14 厘米。雌雄同色。喙略短而较厚，呈黑色，下喙基部粉色。头、上体及两翼棕褐色，头顶具黑色纵纹，上体具显著的黑色斑点。尾羽褐色，较长，为圆尾。颏、喉及下体余部大致为白色，胸侧、两胁及尾下覆羽淡褐色。跗跖褐色或粉色。

分布与生境 中国有限分布于东部地区。一般活动于湿地的苇丛和草地中。在北京为甚罕见旅鸟。

习性与生态 非繁殖期性羞怯，但繁殖期常在芦苇中上层相对暴露处活动，具特征性的空中翻飞炫耀行为。一般营巢于苇丛中；窝卵数 5~6 枚。

Morphological characteristics A small passerine (13-14 cm). Sexes alike. Bill relatively short and thick, mostly black, with pink at base of lower mandible; head, upperparts and wings largely mid-brown; crown and mantle heavily streaked black; brown tail long and rounded; chin, throat and remainder of underparts largely white; breast-sides, flanks and undertail-coverts pale brown; tarsi brown or pink.

Distribution and habitat Restricted areas in eastern China. Rare passage migrant in Beijing. Occurs in extensive reedbeds and wet grassland at wetlands.

Behaviour and ecology Shy in non-breeding season, but often perches on top of vegetation in breeding season. Tail often fanned in flight. Gives unique aerial display flight in breeding territory. Nests in reeds. Clutch size: 5-6.

IUCN 红色名录 近危 (NT) 2016.10.1

崖沙燕	*Riparia riparia*	Sand Martin	燕科 Hirundinidae

北京市二级保护野生动物 | IUCN 红色名录　无危 (LC) 2016.10.1

形态特征　小型鸣禽，体长 11~14 厘米。雌雄同色。成鸟喙黑色；头、上体及两翼皆为灰褐色；尾羽浅叉形；下体白色，胸具显著的灰褐色横带；跗跖褐色。幼鸟羽色大致似成鸟，但上体各羽具淡棕色羽缘，且颏、喉为淡皮黄色。

分布与生境　中国广泛分布。主要活动于接近水域的沙滩、沙丘或河床附近。在北京为区域性常见旅鸟。

习性与生态　喜集群活动。在河岸的沙质悬崖上凿洞为巢；窝卵数 4~6 枚。

Morphological characteristics A small passerine (11-14 cm). Sexes alike. Adult has black bill; plain greyish-brown from head to upperparts, wings and forked tail; underparts mostly white with distinctive greyish-brown breast band; tarsi brown. Juvenile resembles adult but has pale brown fringes on upperparts and pale buff chin/throat.

Distribution and habitat Widespread in China. Locally common passage migrant in Beijing. Usually seen around riverbanks and other sandy habitats near water.

Behaviour and ecology Usually in flocks. Excavates tunnel in vertical sand banks for nest. Clutch size: 4-6.

家燕	*Hirundo rustica*	Barn Swallow	燕科 Hirundinidae

形态特征　中型鸣禽，体长 15~19 厘米。雌雄相似。成鸟喙黑色；前额栗红色，头及上体深蓝色，具金属光泽；翼上覆羽与上体同色，飞羽黑褐色；尾羽深叉形，最外侧尾羽最长、中央尾羽最短，所有尾羽均具白色次端斑；颏、喉及上胸栗红色，其下有一黑色胸带，下体余部白色，但 *tytleri* 亚种下体为橙红色。幼鸟似成鸟，但尾叉较浅。

分布与生境　广泛分布于中国各省。活动于较低海拔的城市、郊野中各种开阔生境，常聚集在屋舍和湿地附近。在北京为甚常见夏候鸟和旅鸟。

习性与生态　喜集群活动，特别是迁徙时常集大群。营巢于屋舍内外的顶棚、墙壁上；为泥制碗状巢；窝卵数 4~5 枚。

Morphological characteristics　A medium-sized passerine (15-19 cm). Sexes alike. Adult has black bill; forehead red; glossy dark blue from head to upperparts and upperwing-coverts; flight feathers blackish-brown; tail deeply forked with white subterminal band; chin and throat red, with a distinctive broad black border below; remainder of underparts white. Subspecies *tytleri* has orange-red underparts. Juvenile similar to adult but has slightly or moderately forked tail.

Distribution and habitat　Widespread in China. Abundant summer breeder and passage migrant in Beijing. Seen in both urban and lowland rural areas. Usually near wetlands or buildings.

Behaviour and ecology　Forms very large flocks on migration. Makes cup-like nest under roof or eaves of buildings. Nest built with mud pellets. Clutch size: 4-5.

北京市二级保护野生动物 | IUCN 红色名录　无危 (LC) 2018.11.5

幼鸟 Juvenile

岩燕 | *Ptyonoprogne rupestris* | Eurasian Crag Martin | 燕科 Hirundinidae

北京市二级保护野生动物

IUCN 红色名录　无危 (LC) 2016.10.1

形态特征　小型鸣禽，体长 13~16 厘米。雌雄同色。喙黑色。头、上体、两翼及尾羽褐色，尾呈浅凹形，具白色次端斑。颏、喉、胸淡皮黄白色或淡灰褐色，腹部及尾下覆羽棕褐色。跗跖甚短，淡黄褐色。与崖沙燕相似，但本种无褐色胸带。

分布与生境　广泛分布于中国除东北和东南外的大部分地区。一般活动于山地和河谷，特别是多悬崖峭壁之处。在北京为不常见留鸟和旅鸟。

习性与生态　喜集小群于山间轻快地飞行。营巢于岩壁缝隙中；窝卵数 3~5 枚。

Morphological characteristics A small passerine (13-16 cm). Sexes alike. Bill black; plain brown from head to upperparts, wings and tail; tail is slightly forked with white spots near tip; chin, throat and breast buff-white or pale greyish-brown; belly and undertail-coverts brown; tarsi very short and brown. Resembles Sand Martin but lacks dark band on breast.

Distribution and habitat Widespread in China, except for the northeast and southeast. Uncommon resident and passage migrant in Beijing. Usually seen in montane areas and river valleys, especially those with cliffs and crags.

Behaviour and ecology Usually in small flocks. Flight light and agile. Nests in crevices on cliffs. Clutch size: 3-5.

毛脚燕	*Delichon urbicum*	Common House Martin	燕科 Hirundinidae

形态特征 小型鸣禽，体长 13~15 厘米。雌雄同色。喙黑色。头、后颈至背部深蓝色，具显著的金属光泽。两翼大致为黑褐色，腰及尾上覆羽白色。尾羽黑褐色，呈叉形。下体全为白色，翼下覆羽灰色。跗跖被白色绒羽。

分布与生境 分布于中国东部及西北地区。一般活动于近湿地的各种开阔生境。在北京为不常见旅鸟。

习性与生态 多集群活动。营巢于岩壁、洞穴或人类建筑物上；巢的主要材料为泥球，开口较小，一般位于巢的顶部或侧顶部；窝卵数 4~6 枚。

Morphological characteristics A small passerine (13-15 cm). Sexes alike. Head, hindneck and mantle glossy dark blue; wings largely blackish-brown; rump and uppertail-coverts white; forked tail blackish-brown; all-white below; underwing-coverts grey; tarsi feathered white.

Distribution and habitat Northwest and eastern China. Uncommon passage migrant in Beijing. Usually occurs in open country near wetlands.

Behaviour and ecology In flocks. Nests on cliffs, in caves or under eaves of buildings. Nest is made of mud pellets, with a small entrance hole at the top. Clutch size: 4-6.

北京市二级保护野生动物 | IUCN 红色名录 无危 (LC) 2016.10.1

烟腹毛脚燕	*Delichon dasypus*	Asian House Martin	燕科 Hirundinidae

北京市二级保护野生动物

IUCN 红色名录　无危 (LC) 2016.10.1

形态特征　小型鸣禽，体长 12~13 厘米。雌雄同色。喙黑色。头顶至背部为金属蓝色。腰白色，具不甚显著的深褐色羽干纹，两翼和尾羽黑褐色，尾呈浅叉形。下体为淡烟灰色，翼下覆羽黑灰色。跗跖被绒羽。与毛脚燕相似，但本种尾叉更浅，接近方形，且下体和翼下覆羽羽色较深。

分布与生境　中国广泛分布（除西北地区外）。在北京为不常见旅鸟和夏候鸟，一般活动于山谷等有悬崖峭壁之处。

习性与生态　常成群在山谷间轻快地飞行。营巢于石壁缝隙中；巢材和形状与毛脚燕相似；窝卵数 3~5 枚。

Morphological characteristics　A small passerine (12-13 cm). Sexes alike. Bill black; glossy steel blue from crown to mantle; rump white with narrow dark shafts; wings and slightly forked tail blackish-brown; underparts largely pale grey-white; underwing-coverts blackish-grey; tarsi feathered white. Closely resembles Common House Martin but has more deeply forked tail, darker underparts and contrastingly dark underwing-coverts.

Distribution and habitat　Widespread in China except for the northwest. Uncommon passage migrant and summer breeder in Beijing. Usually seen in montane areas with cliffs and crags.

Behaviour and ecology　In flocks. Flight fast and agile. Nest resembles that of Common House Martin. Clutch size: 3-5.

金腰燕	*Cecropis daurica*	Red-rumped Swallow	燕科 Hirundinidae

形态特征 中型鸣禽，体长16~20厘米。雌雄相似。喙黑色。头顶及背部深蓝色，具金属光泽。腰砖红色或橙黄色。尾羽黑褐色，较长，呈深叉形。头两侧淡棕色，具深色羽干纹。后颈栗棕色，喉、胸、腹部淡棕白色，具清晰的深色纵纹，尾下覆羽蓝黑色。跗跖较短，黑色。

分布与生境 广泛分布于中国各省。活动于各种开阔区域及城市中。在北京为甚常见夏候鸟和旅鸟。

习性与生态 繁殖期和非繁殖期皆喜集群活动。营巢于屋舍、桥梁等建筑物上；巢为泥制，呈瓶状，开口较小；窝卵数4~6枚。

Morphological characteristics A medium-sized passerine (16-20 cm). Sexes alike. Bill black; crown and upperparts glossy dark blue; rump orange-red or brick-red; blackish tail long and deeply forked; head-sides pale brown with dark shafts; hindneck rufous-brown; off-white from throat and breast to belly, heavily streaked dark; undertail-coverts blackish-blue; tarsi short and black.

Distribution and habitat Widespread in China. Abundant summer breeder and passage migrant in Beijing. Found in both rural open country and urban areas.

Behaviour and ecology In flocks in both breeding and non-breeding seasons. Nests under roof or eaves of human buildings. Nest bottle-shaped, consists of mud pellets with a small tunnel entrance. Clutch size: 4-6.

北京市二级保护野生动物 | IUCN 红色名录 无危 (LC) 2016.10.1

黄额燕	*Petrochelidon fluvicola*	Streak-throated Swallow	燕科 Hirundinidae

IUCN 红色名录　无危 (LC) 2016.10.1

形态特征　小型鸣禽，体长约 12 厘米。雌雄相似。喙黑色。头顶橙棕色。上体蓝黑色，略具金属光泽。腰和尾上覆羽淡棕色。尾羽黑褐色，呈浅叉形。颊、颏、喉至胸淡黄白色，密布黑色短纵纹，下体余部白色。跗跖黑色。

分布与生境　主要分布于南亚，中国仅见于北京。在北京为甚罕见迷鸟，仅于 2014 年 5 月在昌平沙河水库有一次记载。

习性与生态　喜集群活动或与其他燕类混群，飞行时常发出似"chi chi"的颤音。

Morphological characteristics　A small passerine (12 cm). Sexes alike. Bill black; crown orange-brown; upperparts blackish-blue, slightly glossed; rump and uppertail-coverts pale brown; tail blackish-brown, slightly forked; cheek, chin, throat and breast buff-white, heavily streaked black; remainder of underparts white; tarsi black.

Distribution and habitat　Mainly South Asia. Very rare vagrant in Beijing with only one record at Shahe Reservoir, Changping District in May 2014. This is also the only record of Streak-throated Swallow in China.

Behaviour and ecology　Usually in flocks or joins mixed flocks of other hirundines. Gives a flight call "chi chi".

白头鹎	*Pycnonotus sinensis*	Light-vented Bulbul	鹎科 Pycnonotidae

形态特征 中型鸣禽，体长 18~22 厘米。雌雄同色。喙黑色。头顶、眼先及颊黑色，眼后至枕部白色，耳羽后方白色或灰白色。上体灰绿色，两翼和尾羽褐色，具鲜艳的黄绿色羽缘。颏、喉白色，胸灰褐色，腹部及尾下覆羽白色。跗跖黑色。

分布与生境 中国见于华北及其以南地区。一般活动于较低海拔的开阔林地和灌丛。北京为常见留鸟，在北京城区（如公园和高校）和郊区皆较常见。

习性与生态 性活跃而擅鸣，常集 3~10 只的小群。一般营巢于灌丛或树上；巢为杯状，由细树枝、纤维和草等编制而成；窝卵数 3~5 枚。

Morphological characteristics A medium-sized passerine (18-22 cm). Sexes alike. Bill black; crown, lores and cheek black; white patch from rear of eye to nape; ear-coverts white or pale greyish-white; upperparts greyish-green; wings and tail brown, with distinctive yellowish-green fringes; chin and throat white; breast greyish-brown; belly and undertail-coverts white; tarsi black.

Distribution and habitat North, central and southern China. Common resident in Beijing. Occurs in lowland open woodlands and scrub. Common in urban Beijing (e.g. parks and universities) and in rural areas.

Behaviour and ecology Active and vocal. Usually in small flocks (3-10 individuals). The nest is a shallow cup of twigs, grass and coiled fibres. Usually nests in shrubs and trees. Clutch size: 3-5.

北京市二级保护野生动物 | IUCN 红色名录 无危 (LC) 2018.8.9

成鸟 Adult

幼鸟 Juvenile

栗耳短脚鹎	*Hypsipetes amaurotis*	Brown-eared Bulbul	鹎科 Pycnonotidae

IUCN 红色名录　无危 (LC) 2018.8.9

形态特征　中型鸣禽，体长 27~28 厘米。雌雄同色。喙较长，黑色。头、颈灰色，具较短的白色纵纹，耳羽栗棕色并向颈侧略有延伸。上体、两翼和尾羽灰色或灰褐色，下体灰色，具不甚清晰的白色斑点或短纵纹。尾下覆羽黑色，羽缘白色。跗跖灰褐色。

分布与生境　中国主要见于东部沿海地区。在北京为罕见冬候鸟，见于较低海拔的林地和林缘，亦见于植被丰富的城市公园，如圆明园、北京植物园、奥林匹克森林公园等。

习性与生态　单只、成对或集 10 只以内的小群活动。性活泼，鸣声响亮而嘈杂。营巢于茂密的树上；窝卵数 4~5 枚。

Morphological characteristics　A medium-sized passerine (27-28 cm). Sexes alike. Bill long and black; head and neck grey, with short white streaks; brown from ear-coverts to upper neck-sides; upperparts, wings and tail grey or greyish-brown; underparts grey with indistinct white spots or streaks; undertail-coverts black, fringed white; tarsi greyish-brown.

Distribution and habitat　Mainly along coastal areas in eastern China. Rare winter visitor in Beijing. Usually seen in lowland woodland or woodland edge. Also seen in well-vegetated urban parks such as the Old Summer Palace, the Botanical Gardens and the Olympic Forest Park.

Behaviour and ecology　Alone, in pairs or in small flocks (usually less than 10 individuals). Active and vocal. Call is loud and shrill. Nests in dense trees. Clutch size: 4-5.

VI 鸣禽

叽喳柳莺	*Phylloscopus collybita*	Common Chiffchaff	柳莺科 Phylloscopidae

形态特征 小型鸣禽，体长 11~13 厘米。雌雄相似。喙较细，呈黑色，部分个体下喙基部粉色。眉纹淡皮黄色，贯眼纹近黑色。头顶、上体和两翼橄榄褐色（新羽）或灰褐色（旧羽），无翼斑，初级飞羽突出次级飞羽的部分不显著。颏、喉白色，下体余部淡黄白色。跗跖黑褐色。

分布与生境 中国主要见于西部，于东部仅有零星记载。活动于海拔 2000 米以下的林地、灌丛和苇丛。在北京为甚罕见迷鸟。

习性与生态 非繁殖期常单独在浓密的植被中活动。鸣声极具特点，为节奏轻快的“chiff-chaff chiff-chaff”声。营巢于灌丛或草丛中近地面处；窝卵数 4~7 枚。

Morphological characteristics A small passerine (11-13 cm). Sexes alike. Bill fine and black; some individuals are pink at base of lower mandible; supercilium pale buff-white; eye-stripe blackish; crown, upperparts and wings olive-brown (fresh) or greyish-brown (worn), lacking wingbar; short primary projection; chin and throat white; remainder of underparts off-white; tarsi blackish-brown.

Distribution and habitat Widespread in western China. Scarce in eastern areas. Very rare vagrant in Beijing. Seen in forests, scrub and reedbeds below 2,000 m.

Behaviour and ecology In non-breeding season usually skulks in dense vegetation. Song fast and repetitive "chiff-chaff chiff-chaff". Nests close to the ground in shrubs and grass. Clutch size: 4-7.

IUCN 红色名录 无危 (LC) 2016.10.1

林柳莺	*Phylloscopus sibilatrix*	Wood Warbler	柳莺科 Phylloscopidae

IUCN 红色名录　无危 (LC) 2016.10.1

形态特征　小型鸣禽，体长 11~13 厘米。雌雄相似。上喙黑褐色，下喙粉色。眉纹柠檬黄色，贯眼纹深褐色。头顶和上体为鲜艳的黄绿色。两翼黑褐色，具黄绿色羽缘，无翼斑，两翼较长，初级飞羽显著突出于次级飞羽。尾羽黑褐色，具较窄的绿色羽缘。颏、喉及上胸柠檬黄色，下体余部白色。跗跖褐色。

分布与生境　中国主要见于西部地区，东部记录极少。活动于海拔 1500 米以下的各类林地。在北京为甚罕见迷鸟，仅 2018 年 9 月在延庆野鸭湖有一次记载。

习性与生态　性活跃，常单独或成对活动。鸣声为响亮而连续的“chi chi chi chi……”声，节奏逐渐加快直至一连串模糊的颤音。营巢于隐蔽的林下地面；窝卵数 5~7 枚。

Morphological characteristics　A small passerine (11-13 cm). Sexes alike. Upper mandible blackish; lower mandible pink; supercilium long, lemon-yellow; eye-stripe dark brown; crown and upperparts yellowish-green; wings blackish-brown, with yellowish-green fringes and without wingbars; long primary projection; tail blackish-brown, outer vane narrowly fringed green; chin, throat and upper breast lemon-yellow; remaining underparts white; tarsi brown.

Distribution and habitat　Mainly western China. Rare in eastern China. Occurs in woodlands below 1,500 m. Very rare vagrant in Beijing with only one record at Yeyahu, Yanqing District in September 2018.

Behaviour and ecology　Active. Usually alone or in pairs. Song repetitive notes "chi chi chi chi…", accelerating to a trill. Nests on the ground in woodland. Clutch size: 5-7.

褐柳莺	*Phylloscopus fuscatus*	Dusky Warbler	柳莺科 Phylloscopidae

形态特征 小型鸣禽，体长 11~12 厘米。雌雄相似。喙尖细，呈黑灰色，下喙基部粉色。眉纹窄而长，棕白色，甚为清晰。贯眼纹暗褐色，颊棕白色。头顶、上体及两翼褐色，无翼斑，尾羽暗褐色。颏、喉白色或棕白色，下体余部淡灰褐色，两胁和尾下覆羽褐色较浓。跗跖褐色。

分布与生境 广泛分布于中国各省。活动于平原至山区的开阔林地、灌丛，亦见于城区公园、高校等林地。在北京为区域性常见旅鸟。

习性与生态 常在浓密的林地和灌丛下层及地面活动，边在植被中穿梭，边发出单调似"da da da"的鸣声。于灌丛、高草丛中营巢；窝卵数 4~6 枚。

Morphological characteristics A small passerine (11-12 cm). Sexes alike. Bill fine, blackish-grey with pink lower mandible base; supercilium narrow, long and off-white; eye-stripe dark brown; cheek off-white; plain brown from crown to upperparts and wings; lacks wingbar; tail dark brown; chin and throat white or off-white; remainder of underparts pale greyish-brown, darker on flanks and undertail-coverts; tarsi brown.

Distribution and habitat Widespread in China. Locally common passage migrant in Beijing. Occurs in lowland and montane woodland and scrub; seen also in woodlands in urban parks and university campuses.

Behaviour and ecology Usually skulks in dense, low vegetation, sometimes on the ground. Call a dry "da da da". Nests in shrubs and tall grass. Clutch size: 4-6.

北京市二级保护野生动物		IUCN 红色名录 无危 (LC) 2016.10.1

棕腹柳莺 *Phylloscopus subaffinis* Buff-throated Warbler 柳莺科 Phylloscopidae

IUCN 红色名录　无危 (LC) 2016.10.1

形态特征　小型鸣禽，体长 10~12 厘米。雌雄相似。喙黑色，下喙基部淡黄褐色。眉纹淡黄色，贯眼纹暗褐色。背部橄榄褐色，两翼和尾羽暗褐色，各羽皆具显著的橄榄绿色羽缘。颏、喉淡黄色，下体余部淡棕黄色或淡皮黄色。跗跖褐色。

分布与生境　中国主要见于中部及南方地区。主要分布于海拔 900~3600 米的山地森林，非繁殖期亦见于海拔较低处。在北京为甚罕见迷鸟，仅于 2014 年 9 月在温榆河有一笔记载。

习性与生态　多单独或成对在植被中下层活动，非繁殖期行为隐蔽。鸣声为单调重复的“di di di”声。

Morphological characteristics A small passerine (10-12 cm). Sexes alike. Bill fine, mostly black with pale brownish-yellow at base of lower mandible; supercilium pale yellow; eye-stripe dark brown; mantle dark olive-brown; wings and tail dark brown with distinctive olive-green fringes; chin and throat pale yellow; underparts pale brownish-yellow or pale buff; tarsi brown.

Distribution and habitat Central and southern China. Breeds in montane forest at 900-3600 m; descends in winter. Very rare vagrant in Beijing with only one record at the Wenyu River in September 2014.

Behaviour and ecology Usually alone or in pairs. Shy in non-breeding season. Skulks in dense low vegetation. Song a soft repetitive "di di di".

棕眉柳莺 | *Phylloscopus armandii* | Yellow-streaked Warbler | 柳莺科 Phylloscopidae

形态特征 小型鸣禽，体长 11~13 厘米。雌雄相似。喙黑灰色，下喙基部黄色。眉纹棕白色，较长且宽，甚为清晰。贯眼纹褐色，颊淡灰褐色。头顶至背部褐色，腰呈橄榄褐色，两翼和尾羽暗褐色，羽缘为橄榄褐色。下体近白色，喉、胸、腹具模糊的黄色纵纹，两胁淡黄绿色，尾下覆羽皮黄色。跗跖褐色。

分布与生境 分布于中国华北、中部及西南地区。在北京为区域性常见夏候鸟，活动于西、北部山区海拔 1200 米以上的森林和灌丛。

习性与生态 常单独或成对活动。鸣叫声为尖细的"zi zi"声；鸣啭响亮而有节奏，似"zi zi di dilidili"声。营巢于林地灌丛；窝卵数 4~5 枚。

Morphological characteristics A small passerine (11-13 cm). Sexes alike. Bill blackish-grey with yellow at base of lower mandible; supercilium broad, long and off-white; eye-stripe brown; cheek pale greyish-brown; plain brown from crown to mantle; rump olive-brown; wings and tail dark brown with olive-brown fringes; underparts largely off-white; throat, breast and belly have indistinct pale yellow streaks; flanks pale greenish-yellow; undertail-coverts buff; tarsi brown.

Distribution and habitat North, central and southwest China. Locally common summer breeder in Beijing. Usually occurs in woodland and scrub above 1,200 m in the north and west of Beijing.

Behaviour and ecology Solitary or in pairs. Call a short "zi zi". Song a loud, melodious "zi zi di dilidili". Nests in scrub. Clutch size: 4-5.

北京市二级保护野生动物 | IUCN 红色名录 无危 (LC) 2016.10.1

巨嘴柳莺	*Phylloscopus schwarzi*	Radde's Warbler	柳莺科 Phylloscopidae

北京市二级保护野生动物

IUCN 红色名录　无危 (LC) 2016.10.1

形态特征　小型鸣禽，体长 11~14 厘米。雌雄相似。喙较为粗钝，喙峰暗褐色，余部粉色。眉纹较长，呈皮黄色，前端模糊，但自眼后方开始较为清晰。贯眼纹橄榄褐色。头顶、上体及尾羽橄榄褐色，两翼具橄榄绿色羽缘。下体大致为棕白色，两胁淡褐色，尾下覆羽皮黄色。跗跖褐色。

分布与生境　广布于除青藏高原外的中国大部分地区。多见于丘陵至平原地区的林地和灌丛。在北京为区域性常见旅鸟。

习性与生态　行为隐秘，常在浓密的植被下层活动。鸣叫声为单调的“cha cha cha”声。营巢于灌丛或草丛中；窝卵数 4~6 枚。

Morphological characteristics A small passerine (11-14 cm). Sexes alike. Bill relatively short and blunt; mostly pink but culmen dark brown; supercilium long, buff and indistinct before eye; eye-stripe olive-brown; plain olive-brown from crown to upperparts and tail; wings fringed olive-green; underparts largely off-white; flanks pale brown; undertail-coverts buff; tarsi brown.

Distribution and habitat Widespread in China except for the Tibetan Plateau. Locally common passage migrant in Beijing. Seen in woodland and scrubby areas in lowlands and low hills.

Behaviour and ecology Usually shy. Skulks in dense vegetation in non-breeding season. Call a repetitive "cha cha cha". Nests in shrubs or on the ground in grass. Clutch size: 4-6.

灰喉柳莺	*Phylloscopus maculipennis*	Ashy-throated Warbler	柳莺科 Phylloscopidae

形态特征 小型鸣禽，体长 8~10 厘米。雌雄相似。喙黑色。具模糊的白色顶冠纹、深灰色侧冠纹和长而清晰的近白色眉纹，贯眼纹黑灰色，颊浅灰色。上体橄榄绿色，两翼暗褐色，具橄榄绿羽缘，大覆羽和中覆羽具柠檬黄色端斑，形成两道翼斑。腰和尾上覆羽淡黄色，尾羽大致为暗褐色，羽缘橄榄绿色，但最外侧三对尾羽几乎全为白色。颏、喉、颊及上胸灰色，下体余部淡黄色。跗跖褐色。

分布与生境 有限分布于中国西南地区。一般活动于海拔 2000~3500 米的森林和灌丛，非繁殖期向海拔较低处迁移。北京为甚罕见迷鸟，仅在 2019 年 5 月于百花山景区海拔约 2000 米处有一次记载。

习性与生态 非繁殖期多集小群活动，性活跃，鸣叫声为单调的“zi zi”声。

Morphological characteristics A small passerine (8-10 cm). Sexes alike. Bill black; indistinct white median crown-stripe, dark grey lateral crown-stripe and long, off-white supercilium; eye-stripe blackish-grey; cheek pale grey; upperparts olive-green; wings dark brown, with olive-green fringes; greater coverts and median coverts have lemon-yellow tips, forming two prominent wingbars; rump and uppertail-coverts pale yellow; tail largely dark brown, fringed olive-green; tail-sides white; chin, throat and upper breast plain grey; remaining underparts pale yellow; tarsi brown.

Distribution and habitat Restricted in the southwest of China. Usually occurs in woodland and scrubby habitats at 2,000-3,500 m. Descends in non-breeding season. Very rare vagrant in Beijing with only one record at Baihuashan Scenic Area (ca. 2,000 m) in May 2019.

Behaviour and ecology Usually in small flocks in non-breeding season. Active. Call a soft "zi zi".

IUCN 红色名录 无危 (LC) 2016.10.1

云南柳莺	*Phylloscopus yunnanensis*	Chinese Leaf Warbler	柳莺科 Phylloscopidae

北京市二级保护野生动物		IUCN 红色名录 无危 (LC) 2016.10.1

形态特征 小型鸣禽，体长 9~10 厘米。雌雄相似。喙黑色，下喙基部褐色。具清晰的灰白色顶冠纹、暗灰色侧冠纹和长且宽的皮黄色眉纹，贯眼纹暗褐色。上体灰绿色，具两道淡黄白色翼斑，两道翼斑之间的区域（即大覆羽基部）为黑褐色，腰淡黄色。下体大致为白色。跗跖褐色。

分布与生境 中国见于华北、中部至西南地区。活动于海拔 1000 米以上的针叶林、阔叶林和混交林中。在北京为区域性常见夏候鸟。

习性与生态 非繁殖期集小群，繁殖期常立于松树顶端鸣啭，鸣声为清脆而有节奏的"si-ri-di si-ri-di si-ri-di"。营巢于林下地面或近地面之处；窝卵数 4 枚左右。

Morphological characteristics A small passerine (9-10 cm). Sexes alike. Bill mostly black, with brown at base of lower mandible; distinctive greyish-white median crown-stripe, dark greyish lateral crown-stripe and long, broad yellowish-buff supercilium; eye-stripe dark brown; upperparts greyish-green; two prominent pale yellowish-white wingbars separated by a blackish-brown patch (i.e. at the base of greater coverts); rump pale yellow; underparts off-white; tarsi brown.

Distribution and habitat North, central and southwest China. Locally common summer breeder in Beijing. Usually found in deciduous forests, coniferous forests and broadleaf and mixed forests above 1,000 m.

Behaviour and ecology In small flocks in non-breeding season. In breeding season, often sings on top of conifer trees. Song a repetitive "si-ri-di si-ri-di si-ri-di". Nests on the ground or in low vegetation in forests. Clutch size: 4.

黄腰柳莺	*Phylloscopus proregulus*	Pallas's Leaf Warbler	柳莺科 Phylloscopidae

形态特征 小型鸣禽，体长 8~10.5 厘米。雌雄相似。喙黑色。顶冠纹黄色，侧冠纹暗绿色。眉纹宽阔，呈柠檬黄色，甚为鲜艳，贯眼纹暗绿色。上体橄榄绿色，具两道黄色翼斑，飞羽和尾羽黑褐色，具橄榄绿色羽缘，三级飞羽端部白色，腰柠檬黄色。下体白色或灰白色，尾下覆羽略沾淡黄色。跗跖褐色。

分布与生境 广布于中国各省。见于低山至平原（包括城区）的各类林地和灌丛，在北京为常见旅鸟，是春季（4~5 月）和秋季（10 月）北京低海拔最常见的过境柳莺之一。

习性与生态 极活跃，迁徙时常集小群在树冠层不断跳跃，亦常悬停；并发出轻柔的鼻音“zu-wi”声。一般营巢于针叶林树上；窝卵数 5~6 枚。

Morphological characteristics A small passerine (8-10.5 cm). Sexes alike. Bill black. Distinctive yellow median crown-stripe, deep green lateral crown-stripe and broad lemon-yellow supercilium; upperparts and wings largely olive-green, with two prominent yellow wingbars; flight feathers and tail largely blackish-brown with olive-green fringes; tertials tipped white; rump lemon-yellow; underparts white or pale greyish-white; undertail-coverts sometimes have pale yellow wash; tarsi brown.

Distribution and habitat Widespread in China. Common passage migrant in Beijing. Usually seen in lowland and low-elevation woodlands and scrub; one of the commonest warbler species in lowland Beijing in April, May and October.

Behaviour and ecology Active and energetic. Often hover-gleans in canopy. Forms small flocks on migration. Call a soft and nasal "zu-wei". Usually nests in conifer trees. Clutch size: 5-6.

北京市二级保护野生动物　　IUCN 红色名录　无危 (LC) 2016.10.1

黄眉柳莺	*Phylloscopus inornatus*	Yellow-browed Warbler	柳莺科 Phylloscopidae

北京市二级保护野生动物		IUCN 红色名录 无危 (LC) 2018.11.21

形态特征 小型鸣禽，体长 9~11 厘米。雌雄相似。喙黑灰色，下喙基部褐色；头顶暗橄榄绿色，部分个体具模糊的浅色顶冠纹，具较长的淡黄白色眉纹和暗绿色贯眼纹。上体橄榄绿色（新羽绿色较鲜艳，旧羽则偏灰绿色），翼上具两道清晰的淡黄白色翼斑，飞羽黑褐色，羽缘橄榄绿色，三侧飞羽端部白色，腰与上体同色。下体灰白色。跗跖褐色。

分布与生境 中国除西部外广泛分布。见于城区和郊区各类林地，在北京为常见旅鸟，是春季（4~5 月）和秋季（9 月）北京低海拔最常见的过境柳莺之一。

习性与生态 迁徙时集松散的小群在树冠层活动，并发出音调上挑的“zi-wi”鸣叫声。营巢于树上或地上；窝卵数 5~6 枚。

Morphological characteristics A small passerine (9-11 cm). Sexes alike. Bill blackish-grey with brown at base of lower mandible; crown deep olive-green; some individuals have indistinct pale median crown-stripe; supercilium long, pale yellowish-white; eye-stripe dark green; upperparts olive-green (fresh) or greyish-green (worn); two prominent pale creamy wingbars; remiges blackish-brown, fringed olive-green; tertials have white tips; rump concolourous with upperparts; underparts whitish; tarsi brown.

Distribution and habitat Widespread except for western China. Common passage migrant in Beijing. Found in woodlands in both urban and rural areas. It is one of the commonest warblers in lowland areas of Beijing in April, May and September.

Behaviour and ecology Forms small, loose flocks on migration. Usually active in canopy. Call a high-pitched "zi-wi". Nests on the ground or in dense trees. Clutch size: 5-6.

淡眉柳莺 *Phylloscopus humei* Hume's Leaf Warbler 柳莺科 Phylloscopidae

形态特征 小型鸣禽，体长 10~11 厘米。雌雄相似。喙黑色。头顶暗灰绿色，眉纹淡黄白色，贯眼纹暗绿色。上体灰绿色，两翼黑褐色，具狭窄的绿色边缘和两道淡皮黄色翼斑（由中覆羽浅色端部形成的翼斑常不甚清晰），三级飞羽具狭窄的白色尖端，腰与上体同色。下体灰白色或淡皮黄色。跗跖黑褐色。

分布与生境 分布于中国西部、西南、中部至华北地区。在北京为不常见夏候鸟，繁殖于北京山区海拔 1400 米以上的林地，迁徙时亦见于较低海拔地区。

习性与生态 多单独或成对活动，鸣声为拖长的降调“zi……”声。营巢于浓密的灌丛和矮树上；窝卵数 4~5 枚。

Morphological characteristics A small passerine (10-11 cm). Sexes alike. Bill black; crown dark greyish-green; supercilium yellowish-white; eye-stripe dark green; upperparts greyish-green; wings largely blackish-brown with narrow green fringes and two pale buff wingbars (note the median covert wingbar, which is formed by pale tips to the median coverts, is often less distinctive); tertials have very narrow whitish tips; rump concolourous with upperparts; underparts pale buff-white; tarsi blackish-brown.

Distribution and habitat Mainly west, southwest, central and north China. Uncommon and local summer breeder in Beijing. Breeds in forest in montane areas of Beijing (above 1,400 m).

Behaviour and ecology Usually alone or in pairs. Gives drawn-out, insect-like call "zi...". Nests in dense shrubs or trees. Clutch size: 4-5.

IUCN 红色名录 无危 (LC) 2016.10.1

极北柳莺	*Phylloscopus borealis*	Arctic Warbler	柳莺科 Phylloscopidae

北京市二级保护野生动物 IUCN 红色名录 无危 (LC) 2016.10.1

形态特征 小型鸣禽，体长 11~13 厘米。雌雄相似。喙较长，呈黑灰色，下喙基部黄色。头顶暗灰绿色，眉纹较细，仅向前延伸至眼先上方而不至喙基，呈淡黄白色。上体及翼上覆羽大致为橄榄绿色或灰绿色，翼上通常仅具有一道近白色翼斑（由大覆羽浅色端部形成），有时有极不清晰的第二道翼斑。诸飞羽黑褐色，具狭窄的橄榄绿色外翈羽缘，三级飞羽无浅色端斑，腰与上体同色。下体白色。跗跖褐色。

分布与生境 除青藏高原外广泛分布于中国各省。多见于低山和平原的各类林地。于北京为区域性常见旅鸟。

习性与生态 迁徙季常集小群或与其他柳莺混群活动，多在树冠的上层和下层活动，鸣声为响亮的一长串“zi bi-bi-bi-bi”声。营巢于林地地面；窝卵数 5~6 枚。

Morphological characteristics A small passerine (11-13 cm). Sexes alike. Bill long, blackish-grey with yellow at the base of lower mandible; crown dark greyish-green; narrow pale buff-white supercilium reaches above lores; upperparts and upperwing-coverts mostly olive-green or greyish-green; usually has only one whitish wingbar (i.e. the greater coverts wingbar); sometimes also has an indistinct median coverts wingbar; flight feathers largely blackish with olive-green fringes; tertials lack pale tips; rump concolourous with upperparts; underparts white; tarsi brown.

Distribution and habitat Widespread in China except for the Tibetan Plateau. Locally common passage migrant in Beijing. Usually occurs in forests in lowlands and low hills on migration.

Behaviour and ecology In small flocks or joins mixed flocks with other warblers in lower and upper canopy. Call a series of loud and hard notes "zi bi-bi-bi-bi". Usually nests on the ground in forests. Clutch size: 5-6.

双斑绿柳莺	*Phylloscopus plumbeitarsus*	Two-barred Warbler	柳莺科 Phylloscopidae

形态特征 小型鸣禽，体长 11~12 厘米。雌雄相似。喙较长且粗厚。眉纹淡黄白色，较长，向前延伸至喙基且清晰，贯眼纹暗绿色。头顶及上体橄榄绿色，大覆羽和中覆羽端部黄白色，形成两道清晰的翼斑。飞羽黑褐色，具宽阔的橄榄绿色羽缘。尾羽暗褐色，羽缘橄榄绿色。下体白色。跗跖褐色。

分布与生境 中国除青藏高原外广泛分布。主要见于低山和平原的针叶林和针阔混交林中。在北京为不常见旅鸟。

习性与生态 性活跃，迁徙季节常集小群或与其他柳莺混群活动。鸣声通常为三声一度的"chi-li-li"声，但迁徙过境时通常较少发声。营巢于森林中近溪流的地面上；窝卵数 5~6 枚。

Morphological characteristics A small passerine (11-12 cm). Sexes alike. Bill long and thick; pale buff-white supercilium long (reaches bill) and prominent; eye-stripe dark green; olive-green from crown to upperparts; greater and median coverts have yellowish-white tips, forming two prominent wingbars; remiges and tail blackish-brown, fringed green; underparts white; tarsi brown.

Distribution and habitat Widespread in China, except for the Tibetan Plateau. Uncommon passage migrant in Beijing. Mainly seen in mixed and coniferous forest in lowlands and low hills.

Behaviour and ecology Active. Usually forms small, loose flocks or joins mixed flocks with other warblers. Call a three-note "chi-li-li", but less vocal than other warblers on migration. Nests on the ground near streams in forests. Clutch size: 5-6.

北京市二级保护野生动物 | IUCN 红色名录 无危 (LC) 2016.10.1

淡脚柳莺	*Phylloscopus tenellipes*	Pale-legged Leaf Warbler	柳莺科 Phylloscopidae

IUCN 红色名录　无危 (LC) 2016.10.1

形态特征　小型鸣禽，体长 11~12 厘米。雌雄相似。喙黑灰色，下喙基部黄褐色。头顶暗灰褐色，近白色的眉纹甚长且较宽阔，贯眼纹暗绿色。上体暗橄榄绿色或橄榄褐色，具一道或两道较细的淡黄色翼斑，有的个体无翼斑。下体大致为灰白色，部分个体尾下覆羽淡黄白色。跗跖粉灰色。

分布与生境　分布于中国东部及南部区域，迁徙过境时在沿海地区记录较为集中。在北京为不常见旅鸟，多见于平原各类林地。

习性与生态　迁徙时多单独或集松散的小群活动，一般见于林地和灌丛的中下层和地面。鸣声为一连串的快节奏“ji ji ji”声。营巢于林间近水源的地面上；窝卵数 4~6 枚。

Morphological characteristics A small passerine (11-12 cm). Sexes alike. Bill blackish-grey with yellowish-brown at base of lower mandible; crown deep greyish-brown; whitish supercilium long and broad; eye-stripe dark green; upperparts dark olive-green or dark olive-brown; usually shows one or two narrow, pale buff-white wingbars but some individuals lack wingbars altogether; underparts pale greyish-white; sometimes vent and undertail-coverts pale yellowish-white; tarsi greyish-pink.

Distribution and habitat Eastern and southern China. Commoner in coastal areas on migration. Uncommon passage migrant in Beijing. Usually occurs in lowland woodlands.

Behaviour and ecology Usually alone or in small, loose flocks on migration. Feeds in low canopy, in shrubs or on the ground. Song a repetitive "ji ji ji". Nests on the ground near water in forests. Clutch size: 4-6.

乌嘴柳莺	*Phylloscopus magnirostris*	Large-billed Leaf Warbler	柳莺科 Phylloscopidae

形态特征 小型鸣禽，体长 11~13 厘米。雌雄相似。喙黑色，下喙基部黄色。头顶暗灰绿色，无顶冠纹，具淡黄白色眉纹和较宽的暗色贯眼纹。上体为橄榄绿色，翼上具两道（新羽）或一道（旧羽）较细的皮黄色翼斑。下体白色。跗跖褐色。

分布与生境 中国主要见于西南部至中部。繁殖于海拔 1500~4000 米的林地，非繁殖期见于较低海拔至平原地区。罕见于北京，目前仅在北京大学和门头沟区东灵山有过记载，居留型尚不明确。

习性与生态 通常单只或成对在树冠层活动。鸣声为响亮而单调的“di di-di di-di”，五声一度。营巢于林间地面；窝卵数 4~6 枚。

Morphological characteristics A small passerine (11-13 cm). Sexes alike. Bill long and blackish with yellow at base of lower mandible; crown dark greyish-green, lacking median crown-stripe; supercilium pale yellowish-white; broad, dark eye-stripe; upperparts and wings olive-green, with one (worn) or two (fresh) narrow yellowish-white wingbars; underparts largely white; tarsi brown.

Distribution and habitat From southwest to central China. Breeds in forests at 1,500-4,000 m. Descends in winter. Very rare in Beijing with only two records from Peking University campus and Donglingshan, Mentougou District. Residential status: unclear.

Behaviour and ecology Alone or in pairs. Feeds in canopy. Gives loud and repetitive five-note song "di di-di di-di". Usually nests on the ground in woodlands. Clutch size: 4-6.

IUCN 红色名录 无危 (LC) 2018.8.9

冕柳莺 | *Phylloscopus coronatus* | Eastern Crowned Warbler | 柳莺科 Phylloscopidae

北京市二级保护野生动物 | IUCN 红色名录 无危 (LC) 2016.10.1

形态特征 小型鸣禽，体长 11~13 厘米。雌雄相似。喙较长且粗壮，上喙黑灰色，下喙黄色。头顶暗绿色，具模糊的白色顶冠纹，眉纹淡黄白色，贯眼纹暗绿色。上体和两翼橄榄绿色，两翼具一道较细的淡黄色翼斑。尾羽黑褐色，外翈羽缘呈黄绿色。下体近白色，尾下覆羽淡黄色。跗跖褐色。

分布与生境 分布于中国东部及南部地区。在北京为区域性常见夏候鸟及旅鸟，繁殖于北京西部和北部海拔 1100~1800 米的山地森林和灌丛，迁徙时亦见于平原林地。

习性与生态 多在林冠层活动。鸣声为响亮的三声一度“驾 - 驾 - 吉”。营巢于林地近溪流处的地面或近地面的矮树、灌丛中；窝卵数 4~7 枚。

Morphological characteristics A small passerine (11-13 cm). Sexes alike. Bill relatively large with blackish-grey upper mandible and yellow lower mandible; crown dark green, with faint white median crown-stripe; supercilium pale yellowish-white; eye-stripe dark green; upperparts and wings olive-green with one narrow pale yellow wingbar; tail blackish-brown with yellowish-green outer vanes; underparts off-white; undertail-coverts pale yellow; tarsi brown.

Distribution and habitat Widespread in eastern and southern China. Locally common summer breeder and passage migrant in Beijing. Breeds in woodland and scrub at 1,100-1,800 m in the north and west of Beijing. Seen also in lowland forests on migration.

Behaviour and ecology Usually feeds in canopy. Gives loud three-note repetitive song "jia-jia-ji". Nests on the ground or in dense low vegetation near the streams in forests. Clutch size: 4-7.

冠纹柳莺 | *Phylloscopus claudiae* | Claudia's Leaf Warbler | 柳莺科 Phylloscopidae

形态特征 小型鸣禽，体长 10~12 厘米。雌雄相似。喙较长，上喙黑灰色，下喙黄色或粉色。头顶暗绿色，具较长的黄白色顶冠纹和长而宽阔的淡黄色眉纹，贯眼纹深色。上体及两翼橄榄绿色，具两道清晰的淡黄色翼斑。下体白色。跗跖褐色。

分布与生境 分布于中国华北、中部至西南地区。在北京为区域性常见夏候鸟和旅鸟，繁殖于北京西部及北部海拔 1200 米以上的林地，非繁殖期迁至较低海拔处。

习性与生态 多在林冠层活动，但有时亦下至灌丛。鸣声为响亮的颤音，节奏似山雀。营巢于林间地面或靠近地面的树洞中；窝卵数 4~5 枚。

Morphological characteristics A small passerine (10-12 cm). Sexes alike. Bill relatively long with blackish-grey upper mandible and yellow or pink lower mandible; crown dark green with long yellow-white median crown-stripe; supercilium long and broad, pale yellowish; eye-stripe dark; upperparts and wings olive-green with two prominent pale yellow wingbars; underparts white; tarsi brown.

Distribution and habitat North, central and southwest China. Locally common summer breeder and passage migrant in Beijing. Breeds in forests above 1,200 m in the north and west of Beijing. Descends in non-breeding season.

Behaviour and ecology Usually feeds in canopy, sometimes also in shrubs. Song a series of loud, tit-like trills. Nests on the ground or in tree holes close to the ground in forest. Clutch size: 4-5.

北京市二级保护野生动物 | IUCN 红色名录 无危 (LC) 2016.10.1

黑眉柳莺	*Phylloscopus ricketti*	Sulphur-breasted Warbler	柳莺科 Phylloscopidae

IUCN 红色名录　无危 (LC) 2016.10.1

形态特征　小型鸣禽，体长 10~11 厘米。雌雄相似。喙较短，上喙黑灰色，下喙橙黄色。头部羽色甚为鲜艳且对比强烈，具柠檬黄色的顶冠纹和眉纹，以及黑色的侧冠纹和贯眼纹，颊淡黄色。上体和两翼黄绿色或鲜艳的橄榄绿色，具一道或两道黄色翼斑。下体柠檬黄色。跗跖褐色。

分布与生境　主要分布于中国南部地区。主要见于低山和平原的阔叶林和混交林中。在北京为甚罕见迷鸟。

习性与生态　性活跃，常在林冠层活动。鸣声响亮而尖细，似“zi zi bilibibili”声。通常营巢于林下地面的洞穴中；窝卵数 5~6 枚。

Morphological characteristics　A small passerine (10-11 cm). Sexes alike. Bill short with blackish-grey upper mandible and orange-yellow lower mandible; prominent, long lemon-yellow median crown-stripe and supercilium, contrasting with the black lateral crown-stripe and eye-stripe; cheek pale yellow; upperparts and wings yellow-green or bright olive-green with one or two yellowish wingbars; underparts lemon-yellow; tarsi brown.

Distribution and habitat　Mainly southern China. Occurs in mixed or broadleaved forests. Very rare vagrant in Beijing.

Behaviour and ecology　Active in canopy. Gives high-pitched song "zi zi bilibibili". Usually nests in roles in the ground in forests. Clutch size: 5-6.

比氏鹟莺 | *Seicercus valentini* | Bianchi's Warbler | 柳莺科 Phylloscopidae

形态特征 小型鸣禽，体长 10~11 厘米。雌雄相似。上喙黑灰色，下喙黄色。头顶具灰色的顶冠纹和黑色的侧冠纹，头两侧灰绿色，具黄色眼圈。上体及两翼内侧翼上覆羽橄榄绿色，大覆羽和各级飞羽近黑色，具宽阔的绿色羽缘，大覆羽具不显著的淡黄色端斑，形成一道翼斑。尾羽暗褐色，羽缘绿色，最外侧两对尾羽大部分为白色。下体大致为柠檬黄色。跗跖淡褐色。

分布与生境 分布于中国秦岭–淮河以南地区。繁殖期主要活动于海拔 1800 米以上的阔叶林中，非繁殖期亦见于较低海拔甚至平原地区。在北京为甚罕见迷鸟，仅于 2019 年 5 月在温榆河有一笔记载。

习性与生态 性活跃。鸣啭为节奏清晰的“chi-wi chi-wi”。

Morphological characteristics A small passerine (10-11 cm). Sexes alike. Upper mandible blackish-grey; lower mandible yellow; distinctive grey median crown-stripe and black lateral crown-stripe; head-sides greyish-green with yellow eye-ring; upperparts and inner upperwing-coverts olive-green; greater coverts and remiges blackish, with prominent green fringes; narrow pale yellow wingbar, formed by yellow tips to the greater coverts; tail dark brown, fringed green; two pairs of outer rectrices largely white; underparts mostly lemon-yellow; tarsi pale brown.

Distribution and habitat Mainly south of the Qinling–Huaihe Line in China. Breeds in broadleaved forests above 1,800 m. Descends in non-breeding season. Very rare vagrant in Beijing with only one record at the Wenyu River in May 2019.

Behaviour and ecology Active. Song a series of repetitive "chi-wi chi-wi".

北京市二级保护野生动物 | | IUCN 红色名录 无危 (LC) 2016.10.1

淡尾鹟莺	*Seicercus soror*	Plain-tailed Warbler	柳莺科 Phylloscopidae

北京市二级保护野生动物 *

IUCN 红色名录　无危 (LC) 2016.10.1

形态特征　小型鸣禽，体长 10~11 厘米。雌雄相似。上喙黑灰色，下喙黄色。顶冠纹灰色，侧冠纹黑色，眼圈淡黄色。头两侧及上体和两翼皆为橄榄绿色，尾羽黑褐色，羽缘绿色，外侧尾羽部分白色。下体柠檬黄色。跗跖浅褐色。甚似比氏鹟莺，但本种无翼斑，且外侧尾羽白色区域较小。

分布与生境　分布于中国华北、中部及南方地区。在北京为不常见夏候鸟，多见于门头沟东灵山、小龙门林场区域海拔 700~1500 米的中、低山阔叶林。

习性与生态　非繁殖期喜集群活动或与其他柳莺、绣眼等小型鸣禽混群。营巢于林下灌丛、草丛的地面或近地面处；窝卵数 3~4 枚。

● * 注：《北京市二级保护野生动物名录》未收录（更新 / 接受）淡尾鹟莺和比氏鹟莺的物种划分，故此处沿用比氏鹟莺的数据。

Morphological characteristics　A small passerine (10-11 cm). Sexes alike. Upper mandible blackish-grey; lower mandible yellow; median crown-stripe grey; lateral crown-stripe black; eye-ring yellow; olive-green from head-sides to upperparts and wings; tail blackish, fringed green; outer rectrices partly white; underparts lemon-yellow; tarsi pale brown. Closely resembles Bianchi's Warbler but has less white on outer rectrices and lacks wingbar.

Distribution and habitat　North, central and southern China. Uncommon summer breeder in Beijing. Usually occurs in broadleaved forest at 700-1,500 m, including at Donglingshan and Xiaolongmen in Mentougou District.

Behaviour and ecology　In non-breeding season tends to form small flocks or joins mixed flocks with other warblers or white-eyes. Nests in low shrubs and grass or on the ground in forests. Clutch size: 3-4.

栗头鹟莺	*Seicercus castaniceps*	Chestnut-crowned Warbler	柳莺科 Phylloscopidae

形态特征 小型鸣禽，体长 9~10 厘米。雌雄相似。上喙黑灰色，下喙橙黄色。头顶栗色，具不甚清晰的黑褐色侧冠纹，眼圈白色。头两侧、颈部及上背皆为灰色，下背橄榄绿色。两翼黑褐色，各羽具绿色羽缘，大覆羽和中覆羽具黄色端斑，形成两道清晰的翼斑。腰柠檬黄色，尾羽暗褐色，羽缘绿色，外侧尾羽内翈白色。颏、喉及上胸灰色，下胸、腹至尾下覆羽柠檬黄色。跗跖褐色。

分布与生境 主要分布于中国秦岭－淮河以南。主要活动于 2500 米以下的阔叶林林缘、竹林和灌丛。在北京为甚罕见迷鸟，于 2018 年 4 月在延庆官厅水库马场曾有记载。

习性与生态 性甚活跃，常与其他莺类混群在林冠层不同高度活动。营巢于林下地面；窝卵数 4~5 枚。

Morphological characteristics A small passerine (9-10 cm). Sexes alike. Upper mandible blackish-grey; lower mandible orange-yellow; crown chestnut, with indistinct blackish lateral crown-stripe; eye-ring white; grey from head-sides and neck to mantle; lower back olive-green; wings largely blackish-brown, with green fringes; yellow tips to the greater and median coverts form two prominent wingbars; rump lemon-yellow; tail dark brown, fringed green; inner vanes of outer rectrices white; chin, throat and upper breast grey; remainder of underparts lemon-yellow; tarsi brown.

Distribution and habitat Widespread south of the Qinling-Huaihe Line. Mainly in broadleaved forest edge, bamboo and shrub up to 2,500 m. Very rare vagrant in Beijing with only one record at Machang (Guanting Reservoir), Yanqing District in April 2018.

Behaviour and ecology Active and energetic. Often joins mixed flocks with other warbler species at different canopy levels. Nests on the ground in forests. Clutch size: 4-5.

IUCN 红色名录　无危 (LC) 2016.10.1

棕脸鹟莺	*Abroscopus albogularis*	Rufous-faced Warbler	树莺科 Cettiidae

IUCN 红色名录 无危 (LC) 2016.10.1

形态特征 小型鸣禽，体长 9~10 厘米。雌雄相似。喙峰黑灰色，余部粉色。头部大致为红棕色，具甚长的黑色侧冠纹，一直延伸至枕部。上体橄榄绿色，两翼黑褐色，羽缘绿色，无翼斑。下体大致为白色，仅喉为黑色。跗跖褐色。

分布与生境 主要分布于中国秦岭－淮河以南。主要活动于海拔 2500 米以下的竹林和阔叶林中。在北京为甚罕见迷鸟，在城市公园和门头沟小龙门有零星记录。

习性与生态 性活跃，常在林冠层集小群活动。鸣声为清脆似虫鸣的"li-li-li-li-li"声。通常营巢于竹子洞中；窝卵数 3~6 枚。

Morphological characteristics A small passerine (9-10 cm). Sexes alike. Bill pink, with blackish-grey culmen; head largely rufous-brown, with long black lateral crown-stripe; upperparts olive-green; wings largely blackish-brown, fringed green; lacks wingbar; distinctive black throat; remainder of underparts whitish; tarsi brown.

Distribution and habitat Mainly south of the Qinling-Huaihe Line. Occurs in bamboo and broadleaved forest up to 2,500 m. Very rare vagrant in Beijing. Scarce in a few urban parks and the Xiaolongmen Forest Park, Mentougou District.

Behaviour and ecology Very active and energetic. Often feeds in flocks in canopy. Song high-pitched, insect-like whistles "li-li-li-li-li". Usually nests in bamboo holes. Clutch size: 3-6.

远东树莺 | *Horornis canturians* | Manchurian Bush Warbler | 树莺科 Cettiidae

形态特征 中型鸣禽，体长 14~18 厘米。雌雄羽色相似，但雌鸟体型显著小于雄鸟。喙较长而健壮，呈黑灰色，下喙基部淡褐色。头顶红棕色，眉纹灰白色，贯眼纹暗褐色，头两侧及上体余部皆为褐色。颏、喉白色，胸、两胁、尾下覆羽为皮黄色，腹部中央灰白色。跗跖淡褐色。

分布与生境 广泛分布于中国东部、南部及中部地区。一般活动于海拔 1100 米以下的低山及丘陵的林地与灌丛。在北京为不常见夏候鸟和旅鸟。

习性与生态 一般单独或成对活动，常隐匿于浓密的树冠下层和灌丛中觅食及鸣啭。鸣啭声响亮而清脆，似“wu chi-wi-lu-er”。营巢于灌丛下部；窝卵数 4~5 枚。

Morphological characteristics A medium-sized passerine (14-18 cm). Sexes have similar plumage, but female is smaller. Bill relatively long and strong, largely blackish-grey with pale brown at base of lower mandible; crown rufous-brown; supercilium greyish-white and eye-stripe dark brown; head-sides and upperparts mostly brown; chin and throat white; breast, flanks and undertail-coverts buff; central belly greyish-white; tarsi pale brown.

Distribution and habitat Widespread in eastern, central and southern China. Uncommon summer breeder and passage migrant in Beijing. Usually seen in woodland and scrub in low hills (below 1,100 m).

Behaviour and ecology Usually solitary or in pairs. Skulks in, and sings from, dense shrubs or low canopy. Vocal. Song a series of loud whistles "wu chi-wi-lu-er". Nests in low shrubs. Clutch size: 4-5.

IUCN 红色名录 无危 (LC) 2016.10.1

强脚树莺	*Horornis fortipes*	Brownish-flanked Bush Warbler	树莺科 Cettiidae

IUCN 红色名录　无危 (LC) 2016.10.1

形态特征　小型鸣禽，体长 10~12 厘米。雌雄相似。喙尖细，近黑色，下喙基部黄褐色。眉纹浅灰色，于眼后变得不清晰。上体褐色，两翼黑褐色，羽缘橄榄褐色。下体大致为浅灰色，但两胁和尾下覆羽为淡皮黄色。跗跖淡褐色。

分布与生境　广泛分布于中国南方地区。主要活动于海拔 2000 米以下的林地、灌丛和草丛中。罕见于北京，在夏季于门头沟百花山、小龙门林场、东灵山等地有零星记录，非繁殖期亦偶见于城市公园等平原地区，居留型尚不明确。

习性与生态　常单独或成对藏匿于灌丛、草丛和苇丛中活动。鸣声响亮，三声一度，第一个音节拖长，似“wu shi-wi”。营巢于灌丛和草丛的近地面处；窝卵数 3~5 枚。

Morphological characteristics A small passerine (10-12 cm). Sexes alike. Bill fine, blackish above and yellowish-brown at base of lower mandible; supercilium pale greyish, faint behind eye; upperparts brown; wings blackish-brown, fringed olive-brown; underparts largely pale greyish, but flanks and undertail-coverts pale buff; tarsi pale brown.

Distribution and habitat Mainly in southern China. Occurs in forests, shrubs and grassland up to 2,000 m. Rare in montane areas in the west of Beijing, e.g. Baihuashan Reserve and Xiaolongmen Forest Park. Seen occasionally in lowlands, including urban parks in non-breeding season. Residential status: unclear.

Behaviour and ecology Usually alone or in pairs. Skulks in dense vegetation such as shrubs, grasses and reedbeds. Gives loud, three-note song "wu shi-wi". Nests in extensive low vegetation. Clutch size: 3-5.

鳞头树莺 | *Urosphena squameiceps* | Asian Stubtail | 树莺科 Cettiidae

形态特征 小型鸣禽，体长 9~10 厘米。雌雄相似。体型紧凑。喙黑灰色，喙长而健壮。头顶暗褐色，眉纹甚长，延伸至枕部，呈皮黄色，贯眼纹黑色。上体褐色，尾羽甚短。下体污白色，臀及尾下覆羽皮黄色。跗跖淡粉色。

分布与生境 广泛分布于中国东部及南部地区。在北京为区域性常见夏候鸟和旅鸟，繁殖于西部和北部山区海拔 1000~1500 米的林地，非繁殖期见于较低海拔，甚至平原地区。

习性与生态 多单独或成对活动于植被下层和地面。鸣声尖细似虫鸣，似“si-si-si-si”。营巢于灌丛下层或地面；窝卵数 5~7 枚。

Morphological characteristics A small and compact passerine (9-10 cm). Sexes alike. Bill long and strong, blackish-grey; crown dark brown; long, buff supercilium extends to nape; upperparts brown; tail very short; underparts largely off-white; vent and undertail-coverts buff; tarsi pale pink.

Distribution and habitat Widespread in eastern and southern China. Locally common summer breeder and passage migrant in Beijing. Breeds in montane woodlands (1,000-1,500 m) in the north and west of Beijing. Seen in lowlands in non-breeding season.

Behaviour and ecology Usually solitary or in pairs. Occurs in low vegetation and on forest floor. Song an insect-like "si-si-si-si". Nests in low shrubs or on the ground. Clutch size: 5-7.

北京市二级保护野生动物 | | IUCN 红色名录 无危 (LC) 2016.10.1

北长尾山雀	*Aegithalos caudatus*	Long-tailed Tit	长尾山雀科 Aegithalidae

IUCN 红色名录 无危 (LC) 2016.10.1

形态特征 小型鸣禽，体长 13~16 厘米。雌雄相似。喙短小而黑。虹膜深褐色，有时可见粉黄色眼睑。头及下体雪白色，两胁、下腹至尾下覆羽沾粉红色。上背深黑，下背及腰黑色夹杂粉红色，肩羽粉红色杂以白色。尾细长，可达体长之半；除最外侧三对尾羽外翈雪白色外，尾羽余部纯黑。初级飞羽、外侧次级飞羽黑色；内侧次级飞羽、三级飞羽中央黑色，白色羽缘宽阔，翼收拢时形成黑白两色对比。翼上覆羽深黑，翼下覆羽及腋羽灰白。跗跖黑色。

分布与生境 中国东北地区和新疆北部常见留鸟。在北京为罕见冬候鸟或旅鸟，见于树林、灌丛及园林等环境。

习性与生态 在北京常与银喉长尾山雀混群或单独集成小群。鸣声为平调的尖锐金属音“滋–滋–滋–”，似银喉长尾山雀。

Morphological characteristics A small passerine (13-16 cm). Bill short and black. Iris dark brown; yellow-pink eyelids occasionally visible. Head and underparts snow-white with flanks, lower belly and undertail-coverts tinged pink; mantle deep black; back and rump black mixed with pink; scapulars pink mixed with white. Tail narrow and long, reaching up to half of the total body length; rectrices black, with the outermost three having white outer webs. Remiges have distinct black-and-white pattern when folded, with primaries and outer secondaries black, inner secondaries and tertials black or dark brown, with outer web mostly white; wing-coverts deep black; underwing-coverts and auxiliaries greyish-white. Tarsi black;

Distribution and habitat A common resident in northeast China and northern Xinjiang. Rare in Beijing as winter visitor or migrant, found in forests, scrub and parks.

Behaviour and ecology In Beijing, often with small groups of Silver-throated Tit. Calls are monotone, sharp and metallic "zi-zi-zi", similar to Silver-throated Bushtit.

银喉长尾山雀	*Aegithalos glaucogularis*	Silver-throated Bushtit	长尾山雀科 Aegithalidae

形态特征 体小而圆的鸣禽，体长 13~16 厘米。雌雄相似。喙短小而黑。成鸟头顶及枕部黑色，具白色顶冠纹。颈侧、颊、眼先、额基和颏污白色；喉具黑斑。下体亦为污白色但常沾粉红色。背、腰及肩灰色，腰部具一粉红色窄带但通常不可见。尾细长；尾羽深灰黑色，最外侧三对尾羽外翈污白色。飞羽大部灰黑色，三级飞羽及内侧次级飞羽羽缘灰白。翼上覆羽深黑色，翼下覆羽及腋羽污白色。跗跖黑色。幼鸟颜色暗淡，颊和胸前棕红色。

分布与生境 常见于华北、华东和华中。在北京为常见留鸟，见于树林、灌丛及公园等环境。

习性与生态 常集群出现，有时甚喧闹。鸣声简单，通常为平调的尖锐金属音“滋 - 滋 - 滋 -”，无类似山雀的婉转鸣啭。在树侧枝上筑椭球形巢；窝卵数 5~10 枚。

Morphological characteristics A small and round-bodied passerine (13-16 cm). Bill short and black. Adult has black cap and nape, with a white central crown-stripe; side of neck, cheek, lores, forehead and chin whitish; throat spotted black; underparts whitish, often tinged pink; back, rump and scapulars grey, with a narrow pink band on rump but often invisible. Tail long, reaching up to half of the total body length; outer webs of outer three rectrices whitish, other rectrices deep greyish black. Remiges mostly greyish black, with greyish white edges on tertials and inner secondaries. Wing-coverts deep black; underwing-coverts and auxiliaries whitish. Tarsi black. Juvenile dull-coloured, with rufous cheek and breast.

Distribution and habitat Common in northeast, east and central China. A common resident in Beijing, found in forests, scrub and parks.

Behaviour and ecology Gregarious, quite noisy sometimes. Calls are simple, most often a monotone, sharp and metallic "zi-zi-zi", without the melodious song that Paridae (true tits) possess. Builds oval-shaped nests on tree branches. Clutch size: 5-10.

北京市二级保护野生动物		IUCN 红色名录 无危 (LC) 2016.10.1

白喉林莺	*Sylvia curruca*	Lesser Whitethroat	莺鹛科 Sylviidae

IUCN 红色名录　无危 (LC) 2016.10.1

形态特征　体小而修长的鸣禽，体长 12.5~14 厘米。雌雄相似。头顶及枕部灰色，耳羽及眼先深灰黑色。上体灰褐色；喉白；下体污白，常在胸及两胁有灰或褐色晕染而与喉部白色形成对比。尾较长，外侧尾羽全白或部分白色，其余尾羽黑灰色。翼上灰褐色，无翼斑。跗跖黑色。

分布与生境　见于中国华北和西北地区。在北京为罕见旅鸟及冬候鸟，见于灌丛或稀疏林地。

习性与生态　习性较隐匿，常藏匿于茂密的灌丛。鸣啭复杂婉转，鸣叫常为简单的咂嘴声（似褐柳莺）。

Morphological characteristics　A small and slender passerine (12.5-14 cm). Sexes alike. Cap and nape grey; ear-coverts and lores dark greyish-black; upperparts greyish-brown; underparts whitish; breast and flanks often tinged grey or brown, contrasting with the white throat; tail long and mostly blackish-grey, with the outermost rectrices fully or partially white. Wing feathers greyish-brown, without wing-bars; tarsi black.

Distribution and habitat　Distributed in north and northwest China. Rare in Beijing as winter visitor or migrant, found in scrub or areas with scattered trees.

Behaviour and ecology　Secretive, often staying hidden in thickets. Song is complex and melodious; calls are brief and smacking-like, similar to the Dusky Warbler.

山鹛	*Rhopophilus pekinensis*	Chinese Hill Babbler	莺鹛科 Sylviidae

形态特征 中型鸣禽，体长 15~17 厘米。雌雄相似。喙浅灰色，略下弯。虹膜浅黄灰色。头及上体沙褐色，头顶和背部密布褐黑色纵纹；褐黑色贯眼纹及髭纹显著。喉及上胸白色；下体余部白色夹杂红褐色纵纹，两胁及尾下覆羽染皮黄色。尾灰褐色，甚长，展开呈扇形；外侧尾羽端部白色。跗跖浅褐色。

分布与生境 华北地区留鸟。在北京为常见的山地灌丛留鸟，冬季常游荡至平原地区的灌丛和芦苇丛中活动。

习性与生态 常集小群藏匿于灌丛中；喜鸣叫。鸣啭多变，为不断重复的短句或快节奏且较无规律的长句；拖长的三音节降调声“diù-diù-diù-”似山噪鹛但音调更高且各音符等长，似棕头鸦雀但音符更长；告警声为单调的金属颤音。羽色与环境色接近，通常不易发现；窝卵数 4~6 枚。

Morphological characteristics A medium-sized passerine (15-17 cm). Sexes alike. Head and upperparts sandy brown with heavy brownish-black streaks on crown and back; distinct brownish-black eye-stripe and moustache; throat and upper breast white; rest of underparts white with rufous streaks; flanks and undertail-coverts tinged buff; exceptionally long tail greyish-brown and graduated; tips of outer rectrices white; bill pale grey, slightly down-curved; tarsi pale brown; iris pale yellowish grey.

Distribution and habitat Resident in north China. In Beijing, a common resident found in scrub in the hills and mountains. Often found in scrub, and occasionally reedbeds, in the lowlands during winter.

Behaviour and ecology Often forms small flocks and stays hidden in bushes. Very vocal. Song is variable, with short and repeated sequences or a fast, long and irregular strophe. Call is a prolonged and descending trisyllable "dew-dew-dew", similar to Plain Laughingthrush but higher-pitched and notes evenly spaced, also resembling Vinous-throated Parrotbill, but with more prolonged syllables. Alarm call is a monotonous and metallic trill. Well camouflaged and inconspicuous. Clutch size: 4-6.

北京市二级保护野生动物 | IUCN 红色名录 无危 (LC) 2016.10.1

棕头鸦雀	*Sinosuthora webbiana*	Vinous-throated Parrotbill	莺鹛科 Sylviidae

北京市二级保护野生动物

IUCN 红色名录 无危 (LC) 2018.8.9

形态特征 体小而尾长的鸣禽，体长 11~12.5 厘米。雌雄相似。铅灰色的喙短而小。虹膜深黑褐色。各部位羽色为较均一的红褐色，但头顶及飞羽红色更浓重，而下体颜色较浅。跗跖铅灰色。幼鸟整体羽色更浅。

分布与生境 华北、华中及华东地区留鸟。在北京为常见留鸟，低海拔平原至高山的芦苇和灌丛中均可见到。

习性与生态 常集群活动，喜在芦苇中攀援、穿梭。常用喙剥开芦苇、秸秆，寻找虫卵等为食。鸣声细碎，典型的鸣叫为悦耳、短促的三音节降调声“dù -dù -dù -”，以及略沙哑的“drrr-drrr-drrr-”。在灌丛、竹丛或小树上筑小巧的杯状巢；窝卵数 3~7 枚。

Morphological characteristics A small passerine with long tail (11-12.5 cm). Sexes alike. Bill is small, slaty-grey; iris dark black-brown. Plumage largely uniform reddish-brown, with crown and wing feathers darker, and underparts paler. Tarsi slaty-grey. Juvenile plumage is similar but paler.

Distribution and habitat Resident in north, east and central China. In Beijing, a resident commonly found in reeds and scrubby areas from lowland plains to mountain regions.

Behaviour and ecology Often forms flocks, climbing and hopping in reeds; peels reed stalks for insect larvae. Song is variable, loud but melodious chattering. Calls thin and chattering. Typical calls include a melodious, short and descending trisyllable "tiu-tiu-tiu", and a slightly husky chatter "drrr-drrr-drrr-". Builds small, cup-shaped nest in bush, banboo or young trees. Clutch size: 3-7.

VI 鸣禽

震旦鸦雀 | *Paradoxornis heudei* | Reed Parrotbill | 莺鹛科 Sylviidae

形态特征 中型鸣禽，体长 18~20 厘米。雌雄相似。喙黄，粗壮厚实。虹膜深黑褐色。头灰白色，黑褐色眉纹粗重；眼先黑褐色；后颈及上背棕灰色，过渡至下背、肩及腰部为红棕色。喉和上胸灰白色，下胸及两胁沾红棕色，下体余部棕黄色。尾甚长；中央尾羽灰褐色；其余尾羽黑色，羽端具显著的白色端斑。初级飞羽和外侧次级飞羽浅褐色；内侧次级飞羽及三级飞羽黑褐色，具白色羽缘。

分布与生境 中国东北、华北及华东地区留鸟。在北京，自 2016 年首次记录于奥林匹克森林公园以来，已在十余处不同地点的芦苇丛中记录到该种。

习性与生态 常集群活动，喜在芦苇中攀援、穿梭。常用喙剥开芦苇、秸秆，寻找虫卵等为食。最常见的鸣声为一长串哨音“hû-hû-hû……”，常逐渐加速。营巢于苇丛；窝卵数 2~5 枚。

Morphological characteristics A medium-sized passerine (18-20 cm). Sexes alike. Bill yellow and bulky; iris dark blackish-brown. Head greyish-white, supercilium heavy and blackish-brown; lores-blackish-brown; nape and upper back brownish-grey, gradually transiting to rufous-brown on lower back, scapulars and rump; throat and upper breast greyish-white, lower breast and flanks tinged rufous-brown, remainder of underparts brownish-yellow; tail rather long, central rectrices greyish-brown, rest of tail black, with distinctive white terminal spots; primaries and outer secondaries pale brown, inner secondaries and tertials blackish-brown with white edges.

Distribution and habitat Resident in northeast, north, and east China. In Beijing, it has been recorded in the reed beds of more than 10 different sites, since its first record in 2016 at Olympic Forest Park.

Behaviour and ecology Often forms flocks, climbing and hopping amongst reeds; peels reed stalks for insect larvae. Most common vocalisation is a long series of whistles "huew-huew-huew...", often accelerating at the end. Nests in reedbeds. Clutch size: 2-5.

国家二级保护野生动物 | IUCN 红色名录 近危 (NT) 2016.10.1

红胁绣眼鸟	*Zosterops erythropleurus*	Chestnut-flanked White-eye	绣眼鸟科 Zosteropidae

北京市二级保护野生动物　国家二级保护野生动物　IUCN 红色名录　无危 (LC) 2016.10.1

形态特征　小型鸣禽，体长 10~11.5 厘米。雌雄相似。喙细而尖，黑灰或铅灰色。虹膜黑褐色。头绿黄色，额基和眼先处色深；白色眼圈显著，在眼先处断开。上体大部嫩绿色，腰部及尾上覆羽亮黄色。翼上覆羽嫩绿色，飞羽暗绿色。尾短，尾羽灰绿色。颏、喉及尾下覆羽亮黄色，胁部栗红色（雌鸟胁部颜色常较雄鸟淡），下体余部灰白色。跗跖铅灰色。

分布与生境　迁徙时途径中国东部广大地区，在华东及华南地区偶有越冬记录。在北京为常见旅鸟，常见于树林环境。

习性与生态　常集群活动，有时与暗绿绣眼鸟混群。常在树冠层或高处活动，甚好动。鸣声为单音节“dǜ”，甚似暗绿绣眼鸟但常更饱满圆润。

Morphological characteristics A small passerine (10-11.5 cm). Sexes alike. Bill thin and pointy, blackish-grey or bluish-grey; iris blackish-brown. Head greenish-yellow, forehead and lores darker; distinctive white eye-ring, broken at lores. Upperparts mostly fresh green; rump and uppertail-coverts bright yellow; wing-coverts are fresh green; remiges dark green; tail short, greyish-green; chin, throat and undertail-coverts bright yellow; flanks chestnut-red, paler in females; remainder of underparts greyish-white. Tarsi bluish-grey.

Distribution and habitat Migrant through most of eastern China, with occasional wintering records in east and south China. In Beijing, a common migrant found in forests and wooded areas.

Behaviour and ecology Often in flocks, sometimes mixed with Japanese White-eye. Usually found in the canopy or high in the tree; very active. Call is a single note "tiu", very similar to Japanese White-eye but richer and mellower.

暗绿绣眼鸟	*Zosterops japonicus*	Japanese White-eye	绣眼鸟科 Zosteropidae

形态特征 小型鸣禽，体长 10~11 厘米。雌雄相似。喙细而尖，黑灰或铅灰色。虹膜黑褐色。上体及头部橄榄绿色，额基黄色，眼先黑色；白色眼圈显著，在眼先处断开。翼上覆羽橄榄绿色，飞羽暗绿色。尾短，尾羽暗绿色。颏、喉及尾下覆羽淡黄色，下体余部灰色，胁部无栗红色（区别于红胁绣眼鸟）。跗跖铅灰色。

分布与生境 繁殖于华北以南。在北京为常见旅鸟及山地林区不常见的夏候鸟，主要见于林地环境。

习性与生态 常集群活动，有时与红胁绣眼鸟混群。常在树冠层或高处跳动，甚活跃。鸣声为单音节“dǜ”，甚似红胁绣眼鸟但常更尖细而带颤音。于树上或灌丛中营巢；窝卵数 2~6 枚。

Morphological characteristics A small passerine (10-11 cm). Sexes alike. Bill thin and pointy, blackish-grey or bluish-grey; iris blackish-brown. Upperparts and head are olive-green, forehead yellow and lores black; distinctive white eye-ring, broken at lores. Wing-coverts are olive green; remiges dark green. Tail is short and dark green. Chin, throat and undertail-coverts pale yellow, remainder of underparts grey; no reddish on flanks in contrast to Chestnut-flanked White-eye. Tarsi bluish-grey.

Distribution and habitat Breeds in south of north China. In Beijing, a common migrant found in woodlands, and an uncommon summer breeder in the forests of mountainous regions.

Behaviour and ecology Often in flocks, sometimes mixed with Chestnut-flanked White-eye. Hops in the canopy or high in the tree; very active. Call is a single note "tiu", very similar to Chestnut-flanked White-eye, but sharper and with trills. Nests in trees or bushes. Clutch size: 2-6.

IUCN 红色名录 无危 (LC) 2019.8.15

山噪鹛	*Garrulax davidi*	Plain Laughingthrush	噪鹛科 Leiothrichidae

北京市二级保护野生动物 | IUCN 红色名录 无危 (LC) 2016.10.1

形态特征 中等体型的鸣禽，体长 23~25 厘米。雌雄相似。喙大部浅黄色，喙尖色暗而下弯。体壮而尾长的暗灰褐色鸟，全身羽色纯而均一。眼先及颏部略黑；飞羽及尾羽沾灰色；尾羽末端近黑。跗跖肉褐色。

分布与生境 中国北方地区山地广布的留鸟。在北京为山区常见留鸟，冬季有时游荡至低海拔平原地区越冬，多见于灌丛。

习性与生态 性喧闹，常单独或集小群活动，在浓密的山地灌丛中穿梭。鸣啭复杂多变，以短句为主，音色暗哑，常效鸣鹰鹃、红嘴蓝鹊、宝兴歌鸫等其他鸟类。鸣叫常有多音节降调声“diù……diù……diù……diù……”，似山鹛但首音符明显拖长。营巢于灌丛中；窝卵数 3~6 枚。

Morphological characteristics A medium-sized passerine (23-25 cm). Sexes alike. Bill is mostly pale yellow with darker tip and down-curved. Robust with long tail, overall uniformly dark greyish-brown. Lores and chin slightly blackish; remiges and tail tinged grey; tail tips blackish.; tarsi flesh-brown.

Distribution and habitat Resident in hilly regions of northern China; widespread. In Beijing, a resident commonly found in scrub in hilly regions, sometimes wandering to the lowland plains for winter.

Behaviour and ecology Often seen alone or in small flocks, skulking in dense bushes on hills; very vocal. Song is complex and variable, consisting of short coarse strophes, often mimicking other bird species, such as Large Hawk-cuckoo, Red-billed Blue Magpie, Chinese Thrush, etc. Calls include descending multi-syllabic "dew... dew...dew...dew...", similar to Chinese Hill Babbler, but with the first note prominently prolonged. Nests in bushes. Clutch size: 3-6.

欧亚旋木雀 *Certhia familiaris* Eurasian Treecreeper 旋木雀科 Certhiidae

形态特征 小型鸣禽，体长12.5~14厘米。雌雄相似。喙细长下弯，上喙黑灰色，下喙铅灰或淡黄色。头顶及耳羽棕褐色带有白色斑点；具明显的白色眉纹；眼先褐色。上背棕褐色，下背、腰及尾上覆羽红褐色，杂以白斑。尾羽细长，尖端呈针状，棕色无横斑。翼上深褐色夹杂乳白色斑纹，具两道白色翼斑。下体白色，两胁后部至尾下沾皮黄色。跗跖深褐色。

分布与生境 广布于中国北方大部分地区的留鸟。在北京为罕见冬候鸟和旅鸟，见于成熟的针叶林或针阔混交林，有时记录于城市公园。

习性与生态 常单独活动，以尾羽支撑身体贴于树干旋转向上爬行，取食藏在树皮中的昆虫及虫卵。鸣啭婉转细弱而音调甚高；鸣叫为尖细的颤音“zi……”，似戴菊，但音符较长而缺乏快速重复。

Morphological characteristics A small passerine (12.5-14 cm). Sexes alike. Bill is slender and down-curved; upper mandible blackish-grey; lower mandible bluish-grey or pale yellow. Crown and ear-coverts brown, speckled with white spots; supercilium prominent and white; lores brown; upper mantle brown, lower mantle to uppertail-coverts rufous-brown, mottled with white streaks. Rectrices are prolonged with needle-like tips, uniformly brown and without bars. Wings deep brown, mottled with creamy-white streaks, with two white wing-bars. Underparts white, with buffish tinge to the lower flanks and undertail-coverts. Tarsi dark brown.

Distribution and habitat A resident widespread in northern China. In Beijing, a rare winter visitor and migrant, found in mature coniferous or mixed forests and sometimes in parks.

Behaviour and ecology Usually solitary. Clings onto tree trunks with support of its tail and climbs up in a spiral, foraging for insects and larvae hidden in bark. Song is melodious, weak and rather high-pitched. Calls include sharp and thin trills "zi…", like Goldcrest, but syllables longer and lacks rapid repetition.

IUCN 红色名录 无危 (LC) 2016.10.1

普通鳾	*Sitta europaea*	Eurasian Nuthatch	鳾科 Sittidae

北京市二级保护野生动物

IUCN 红色名录　无危 (LC) 2018.8.9

形态特征　小型鸣禽，体长 11.5~13.5 厘米。雌雄相似。喙强壮，上喙基铅灰，下喙基灰白，喙尖黑灰。头顶及上体灰蓝色，明显的黑色贯眼纹延至颈部，无眉纹。颏及眼下白色。亚种 *amurensis* 颊、喉、颈侧及上胸部白色，下胸、两胁和腹部肉桂色，尾下覆羽带有斑驳的砖红色；亚种 *sinensis* 喉、耳羽及整个下体均为肉桂色。两亚种幼鸟下体污白或浅皮黄色。尾短，灰蓝色。翼上大部灰蓝，初级飞羽外翈灰黑。跗跖褐色。

分布与生境　中国东北、华北、华中及华东地区留鸟。在北京为山地常见留鸟，冬季游荡至低海拔或平原，多见于较成熟的针叶林或混交林。

习性与生态　常紧贴树皮活动，可头部向下爬行，取食松子、昆虫等。鸣啭为响亮的单音节哨音“huí”等；鸣叫为音色特殊的撞击声“twíng，twíng……”等。营巢于树洞中；窝卵数 4~13 枚。

Morphological characteristics　A small passerine (11.5-13.5 cm). Sexes alike. Bill is strong, base of upper mandible bluish-grey, base of lower mandible greyish-white, tip blackish. Crown and upperparts slaty-blue; distinct black eye-stripe extending to neck, no supercilium; chin and cheek white. Race *amurensis* has white cheek, throat, side of neck and upper breast; lower breast, flanks and belly uniformly cinnamon; undertail-coverts greyish white with mottled brick red. Race *sinensis* cinnamon on throat, ear-coverts, and underparts. Tail is stubby and slaty-blue. Wing feathers are mostly slaty-blue, with outer webs of primaries greyish-black. Tarsi brown. Juveniles of both races have whitish or pale buffy underparts.

Distribution and habitat　Resident in northeast, north, east and central China. In Beijing, a common resident found in mature coniferous or mixed forest in mountainous areas, descending to lower regions during winter.

Behaviour and ecology　Forages on pine cones, seeds and insects; often seen moving along the tree bark like a woodpecker, and famous for climbing down tree trunks head first. Song includes a loud whistling of single notes "weeu-", often repeated. Calls include an ascending metal clanking "twing-twing…". Nests in tree holes. Clutch size: 4-13.

黑头鳾 | *Sitta villosa* | Chinese Nuthatch | 鳾科 Sittidae

形态特征 小型鸣禽，体长 10~11 厘米。雌雄相似。喙强壮，上喙基黑灰色，下喙基铅灰色，喙尖灰黑色。上体蓝灰色，头顶辉黑色，有明显的白色眉纹，贯眼纹黑色；飞羽及覆羽大部褐灰色，初级飞羽外翈色深。尾短，中央尾羽蓝灰色，两侧尾羽黑色且具有白色端斑或次端斑。颏、喉白色，下体余部皮黄或污白色。跗跖灰褐色。

分布与生境 中国华北及西北地区留鸟。在北京为常见留鸟，见于低海拔平原及中低海拔山区的针叶林。

习性与生态 常成对或集群活动，行为与普通鳾等其他鳾类似。鸣啭为成串的快速哨音或气泡音“wee-wee-wee-wee”等；鸣叫为枯燥干哑的单音节。营巢于树洞中；窝卵数 4~9 枚。

Morphological characteristics A small passerine (10-11 cm). Sexes alike. Bill is strong; base of upper mandible blackish-grey, base of lower mandible bluish-grey, tip greyish-black. Upperparts bluish-grey; crown glossy black; distinctive white supercilium and black eye-stripe. Wing feathers are mostly brownish-grey, with darker outer webs of primaries. Tail is stubby; central rectrices bluish-grey, outer rectrices black with white tips or subterminal areas. Chin and throat white; remainder of underparts buffy or whitish. Tarsi greyish-brown.

Distribution and habitat Resident in north and northwest China. In Beijing, a common resident found in lowland coniferous forests and low to medium-high mountainous regions.

Behaviour and ecology Often in pairs or small flocks. Behaviour resembles Eurasian Nuthatch and other typical nuthatches. Song includes a series of rapid whistling or bubbling "wee-wee-wee-wee". Calls are coarse single notes. Nests in tree holes. Clutch size: 4-9.

北京市二级保护野生动物 | | IUCN 红色名录 无危 (LC) 2016.10.1

红翅旋壁雀 | *Tichodroma muraria* | Wallcreeper | 鳾科 Sittidae

北京市一级保护野生动物 | IUCN 红色名录　无危 (LC) 2018.8.9

形态特征　中型鸣禽，体长 14~18 厘米。喙细长，略下弯，为黑褐色。头顶和上体烟灰色。雄鸟繁殖羽颏、喉黑色，胸及两胁深灰色；尾下覆羽具黑灰和灰白色横斑。尾羽深灰黑色，内侧尾羽羽端烟灰色，外侧尾羽羽端白色。大覆羽外侧及中覆羽、小覆羽玫瑰红色，大覆羽内侧黑褐色；飞羽黑色，次级飞羽及内侧初级飞羽外翈基部玫瑰红色；外侧初级飞羽仅具 1 或 2 枚白圆斑。雌鸟及雄鸟非繁殖羽颏及喉污白色。

分布与生境　中国中西部地区留鸟或夏候鸟，华北及华东地区冬候鸟。在北京为罕见但记录较稳定的冬候鸟（十渡），偶见于近郊平原地区。在山区崖壁等多岩环境生活。

习性与生态　较少集群，常展翼再快速收拢，攀附于崖壁上觅食石缝中的昆虫。

Morphological characteristics A medium-sized passerine (14-18 cm). Bill is slender, slightly down-curved and blackish-brown. Smoky-grey on the crown and upperparts. Breeding males have black on chin and throat, and deep grey on breast and flanks; undertail-coverts streaked with blackish-grey and greyish-white bars; deep greyish-black on the tail, with smoky-grey on the tips of inner rectrices, and white on the tips of outer rectrices. Rosy on median, lesser and outer greater coverts; blackish-brown on inner greater coverts. Remiges mostly black, with the base of outer webs of secondaries and inner primaries rosy-tinged, with one or two round white patches on the outer primaries. Females and non-breeding adults have sullied-white chin and throat.

Distribution and habitat A resident or summer breeder in central-west China; winter visitor in north and east China. In Beijing, a rare but repeatedly-seen winter visitor at Shidu, occasionally seen elsewhere. Found on rocky cliffs of montane regions.

Behaviour and ecology Usually solitary. Often expands and folds its wings rapidly. Forages for insects in rock crevices by climbing vertically on cliff walls.

鹪鹩	*Troglodytes troglodytes*	Eurasian Wren	鹪鹩科 Troglodytidae

形态特征 体态短圆的小型鸣禽，体长 9~10 厘米。雌雄相似。喙短，略下弯，深灰黑色。通体深褐色。头和枕部棕色，具乳白或皮黄色眉纹及黑褐色贯眼纹。背及腰部具不明显黑色横斑。尾短，尾羽棕色，具黑色横纹。两翼褐色杂以乳白色斑点。下体棕褐，常较上体颜色略淡。胁、腹部及尾下覆羽具黑褐色横斑。跗跖褐色。

分布与生境 中国东北、西北、华北、华中至西南地区夏候鸟，在华北、华东及西藏有冬候鸟。在北京高海拔山区为区域性常见夏候鸟，见于林下潮湿的多石环境；在平原地区为旅鸟及冬候鸟，多见于水边。

习性与生态 多活动于地面，以昆虫为食。好动，常弹尾，在石缝或茂密的植被中穿梭。鸣啭为尖细高频的快速颤音组成的长句，婉转悦耳；鸣叫为干涩的“chi”声。通常营巢于岩缝、岩洞或树洞中；窝卵数 3~9 枚。

Morphological characteristics A small passerine (9-10 cm). Sexes alike. Bill short and slightly down-curved, deep greyish-black. Adult deep brown overall; creamy-white or buffish supercilium and blackish-brown eye-stripe; inconspicuous black streaks on the mantle to rump; tail short and brown, with black bars. Wings brown, fringed with creamy white spots. Underparts buffish-brown, usually slightly paler than the upperparts; flanks, belly and undertail-coverts are streaked with blackish-brown bars; tarsi brown.

Distribution and habitat In China, a summer breeder to northwest, northeast, north, central to southwest China. Winters in Tibet and north to east China. In Beijing, a locally common summer breeder in moist woods with mossy stones in the mountains; also a migrant and winter visitor to lowland Beijing, usually near water.

Behaviour and ecology Often seen on the ground, foraging for insects. Very active; often flicks tail and flies in short, quick bursts along rock crevices or in dense vegetation. Song consists of fast, long, high-pitched strophes including melodious trills. Call is a single dry and almost scolding "chi". Nests in cracks, crevices or holes in rocks or trees. Clutch size: 3-9.

IUCN 红色名录 无危 (LC) 2018.8.9

褐河乌	*Cinclus pallasii*	Brown Dipper	河乌科 Cinclidae

北京市一级保护野生动物 | IUCN 红色名录　无危 (LC) 2016.10.1

形态特征　中型鸣禽，体长 21~23 厘米。雌雄相似。喙强壮，近黑色。成鸟全身大部为均匀的深棕色，翼上和尾端颜色更深。幼鸟色浅，上体密布浅色点斑；下体皮黄，杂以黑褐色鳞状斑。跗跖深灰色。

分布与生境　广布于中国大部分地区（除荒漠及草原外），为北京西部和北部山区的不常见留鸟，生活于溪流或急流环境。

习性与生态　常单独或成对活动。擅游泳及潜水捕食昆虫、软体动物及小型鱼虾。鸣叫为响亮的颤音“drrrrreeee”。营巢于近水岩石隙中；窝卵数 3~6 枚。

Morphological characteristics A medium-sized passerine (21-23 cm). Sexes alike. Bill strong and blackish. Adult plumage is uniformly deep brown, with darker wing-coverts and tail tip. Tarsi deep grey. Juvenile similar to adult but paler, with upperparts densely covered with pale spots; underparts buff, scattered with black-brown scaly spots.

Distribution and habitat Widespread in most of China, except for desert and steppe regions. In Beijing, an uncommon resident in west and north montane areas. Found along fast-moving river valleys and creeks.

Behaviour and ecology Tends to be encountered singly or in pairs. Swims or dives for insects, molluscs, small shrimps and fish. Call is a loud trill "drrrrreeee". Nests in crevices near water. Clutch size: 3-6.

八哥	*Acridotheres cristatellus*	Crested Myna	椋鸟科 Sturnidae

北京市二级保护野生动物　　IUCN 红色名录　无危 (LC) 2016.10.1

形态特征　中型鸣禽，体长 25~27 厘米。雌雄相似。喙象牙白色，下喙基粉红色，喙尖略带弯钩。虹膜亮橙色。头黑色，额基处具明显的簇状羽。周身为均匀的深灰黑色，尾下覆羽具白斑。尾较短，尾羽黑色，外侧尾羽羽端白色。翼上深黑色，初级飞羽基部及初级覆羽白色，形成醒目的白色“腕斑”。翼下覆羽白斑亦明显，飞行时清晰可见。跗跖暗黄色。

分布与生境　华北及其以南的低海拔地区留鸟，较适应城市园林等环境。在北京曾罕见，近年来数量增加，在城市公园及近郊平原地区有时可见大群，为区域性常见留鸟。

习性与生态　通常集群或成对出现，偶尔单独活动。在地面上掘取昆虫、蚯蚓等为食；冬季亦食果及种子。鸣啭喧闹而复杂多变，擅模仿；鸣叫为干涩的颤音“zhi-zhi……”声。营巢于树洞；窝卵数 1~6 枚。

Morphological characteristics A medium-sized passerine (25-27 cm). Sexes alike. Bill ivory-white, with pink on the base of the lower mandible, slightly hooked at the tip; iris bright orange. Head black, with prominent tufted feathers at the base of forehead; uniformly dark greyish-black overall, with white streaks on the undertail. Tail short and black, with white tips on the outer rectrices. Jet-black wings, with white patches on primary coverts and at the base of primaries, forming a prominent white carpal patch. White patches on underwing also striking, especially in flight. Tarsi darkish yellow.

Distribution and habitat In China, a resident in lowland southern and north China, often inhabiting urban environments. Formally rare in Beijing, but the local population has significantly increased in the past few decades. Large flocks can be seen in some urban parks and suburban plains.

Behaviour and ecology Usually in flocks or pairs, occasionally single. Feeds for insects and worms on the ground; eats fruit and seeds in winter. Song is loud and variable, well-known for mimicking. Call is a dry trill "drri-drri...". Nests in tree holes. Clutch size: 1-6.

丝光椋鸟	*Spodiopsar sericeus*	Silky Starling	椋鸟科 Sturnidae

形态特征 中等体型的鸣禽，体长 21~24 厘米。喙橘红或深红色，喙尖黑。虹膜近黑色。雄鸟头及枕部白色，羽呈丝状。肩、背部青灰色，腰及尾上覆羽浅灰色。后颈至上胸具深灰色领环。下体余部灰白色。尾羽及翼上暗绿色，带金属光泽；初级飞羽基部大面积白斑。雌鸟羽色暗淡，头、上体及下体均为淡灰褐色。跗跖橙黄色。

分布与生境 华北及华北以南的低海拔平原地区的广布留鸟，北方地区有部分繁殖种群越冬于南方。在北京曾经罕见，现为城市公园常见的夏候鸟。

习性与生态 常集群活动，多见于园林环境。鸣啭复杂多变；鸣叫为干涩的颤音“chi-li-li……”声，亦有似鸦科的粗哑哈气声“khaaaa……”。营巢于树洞；窝卵数最多可达 7 枚。

Morphological characteristics A medium-sized passerine (21-24 cm). Bill orange-red or crimson, with black on the tip; iris blackish. Male has filamentous white feathers on the head and nape; slaty-grey on the scapular, mantle and back, with pale grey on the rump and uppertail-coverts; deep grey collar on hind neck reaching to upper breast; remainder of underparts greyish-white; tail and wings shiny dark green, with large white patches at the primary bases. Tarsi orange yellow. Female duller and overall light greyish-brown.

Distribution and habitat In China, a widespread resident in the lowlands from north to southern China. Some that breed in north China winter in the south. In Beijing, formerly rare but now a common breeder in urban parks.

Behaviour and ecology Often seen in flocks, inhabiting parks and large gardens. Song is complicated and varied; call is a dry trill "chi-li-li...", also also a corvid-like husky "khaaaa…". Nests in tree holes. Clutch size: up to 7.

北京市二级保护野生动物		IUCN 红色名录 无危 (LC) 2018.8.9

灰椋鸟 | *Spodiopsar cineraceus* | White-cheeked Starling | 椋鸟科 Sturnidae

IUCN 红色名录　无危 (LC) 2016.10.1

形态特征　中型鸣禽，体长 22~24 厘米。成年雄鸟头顶及枕部深灰黑色，背、肩部暗灰色，腰白色，尾上覆羽深灰色；眼先灰黑色，额基至眼上方为驳杂的灰白色；眼先下方及耳羽近白；颏、喉、颈侧及上胸部深灰黑色；尾下覆羽污白色，下体余部暗灰色；尾平，尾羽深灰褐色，外侧尾羽具白色端斑；翼上暗灰褐色，无翼斑。成年雌鸟头颈及胸部不如雄鸟色深，耳羽白色面积较小。飞行时轮廓似三角形。

分布与生境　中国北方地区夏候鸟，华北亦有越冬种群。在北京一年四季均可见到，但尚不能确定它们为留鸟还是分属不同居留类型的迁徙种群。常见于树林、公园及农田等环境。

习性与生态　喜集群活动，常在地面觅食，也在树上取食桑葚等浆果。营巢于树洞中；窝卵数 2~10 枚。

Morphological characteristics　A medium-sized passerine (22-24 cm). Adult male has deep greyish-black on the crown and nape, dark grey on the mantle, back and scapulars, white on the rump and deep grey on the undertail-coverts; greyish-black lores, mottled with greyish-white at the base of forehead to above the eyes; nearly white on ear-coverts and under eyes; deep greyish-black on chin, throat, neck-sides and upper breast; undertail sullied white; remainder of underparts darkish grey; tail flat, deep greyish-brown, with white patches on the tips of outermost rectrices; wings dark greyish-brown, without wing patches. Adult female is paler on the head, neck and breast with less white on ear-coverts. Silhouette in flight resembles a triangle.

Distribution and habitat　In China, breeds in north and northeast, with some also wintering in north China. Present in Beijing all year but so far unclear whether some birds are resident or whether birds seen in summer and winter relate to different, migratory, populations. Common in wooded areas, parks and farmland.

Behaviour and ecology　Often seen in flocks, especially outside breeding season, feeding on the ground for insects or in trees for berries such as mulberry. Nests in tree holes. Clutch size: 2-10.

北椋鸟	*Agropsar sturninus*	Daurian Starling	椋鸟科 Sturnidae

形态特征 中型鸣禽，体长 16~19 厘米。雄鸟头颈至下体白灰色，头顶后部具一辉紫色斑；背部为带有金属光泽的暗紫色，肩羽暗紫色过渡至白色，腰白或淡灰色，尾上覆羽灰褐色；尾羽为带有金属光泽的墨绿色；小覆羽暗紫色，中覆羽白色，大覆羽及次级飞羽、三级飞羽墨绿色；次级飞羽基部灰褐色；初级飞羽黑褐色，外翈基部具白斑。雌鸟头顶灰褐色，无紫斑；上体、尾及两翼淡茶褐色或黑褐色，无金属光泽。幼鸟与雌鸟羽色接近。

分布与生境 繁殖于中国华北及东北地区，在北京为罕见旅鸟及夏候鸟，见于开阔的树林环境，如公园及大学校园等。

习性与生态 常单独或集小群活动。营巢于树洞；窝卵数 3~7 枚。

Morphological characteristics A medium-sized passerine (16-19 cm). Adult male is whitish-grey from the head and neck to the underparts, with a shining purple patch to the rear of the crown; metallic dark purple on the mantle and back, transiting to white on the scapulars; rump white or light grey; uppertail-coverts greyish-brown; metallic dark green on the tail; lesser coverts dark purple; median coverts white; greater coverts, secondaries and tertials dark green, with greyish-brown on the base of secondaries; blackish-brown on the primaries, with white patches on the outer webs. Adult female is greyish-brown on the crown with no purple spots; light brown or blackish-brown on the upperpart, tail and wings, without metallic shine. Juvenile similar to female.

Distribution and habitat In China, breeds in north and northeast. In Beijing, a rare migrant and summer breeder, inhabiting open wooded areas, such as parks and university campuses.

Behaviour and ecology Usually seen in pairs or singles. Nests in tree holes. Clutch size: 3-7.

IUCN 红色名录 无危 (LC) 2016.10.1

雄鸟 Male

雌鸟 Female

紫翅椋鸟	*Sturnus vulgaris*	Common Starling	椋鸟科 Sturnidae

IUCN 红色名录　无危 (LC) 2018.10.31

形态特征　中等体型的鸣禽，体长25~27.5厘米。成鸟喙亮黄色（繁殖期）或黑色（非繁殖期）；全身墨绿色，闪耀金属光泽并密布白色点斑；头部墨绿色更深，近黑。雄鸟颈部深辉紫色；飞羽咖啡色。雌鸟整体颜色更暗，金属光泽不如雄鸟明显。幼鸟深灰褐色，在换羽的不同阶段逐渐出现金属墨绿色及白色点斑。成鸟跗跖暗红色，幼鸟跗跖黑色。

分布与生境　中国西北地区留鸟，少部分迁徙至东部。在北京为早春和深秋不常见的旅鸟，见于农田及水库周边等开阔环境。

习性与生态　常单独或集群活动，有时与灰椋鸟或凤头麦鸡混群。在欧洲及北美是甚常见的适应城市环境的鸟类，常集大群。甚喧闹，鸣啭复杂，擅效鸣。

Morphological characteristics A medium-sized passerine (25-27.5 cm). Bill bright yellow (breeding season) or black (non-breeding season). Adult dark and metallic green overall, dotted with small white speckles, and deeper green to black on the head. Male has shining purple on the neck and brown on the remiges. Female duller and less shiny. Juvenile deep greyish-brown overall, transiting to metallic green with white speckles after moulting. Tarsi dark reddish (adult) or black (juvenile).

Distribution and habitat In China, a resident in the northwest, with a small population moving east in winter. In Beijing, an uncommon migrant, seen in open areas such as farmland and reservoir surroundings in early spring and late autumn.

Behaviour and ecology Usually in small flocks, sometimes mixed with White-cheeked Starling or Northern Lapwing. In Europe and North America, a very common bird in urban habitats, often seen in huge flocks. Very vocal and noisy with complicated songs including imitations of other bird species.

橙头地鸫 *Geokichla citrina* Orange-headed Thrush 鸫科 Turdidae

形态特征 中型鸣禽，体长 20~23 厘米。喙灰黑色。成年雄鸟头、颈部橙色，上体余部及尾羽青灰色，背羽染黑色；眼先、颏、喉及颊杏黄色杂以浅橙色，部分亚种眼下具一黑褐色“泪纹”，耳羽后部具一黑色“月牙”斑；肩及覆羽青灰色，中覆羽具宽阔的白色羽缘形成浅色翼斑；飞羽灰黑色；胸、腹及两胁为均匀的橙色，尾下覆羽白色，偶带青灰色晕染。成年雌鸟橙色不如雄鸟明亮，上体褐色。

分布与生境 华中、华东及华南地区留鸟。在北京为罕见迷鸟，见于城市公园。

习性与生态 性机警，常单独或成对活动，在地面以蚯蚓等为食。

Morphological characteristics A medium-sized passerine (20-23 cm). Bill greyish-black. Adult male has orange on the head and neck, slaty-grey on the tail and on remainder of upperparts, with black mottling on the back; apricot lores, chin, throat and cheek, mottled with pale orange; a few subspecies have blackish-brown "tear drop" under eye, and a crescent-shaped black spot rear on the ear-coverts. Scapulars and wing-coverts greyish-blue, with broad white fringes on the median coverts, forming pale wingbars; greyish-black remiges; uniformly orange on the breast, flanks and belly; white undertail, occasionally with a slaty-greyish hue. Adult female less bright, with brown on the upperparts.

Distribution and habitat In China, resident in central, east and south China. Rare vagrant to Beijing's parks.

Behaviour and ecology Often single or in pairs; very shy; feeds on the ground, where it forages for worms etc.

IUCN 红色名录 无危 (LC) 2019.6.13

左雌右雄 Female (left), male (right)

白眉地鸫	*Geokichla sibirica*	Siberian Thrush	鸫科 Turdidae

雌鸟 Female

雄鸟 Male

幼鸟 Juvenile

IUCN 红色名录 无危 (LC) 2016.10.1

形态特征 中型鸣禽，体长 20.5~23 厘米。成年雄鸟具一醒目的白色眉纹，头、上体、尾羽及下体大部青黑色，腹部具不明显的灰白色鳞状斑，尾下覆羽具黑白相间的横斑；初级飞羽及初级覆羽黑色，翼上余部青黑色。成年雌鸟几乎无黑色，头顶及上体橄榄褐色，眉纹及颏乳白，眼先褐色，颊及耳羽白色和橄榄褐色斑驳；下体大部白色，密布深褐色“月牙”斑；两胁及尾下覆羽带棕色晕染；大覆羽及飞羽棕褐；翼上余部橄榄褐色。幼鸟似雌鸟，翼上覆羽多浅色羽端。

分布与生境 繁殖于中国东北地区森林，越冬于中国南部。在北京为罕见旅鸟，见于植被繁茂的树林，于城市公园偶有记录。

习性与生态 性隐匿，多藏于繁密的树林之中，在地面觅食。

Morphological characteristics A medium-sized passerine (20.5-23 cm). Adult male is mostly bluish-black on the head, upperpart, tail and underparts, with a prominent white supercilium; inconspicuous greyish-white scaly spots on the belly, and black-and-white streaks on the undertail-coverts; slaty-grey wings, darker on primary coverts and primaries. Adult female lacks blackish tones; olive-brown on crown and upperparts; creamy-white supercilium and chin; brown lores, cheeks and ear-coverts, mottled with sullied white; malar brown; mostly white on the underparts, speckled with crescent-shaped deep brown spots; brown-tinged on flanks and undertail-coverts; rufous-brown on greater coverts and remiges; olive-brown on the remainder of wings. Juvenile similar to female, with pale tips to the wing-coverts.

Distribution and habitat In China, breeds in forests of northeast, wintering in south. In Beijing, a rare migrant, found in dense woodland and occasionally parks.

Behaviour and ecology Shy and stay well-hidden in dense woodland; feeds on the ground.

虎斑地鸫	*Zoothera aurea*	White's Thrush	鸫科 Turdidae

形态特征 中型鸣禽，体长 28~30 厘米。雌雄相似。头顶及上体亮褐色，下体白。周身密布带有金黄色羽缘的黑色月牙斑。耳羽处有一不规则黑斑，通常呈月牙状或水滴状；眼下黑白驳杂，有时似带一“泪痕”。颏及眼先白。尾长，中央尾羽亮褐色，外侧尾羽褐黑色，最外侧两对尾羽外缘白色。肩及覆羽棕黑两色，带有乳白色羽缘。飞羽黑褐色，翼下形成两组黑白相间的翼带。

分布与生境 繁殖于中国东北、中部及西南山区，越冬于南方地区。在北京为不常见旅鸟，可见于城市及山区林地，偶有越冬记录，可能在山区有繁殖。

习性与生态 常在地面觅食无脊椎动物，繁殖季常藏匿于密林之中高声鸣啭。鸣啭为极富特色的平调长哨音“hu……”，常在不同音调间切换，多闻于黎明甚至深夜。营巢于较低处树枝；窝卵数 4~5 枚。

Morphological characteristics A medium-sized passerine (28-30 cm). Sexes alike. Overall speckled with golden-fringed black crescent-shaped spots; bright brown on crown and upperparts, white on underparts; crescent- or drop-shaped, black spot on the ear-coverts; black and white under eyes, resembling a "tear stain"; chin and lores white; tail long, bright brown on the central rectrices and brownish-black on the outers, with white fringes to the outermost two; bi-coloured (brown and black) scapulars and upperwing-coverts fringed with creamy white edges; remiges blackish-brown. In flight, shows two conspicuous black and white underwing bands.

Distribution and habitat In China, breeds in northeast, central and southwest mountains, wintering to south. In Beijing, an uncommon migrant in urban or montane woodlands, occasional in winter and a possible breeder in the mountains.

Behaviour and ecology Feeds on the ground for invertebrates. Sings from well-hidden perch in the dense forest in the breeding season. Song very distinctive, consisting of plain and prolonged whistles "hu…", alternating between different pitches, often heard pre-dawn and even during the night. Nest is low down in trees. Clutch size: 4-5.

IUCN 红色名录 无危 (LC) 2019.8.8

灰背鸫	*Turdus hortulorum*	Grey-backed Thrush	鸫科 Turdidae

雄鸟 Male

雌鸟 Female

IUCN 红色名录 无危 (LC) 2018.8.9

形态特征 中型鸣禽，体长 20~23 厘米。成年雄鸟喙黄色；头、上体及上胸部石板灰色；下胸、两肋及侧腹部橙色，下体余部白色；尾羽及两翼灰色。成年雌鸟上喙黑褐色，下喙基暗黄色；头部、上体及两翼土褐色，颏、喉及上胸近白，具黑色点斑。未成年雄鸟灰色不如成年雄鸟明显，颏、喉及上胸似雌鸟。跗跖肉褐色。

分布与生境 在中国主要繁殖于东北，越冬于华中、华南和西南地区。在北京主要为不常见旅鸟，偶有越冬记录。近年夏季在北京西部及北部中高海拔山区森林数次发现鸣啭个体和可能正在育雏的个体。过境时见于林地及城市公园等。

习性与生态 过境和繁殖季较隐匿，常活动于近水的林地，以昆虫等为食。鸣啭为多变的短句，带有金属颤音及后缀装饰音。营巢于林间树上较低处；窝卵数 3~5 枚。

Morphological characteristics A medium-sized passerine (20-23 cm). Adult male has yellow bill and slate-grey head, upperparts and upper breast; orange on the lower breast, flanks, and belly; white on the remainder of underparts; grey tail and wings. Adult female dusty-brown head. Bill blackish-brown with yellow on the base of lower mandible. Upperparts and wings brown. Off-white chin, throat and upper breast, speckled with black spots. Immature male has less grey than adult; chin, throat and upper breast resemble adult female. Tarsi fleshy-brown.

Distribution and habitat In China, mostly breeds in northeast, wintering in south, central and southwest China. In Beijing, mostly an uncommon migrant, with a few winter records. In recent years, singing males and adults feeding young have been found in north and west mountainous areas of Beijing. Migrants usually encountered in wooded areas, including parks.

Behaviour and ecology Usually shy on migration and in breeding season, inhabiting woods near water, where it feeds on the ground for insects. Song consists of variable short strophes with metallic trilling, ending with a flourish. Nests low in trees in forest. Clutch size: 3-5.

乌灰鸫 | *Turdus cardis* | Japanese Thrush | 鸫科 Turdidae

形态特征 中型鸣禽，体长 21~22 厘米。成年雄鸟喙全黄，眼圈金黄；头颈至胸部乌黑，上体和两翼青灰色，略带光泽；胁、腹部及尾下覆羽白色，胸侧及两胁具特殊的三角形或心形黑斑。成年雌鸟喙褐色，喙端近黑；头颈、上体、两翼及尾羽橄榄灰色，颏喉至上胸皮黄，具黑色细纹；胸侧及两胁暗橙色，下体余部白，三角形或心形黑斑显著。未成年雄鸟背部、两翼及尾羽黑灰色，较头颈颜色浅。

分布与生境 在中国主要繁殖于长江流域，越冬于华南。在北京及周边地区有数笔迷鸟记录，见于园林环境。

习性与生态 性机警，单独或成对活动，常在地面取食昆虫等。

Morphological characteristics A medium-sized passerine (21-22 cm). Adult males have yellow bill and eye-ring. Jet black on the head and neck to breast; bluish-grey on upperparts and upper wings; white on flanks, belly and undertail-coverts; distinctive triangle or heart-shaped black spots on the breast-sides and flanks. Adult females have brown bill with black tip. Head, neck, upperparts, wings and tail olive-grey; buffish on chin, throat to upper breast, streaked with fine black stripes; dark orange on the breast-sides and flanks and white on the remainder of underparts, speckled by triangle or heart-shaped black spots. Immature males are blackish-grey on the mantle, wings and tail, paler than the head and neck.

Distribution and habitat In China, mostly breeds around the Yangtze River basin and winters in south China. Several vagrants have been documented in Beijing in late autumn and winter, mostly in parks.

Behaviour and ecology Often seen singly or in pairs; very shy; forages on the ground for insects.

IUCN 红色名录 无危 (LC) 2018.8.9

雄鸟 Male

雌鸟 Female

灰翅鸫	*Turdus boulboul*	Grey-winged Blackbird	鸫科 Turdidae

IUCN 红色名录　无危 (LC) 2016.10.1

形态特征　中型鸣禽，体长 27~29 厘米。成年雄鸟喙橙黄色，具金黄色眼圈；周身漆黑色，似乌鸫，但翼上覆羽、次级飞羽和三级飞羽（外翈）烟灰色。成年雌鸟周身暗褐色，上体较下体颜色略灰；翼上覆羽、次级飞羽和三级飞羽（外翈）土褐色，翼上余部近黑。未成年雄鸟似成年雄鸟但黑色较浅，颏、喉部具皮黄色。跗跖暗黄或暗红色。

分布与生境　为中国西南地区留鸟，见于林地及灌丛。在北京为罕见迷鸟或夏候鸟（有待证实），仅在夏季记录于东灵山地区，为鸣啭的雄鸟，不排除存在繁殖个体甚至小种群的可能。

习性与生态　性机警，单独或成对活动。繁殖期在高大的树冠上鸣啭求偶。于树上营杯状巢；窝卵数 2~4 枚。

Morphological characteristics A medium-sized passerine (27-29 cm). Adult male has orange-yellow bill and golden eye-ring. It is pitch-black overall, similar to Chinese Blackbird, but with smoky-grey on upperwing-coverts, secondaries and outer webs of tertials. Adult female is dark brown overall, slightly greyer on the upperparts than underparts; dusty brown on upperwing-coverts, secondaries and the outer webs of tertials; nearly black on the remainder of wing. Immature males similar to adult male but less black, and with buffish on chin and throat. Tarsi dark yellow or dark red.

Distribution and habitat Resident in southwest China, inhabiting woodlands and scrub. In Beijing, a rare vagrant or possibly a rare summer breeder (unproven); only records are of singing males from Donglingshan in summer.

Behaviour and ecology Often seen singly or in pairs; shy; in breeding season, males sing from the canopy of tall trees. Builds cup-shaped nests in trees. Clutch size: 2-4.

乌鸫	*Turdus mandarinus*	Chinese Blackbird	鸫科 Turdidae

形态特征 中型鸣禽，体长 28~29 厘米。喙橙黄色（成鸟）或锈褐色（幼鸟）。眼圈金黄色。成年雄鸟周身黑色，略具光泽。成年雌鸟深褐黑色，幼鸟上体褐色，下体密布棕白色鳞状斑。跗跖黑褐色。

分布与生境 广泛分布于中国中部及南部。在北京为常见留鸟及夏候鸟，可见于城市园林、树林及草地等低海拔平原生境。

习性与生态 喜奔走于草地上捕食蚯蚓或在园林中取食植物果实，繁殖季常站在高树或建筑物上放声鸣啭。甚喧闹，鸣啭复杂多变，富含颤音及重复音节，句长及音色变化丰富，擅效鸣其他鸟类的声音甚至人为噪音；典型鸣叫为单薄的“zsssi……”声。于树上营碗状编织巢；窝卵数 2~6 枚。

Morphological characteristics A medium-sized passerine (28-29 cm). Bill orange (adult) or rusty brown (juvenile). Eye-ring yellow. Shiny pitch-black (adult male) or deep brownish-black (adult female) overall. Tarsi blackish-brown. Juvenile has brown upperparts and scaly brown-and-white underparts.

Distribution and habitat Widely distributed in central and southern China. In Beijing, a common resident and summer breeder, found in urban gardens, parks and other wooded areas in the lowlands.

Behaviour and ecology Prefers to forage on the ground, where it feeds on worms or fruits. Very vocal with complicated and variable songs, enriched with trills and repeated notes. Strophes vary in length and timbre. Famous for mimicking the sounds of other species or even anthropogenic noises. Typical call is a dragged and thin metallic "zssssi...". In breeding season, loud songs often delivered from a high perch, such as a tall tree or building. The nest is a woven bowl in trees. Clutch size: 2-6.

IUCN 红色名录 无危 (LC) 2018.8.9

雄鸟 Male

雌鸟 Female

灰头鸫	*Turdus rubrocanus*	Chestnut Thrush	鸫科 Turdidae

雄鸟 Male

雌鸟 Female

雄鸟 Male

IUCN 红色名录　无危 (LC) 2016.10.1

形态特征　中型鸣禽，体长 23~27 厘米。喙亮黄色。虹膜黑褐色，眼圈金色。成年雄鸟（*gouldii* 亚种）头颈部灰色，上体余部及胸、腹部醒目的栗红色，两翼及尾全黑；尾下覆羽黑色，具白色羽缘。雌鸟头颈部淡褐色，体羽栗色亦较淡。跗跖暗黄色。

分布与生境　分布于中国西北、中部及西南山区，喜繁密的森林。在北京为罕见迷鸟，见于城市园林（奥林匹克森林公园）和西部山区（百花山）。

习性与生态　常在地面捕食无脊椎动物；非繁殖季节在树上取食植物果实。

Morphological characteristics A medium-sized passerine (23-27 cm). Bill bright yellow; iris blackish-brown, circled by a golden eye-ring. Adult male (ssp. *gouldii*) has grey on head and neck; remainder of upperparts, breast and belly rich chestnut; solid black on upper wings and tail; black on undertail-coverts, with white fringes. Female light brown on head and neck, paler chestnut on the underparts. Tarsi dark yellow.

Distribution and habitat Resident in the mountains of northwest, central and southwest China, where it prefers dense forests. In Beijing a rare vagrant in urban parks (Olympic Forest Park) and west mountains (Baihuashan).

Behaviour and ecology Forages for invertebrates on the ground or, outside the breeding season, feeds on fruits in trees.

褐头鸫	*Turdus feae*	Grey-sided Thrush	鸫科 Turdidae

形态特征 中型鸣禽，体长 22~25 厘米。雌雄相似。上喙及喙端黑色，下喙基明黄色。成鸟头、上体深棕色；眼先黑色，眉纹及眼下白色；下体暗灰色，尤以两肋灰色浓重，腹部及尾下覆羽灰色浅，有时近白；两翼大部棕色，初级飞羽褐黑色。幼鸟下体皮黄色，密布深褐色鳞状斑。跗跖暗黄色。

分布与生境 繁殖于中国华北山区，越冬于中南半岛。在北京为罕见夏候鸟和旅鸟，繁殖于中高海拔山区密林，过境时于城市公园偶有记录。

习性与生态 性隐匿，繁殖期于高处枝头高声鸣啭。鸣啭悦耳但较单调，常由数个固定的短促乐音起始，旋即接以尖细的金属颤音，并常在末尾缀以细弱的装饰音，乐句较短，不断重复；鸣叫为低沉的金属颤音"dzii……"。营巢于灌丛中或树上；窝卵数 4~5 枚。

Morphological characteristics A medium-sized passerine (22-25 cm). Sexes alike. Bill largely black with bright yellow at the base of lower mandible. Adult deep brown on the head and upperparts; black lores; white supercilium and under eye; dark grey on underparts, darkest on the flanks but paler, even nearly white occasionally, on the belly and undertail-coverts; upperwing largely brown, with brownish-black on primaries. Juvenile buffish on underparts, speckled with dark brown scaly patches. Tarsi dark yellow.

Distribution and habitat Only known to breed in the mountains of north China; winters in Indochina. Rare summer breeder and migrant to Beijing, breeding in dense forest in medium to high montane areas; occasionally seen in city parks on migration.

Behaviour and ecology Very shy. Males sing loudly from tall tree in the breeding season. Song consist of melodious, short strophes, starting with several musical notes, followed by a thin metallic trilling and ending with a soft and weak flourish. Call is a muffled metallic trilling "dzii...". Nests in bushes or trees. Clutch size: 4-5.

北京市一级保护野生动物 | 国家二级保护野生动物 | IUCN 红色名录 易危 (VU) 2016.10.1

白眉鸫	*Turdus obscurus*	Eyebrowed Thrush	鸫科 Turdidae

雄鸟 Male
雌鸟 Female

IUCN 红色名录　无危 (LC) 2016.10.1

形态特征　中型鸣禽，体长 19~24 厘米。成年雄鸟头部及颈侧深灰，背、腰及尾上覆羽棕褐；眼先黑色，眉纹、眼下及颏白；喉深灰，过渡至上胸、胸侧及两胁为橙褐色；喉及耳羽常夹杂不清晰的白色纵纹；下体余部白色；尾羽深灰褐；两翼大部棕褐色，初级飞羽黑褐色。雌鸟头颈部不灰，亦为棕褐色，颏及喉白。

分布与生境　在中国仅繁殖于黑龙江东北部及内蒙古北部，过境时可见于中部及东部大部分地区，于南部地区越冬。在北京为不常见旅鸟，见于公园等林地环境。

习性与生态　过境或受惊时常在郁闭的树上停留，较机警。常集小群，在地面取食蚯蚓等。鸣叫为金属颤音“dzee……”，较褐头鸫音调高，常在惊飞及夜间迁徙时发出。

Morphological characteristics　A medium-sized passerine (19-24 cm). Adult male deep grey on head and neck-sides; rufous brown on mantle, to rump and uppertail-coverts; black lores; white supercilium, under-eye and chin; deep grey on throat, transiting to orange-brown on upper and side of breast and flanks; throat and ear-coverts are often mottled by inconspicuous white streaks; remainder of underparts white; greyish-brown tail; upper wings brown, with blackish-brown on the primaries. Female brown, instead of grey, on head and neck, white on chin and throat.

Distribution and habitat　In China, breeds in northeast Heilongjiang and north Inner Mongolia. Migrants widely seen in central and eastern China; winters in southern China. In Beijing, an uncommon migrant found in wooded parks.

Behaviour and ecology　On migration or when flushed, often perches on trees with closed-canopy; quite shy. Usually found in small flocks, foraging on the ground for worms. Song consists of melodious strophes with trills and flourishes. Call is a metallic trill "dzeee…", higher-pitched than Grey-sided Thrush, often delivered when flushed or when migrating at night.

白腹鸫	*Turdus pallidus*	Pale Thrush	鸫科 Turdidae

形态特征 中型鸣禽，体长 22~23 厘米。上喙黑色，下喙基黄，端部黑。虹膜深褐色，带橙黄色眼圈。成年雄鸟头深灰色，上体赭褐色，尾灰黑色；眼先黑色，额基沾褐色；喉、胸至两胁晕染橙棕色，腹部灰白，尾下覆羽近白。成年雌鸟头褐色，颏、喉白，杂深褐色纵纹。未成年雄鸟似雌鸟，但头、面部带少量灰色晕染。跗跖暗黄色。

分布与生境 繁殖于中国东北地区，越冬于长江以南大部地区。在北京为不常见旅鸟和罕见冬候鸟。见于城市公园等林地环境。

习性与生态 单独或成对活动，性隐匿。

Morphological characteristics A medium-sized passerine (22-23 cm). Bill black with yellow at the base of lower mandible; iris deep brown, circled by an orange-yellow eye-ring. Adult male deep grey on the head, ochre-brown on upperparts, and greyish black on the tail; black lores with brown tinge on the base of forehead; orange-brown mottled on throat and breast to flanks; greyish white belly, undertail nearly white. Adult female brown head, white chin and throat, speckled with deep brown streaks. Immature male similar to female, but with less grey on head and cheek. Tarsi dark yellow.

Distribution and habitat In China, breeds in northeast and winters mainly south of Yangtze. In Beijing, an uncommon migrant and rare winter visitor, found in wooded areas, including urban parks.

Behaviour and ecology Often seen singly or in pairs; rather shy.

IUCN 红色名录 无危 (LC) 2016.10.1

雄鸟 Male

雌鸟 Female

PASSERINES

赤胸鸫	*Turdus chrysolaus*	Brown-headed Thrush	鸫科 Turdidae

雄鸟 Male

雌鸟 Female

IUCN 红色名录　无危 (LC)　2016.10.1

形态特征　中型鸣禽，体长 23~24 厘米。喙近黑，下喙基黄。虹膜黑色，带金黄色眼圈。成年雄鸟头、上体及两翼锈褐色，喉、胸、两胁为大面积橙棕色；上腹中央及下腹和尾下覆羽白色，与胸及侧腹部颜色分界明显，在下胸部构成一“尖角”。成年雌鸟似雄鸟，但头及上体褐色较淡，颏、喉白色，夹杂褐色纵纹。跗跖褐色。

分布与生境　在中国尚无繁殖记录，为华东沿海地区少见的旅鸟，于华南地区有越冬个体。在北京为甚罕见迷鸟，仅记录于北京大学校园。

习性与生态　性机警，常在茂密的灌丛或林下活动。

Morphological characteristics　A medium-sized passerine (23-24 cm). Bill nearly black with yellow at base of lower mandible; iris black, circled by a golden eye-ring. Adult males rusty brown on head, upperparts and wings; largely orange-brown on throat, breast and flanks; white on central upper belly, lower belly and undertail-coverts, with a well-defined boundary to breast and side belly, constituting a "sharp angle" below the lower breast. Adult female similar but duller head and upperparts; white on chin and throat, mottled with brown streaks. Tarsi brown.

Distribution and habitat　No breeding records in China; a scarce migrant to coastal east China, with some wintering in south. Very rare vagrant to Beijing (only once recorded in Peking University).

Behaviour and ecology　Shy; inhabits thick scrub and woodland understory.

黑喉鸫	*Turdus atrogularis*	Black-throated Thrush	鸫科 Turdidae

形态特征 中型鸣禽，体长 24~27 厘米。成年雄鸟头顶、耳羽至颈后灰黑色；背部至尾上覆羽及两翼灰色；眼先深黑色；眉纹及颊、颏、喉部纯黑色，并延伸至颈侧及上胸部；下胸至尾下覆羽白色，着淡灰色杂斑；尾羽深灰黑色。成年雌鸟似雄鸟，但头、胸部黑色不纯且常呈鳞状黑斑，有时甚至仅为黑灰色晕染；颊及颏、喉常不全黑，下体尤其两胁处带黑灰色细纵纹。未成年雄鸟介于二者之间；未成年雌鸟头及下体黑色甚淡斑纹更显著。

分布与生境 在中国繁殖于新疆，向南或向东迁徙越冬。在北京为不常见的冬候鸟和旅鸟，见于平原及山区的灌丛和树林环境。

习性与生态 常集群活动，在北京数量较少，与其他鸫混群。冬季以浆果等为食。鸣叫似赤颈鸫。

Morphological characteristics A medium-sized passerine (24-27 cm). Adult male greyish-black on crown, ear-coverts and hind neck; grey on wings and mantle to undertail-coverts; dark black lores; uniformly black supercilium, cheek, chin, throat, extending to neck-sides and upper breast; white on lower breast to undertail-coverts, speckled with pale grey; tail deep greyish-black. Adult female similar to male but with less black, more blackish-grey, speckled with scaly black streaks on head and breast; usually not fully black on cheek, chin and throat; fine blackish-grey steaks on the underparts, especially flanks. Immature male intermediate between adult male and female. Immature female is dullest of all ages and sexes, distinguished by limited black on head and throat.

Distribution and habitat In China, breeds in Xinjiang, migrating to south and east in winter. In Beijing, an uncommon winter visitor and migrant inhabiting scrub, forests and parks in lowland and mountains.

Behaviour and ecology Often seen in flocks, mixing with other winter thrushes. In winter, feeds mostly on berries and other fruit. Calls similar to Red-throated Thrush.

IUCN 红色名录 无危 (LC) 2016.10.1

雄鸟 Male

雌鸟 Female

赤颈鸫	*Turdus ruficollis*	Red-throated Thrush	鸫科 Turdidae

IUCN 红色名录　无危 (LC) 2018.8.9

形态特征　中型鸣禽，体长 23~27 厘米。成年雄鸟头顶、耳羽及上体青灰色；眼先深黑色；眉纹及颊、颏、喉部绛红色，并延伸至颈侧及上胸部；下胸至尾下覆羽白色，着淡灰褐色杂斑；中央尾羽深棕色，其余尾羽深橘红色；翼上黑灰色。成年雌鸟似雄鸟，但头、胸部红色较淡而不纯，颊及颏、喉夹杂不清晰的黑色纵纹，下体灰褐色斑纹更显著。未成年雄鸟介于二者之间；未成年雌鸟羽色最为暗淡，头及喉部及尾羽仅可见少量红褐色。

分布与生境　在中国繁殖于新疆西北部地区，越冬于长江以北的大部分地区。在北京为常见的冬候鸟和旅鸟，见于平原及山区的灌丛和树林环境。

习性与生态　常成群活动，与红尾斑鸫、斑鸫及黑喉鸫混群。冬季以浆果等为食。鸣叫为干涩的上扬音“fī-”，以及似太平鸟的颤音“trrrrii”；告警为刺耳的“chuk-chuk-”。

Morphological characteristics　A medium-sized passerine (23-27 cm). Adult male grey on crown, ear-coverts and upperparts; dark black lores; deep red supercilium, chin and throat, spreading to neck-sides and upper breast; white on lower breast and undertail-coverts, speckled with pale greyish-brown; central rectrices deep brown, outer retrices deep orange-red; upper wing blackish-grey. Adult females similar but paler and less monotone on head and breast, also streaked with thin black stripes on cheek, chin and throat, and more speckled underparts. Immature male intermediate between adult male and female. Immature female the dullest of all ages and sexes, with limited reddish-brown on head, throat and tail.

Distribution and habitat　In China, breeds in northwest Xinjiang, wintering mostly north of Yangtze. In Beijing, a common winter visitor and migrants inhabiting scrub and woodland in both lowlands and mountains.

Behaviour and ecology　Often seen in flocks, mixing with Naumann's, Dusky and Black-throated Thrushes. In winter, feeds on berries and other fruits. Call includes a dry, ascending "fi-", and a waxwing-like thin trilling "trrrrrii"; alarm is a harsh "chuk-chuk-".

红尾斑鸫	*Turdus naumanni*	Naumann's Thrush	鸫科 Turdidae

形态特征 中型鸣禽，体长 23~25 厘米。成年雄鸟头顶、耳羽、后颈及上背橄榄灰色，染橙红色；下背、腰部及尾上覆羽橙红色；眼先黑褐色，眉纹、眼下、颏、喉至上胸橙红色，常具不明显黑色下髭纹；下胸至两胁及侧腹部和尾下覆羽大面积暗橙红色，羽缘白色，形成鳞状或矛状斑纹；腹部中央近白；尾羽大部暗橙红色，末端深褐色；翼上大部橄榄灰色，杂以橙红色；初级覆羽及初级飞羽褐色；翼下覆羽及腋羽橙红色。成年雌鸟头及上体灰色和下体橙色皆较雄鸟浅。

分布与生境 中国广泛分布。在北京为常见的冬候鸟和旅鸟，在公园、山区林地、灌丛、草地等植被较稀疏处均可见。

习性与生态 常成群活动，与其他鸫类，尤其是斑鸫及赤颈鸫等混群。在地面及树上取食浆果、种子等。

Morphological characteristics A medium-sized passerine (23-25 cm). Adult male has olive-grey crown, ear-coverts, hind neck, mantle and upper back, mottled with rufous; lower back, rump and uppertail-coverts rufous; lores blackish-brown; rufous supercilium, under-eye, chin and throat to upper breast; some individuals have inconspicuous black sub-malar; lower breast to flanks and side of the belly to undertail-coverts show dark orange-red and with white fringes, forming scaly or dart-shaped streaks; central belly is nearly white; tail mostly dark red, with deep brown tip; upperwing largely olive-grey, mottled by orange-red; primary coverts and primaries brown; underwing-coverts and auxiliaries orange-red. Adult females paler and more brownish on head and upperparts; orange tones on underparts also lighter than male.

Distribution and habitat Widespread in China. In Beijing, a common winter visitor and migrant to parks, montane forests, scrub and grassland with scattered trees or bushes.

Behaviour and ecology Often seen in flocks, mixing with other winter thrushes, especially Dusky and Red-throated Thrush. Forages on the ground for berries and seeds.

北京市二级保护野生动物 | | IUCN 红色名录 无危 (LC) 2016.10.1

PASSERINES

斑鸫	*Turdus eunomus*	Dusky Thrush	鸫科 Turdidae

北京市二级保护野生动物 *

IUCN 红色名录 无危 (LC) 2016.10.1

形态特征 中型鸣禽，体长 23~25 厘米。成年雄鸟头顶、耳羽至颈后深灰黑（新羽）或灰褐（旧羽）；上背黑褐，羽缘皮黄或白色；下背至尾上覆羽棕褐色驳杂，尾羽近黑；眉纹、眼下及颏和颈侧白（新羽）或皮黄（旧羽）；喉白，可带黑褐色细纹；胸、胁及侧腹部黑，羽缘白，形成鳞状或矛状斑；腹中央及尾下覆羽近白；肩及小覆羽黑褐，翼上余部亮棕。雌鸟头及上体偏褐，眉纹皮黄；眼下及颈侧多黑色杂斑；下体黑色不显著；翼上棕色浅。未成年鸟似雌鸟，头及下体更淡。存在与红尾斑鸫杂交的现象。

分布与生境 中国广泛分布。为北京林区常见冬候鸟及旅鸟。

习性与生态 常成群活动，在北京多与红尾斑鸫混群。以浆果、种子等为食。

● * 注：《北京市二级保护野生动物名录》未收录（更新 / 接受）斑鸫和红尾斑鸫的物种划分，故此处沿用红尾斑鸫的数据。

Morphological characteristics A medium-sized passerine (23-25 cm). Adult male has dark greyish-black (fresh) or brown (worn) crown, ear-coverts and hind neck; upperparts largely blackish-brown with buffish or white fringes; uppertail-coverts speckled brown; tail nearly black; white (fresh) or buffish (worn) supercilium, under-eye, chin, and neck-sides; throat white or speckled with fine blackish-brown streaks; black on breast, flanks and side of the belly, with white fringes, forming scaly or dart-shaped patterns; central belly and undertail-coverts nearly white; scapulars and lesser converts blackish-brown; remainder of upperwing bright brown. Female has more brownish head and upperparts; supercilium yellowish; under-eye and side neck mottled with black spots; black on underparts is less obvious than males, as is the brown on upperwing. Immature similar to female but with paler head and underparts. Hybridization between Dusky Thrush and Naumann's Thrush is not rare.

Distribution and habitat Widespread in China. In Beijing, a common winter visitor and migrant to any wooded areas.

Behaviour and ecology Often seen in flocks. In Beijing, usually seen with Naumann's Thrush, feeding on berries and seeds.

田鸫	*Turdus pilaris*	Fieldfare	鸫科 Turdidae

形态特征 中型鸣禽，体长 24~28 厘米。雌雄相似。喙大部黄色，喙端黑。成鸟头顶、耳羽及颈后及下背至尾上覆羽石板灰色；背、肩栗色，带淡灰色羽缘；眉纹及颏、喉污白色或皮黄色，具黑色细纹；颈侧及上胸暗橙色，着黑色不规则点斑；下体余部白，胸侧、两胁至侧腹部密布黑色箭头斑；尾下覆羽具暗褐色斑；尾羽全黑；初级飞羽及初级覆羽深灰色；翼上余部淡栗色，杂以灰色；腋羽及翼下覆羽近白。幼鸟头及上体多锈褐色，下体斑纹粗重。

分布与生境 在中国繁殖于新疆北部，越冬于西北地区。在北京为甚罕见迷鸟，仅在东灵山有一笔冬季记录。栖息于林地、灌丛等环境。

习性与生态 集群活动，在北京出现时与赤颈鸫等其他鸫类混群。冬季取食浆果等。

Morphological characteristics A medium-sized passerine (24-28 cm). Sexes alike. Bill mostly yellowish with black on the tip. Adult has slaty-grey crown, ear-coverts, hind neck, and lower back to uppertail-coverts; scapulars, mantle and upper back chestnut with pale grey fringes; supercilium, chin and throat sullied white or buffish, with fine black stripes; neck-sides and upper breast dark orange, spotted with irregularly shaped black dots; remaining underparts white, covered with dart-shaped black streaks on the breast-sides, flanks and side of the belly; dark brown streaks on undertail-coverts; rectrices fully black; primaries and primary coverts dark grey; remainder of upperwing pale chestnut smudged with grey; auxiliaries and underwing-coverts nearly white. Juvenile more rusty on head and upperparts, and more streaked on underparts.

Distribution and habitat In China, breeds in north Xinjiang, and winters in northwest China. Very rare vagrant in Beijing, with only one record in winter (Donglingshan). Inhabits open woodland and scrub.

Behaviour and ecology Usually in flocks. In Beijing, in mixed flock with Red-throated Thrush and other thrushes. In winter, feeds on berries and other fruit.

IUCN 红色名录 无危 (LC) 2016.10.1

白眉歌鸫

Turdus iliacus

Redwing

鸫科
Turdidae

IUCN 红色名录　近危 (NT) 2016.10.1

形态特征　中型鸣禽，体长 20~24 厘米。雌雄相似。上喙全黑，下喙基部黄色，端部黑色。头顶、耳羽、上体及两翼均匀的橄榄褐色。眼先黑色，眉纹白色，污白色下颊纹延伸至颈侧。髭纹黑色，颏、喉白色具黑色细纹。下体大部白色，胸至两胁及侧腹部密布灰黑色纵纹，胁部至腋羽及翼下覆羽具醒目的锈红色。跗跖肉色。

分布与生境　分布于中国新疆北部。在北京为罕见迷鸟，与其他鸫类混群，见于城市园林或大学校园。

习性与生态　常在地面捕食无脊椎动物; 冬季在树上取食浆果等植物果实。鸣叫为下降的高频颤音“tsiii”，在北京可区别于与之混群的其他鸫类。

Morphological characteristics A medium-sized passerine (20-24 cm). Sexes alike. Bill tip and upper mandible black, contrasting with yellowish base of lower mandible. Uniformly olive-brown on the crown, ear-coverts, upperpart and wings; dark lores; white supercilium, and sullied white sub-malar, extending to neck; malar black, chin and throat white with fine black stripes; underparts mostly white, with greyish-black streaks on the breast, flanks and side of the belly; prominent rusty-red on flanks, axillaries, and underwing-coverts. Tarsi flesh-colour.

Distribution and habitat In China, most records are from north Xinjiang. Rare vagrant to Beijing, usually mixed with other thrushes in parks or university campuses.

Behaviour and ecology Feeds on invertebrates on the ground; in winter on berries and other fruit. Typical call is a descending high-pitched trill "tsiii", distinctive from other thrushes in Beijing.

宝兴歌鸫	*Turdus mupinensis*	Chinese Thrush	鸫科 Turdidae

形态特征 中型鸣禽，体长 22~24 厘米。雌雄相似。上喙灰黑色，下喙基暗黄色，喙端灰黑色。头顶、上体、尾羽及两翼均匀的深橄榄棕色。眼先及面部污白，杂以模糊的褐色斑，耳羽后部具黑月牙斑。颏、喉近白，髭纹黑色。下体白色，除尾下覆羽外密布黑色圆斑，胸侧及两胁偶带皮黄色。中、大覆羽羽端皮黄，形成两道翼斑。跗跖肉色。幼鸟下体圆斑不规则。

分布与生境 在中国中部和西南地区为留鸟，在华北为夏候鸟。在北京繁殖于中高海拔山区阔叶林，为不常见夏候鸟。迁徙时见于平原地区林地和林下草地等环境，冬季在近山平原处有零星记录。

习性与生态 性机警，常隐匿于树冠中鸣啭。鸣啭为悦耳多变的短句，常效鸣。常在地面采食无脊椎动物。营巢于阔叶树上；窝卵数 4~6 枚。

Morphological characteristics A medium-sized passerine (22-24 cm). Sexes alike. Bill greyish-black on the tip and upper mandible, contrasting with the dark yellow base of lower mandible. Uniformly dark olive-brown on crown, upperparts, tail and wings; lores and cheek sullied white and smudged with brown; black crescent at rear of ear-coverts; chin and throat nearly white, contrasting with black malar stripes; underparts white, except for undertail-coverts, which are spotted with small round black marks; some individuals sullied by buff on the breast-sides and flanks; buff tips on median and greater coverts form two wing-bars. Tarsi pinkish. Juvenile has irregularly shaped dark spots on the underparts.

Distribution and habitat Residents in central and southwest China and a summer breeder to north China. In Beijing, an uncommon summer breeder in broad-leaved forests at medium to high elevation. Migrants occur in forests and glades, occasionally in wooded parks in lowland Beijing.

Behaviour and ecology Usually shy; sings from well-hidden perch in the canopy. Song is melodious, consisting of variable short strophes, including mimicry. Feeds on the ground where it forages for invertebrates. Nests in broadleaved trees. Clutch size: 4-6.

北京市二级保护野生动物

IUCN 红色名录　无危 (LC) 2016.10.1

欧亚鸲	*Erithacus rubecula*	European Robin	鹟科 Muscicapidae

形态特征 小型鸣禽，体长 13~15 厘米。雌雄相似。喙黑色。上体橄榄褐色，前额、眼先，颊橙红色。下体色彩鲜艳分明，颏、喉和胸部橙红色，腹中部及尾下覆羽白色，两胁浅橄榄褐色延伸至腹部。耳羽、喉侧及胸侧蓝灰色呈连续带状；跗跖粉色至粉棕色。

分布与生境 分布于中国新疆西部。在北京为罕见迷鸟，见于城市园林灌丛。

习性与生态 常在灌丛之下的地面或低矮枝条间跳动，擅隐蔽；鸣啭时则站在显眼的枝头处。鸣啭婉转多变，常有高音调的金属哨音和颤音；鸣叫亦多变，且长度不一。

Morphological characteristics A small passerine (13-15 cm). Sexes alike. Bill black. Upperparts are olive-brown; forehead, lores and cheek orange-red; underparts, chin, throat and breast orange-red, central belly and undertail-coverts white; flanks pale olive-brown, spreading to sides of belly; ear-coverts, side of throat and side of breast bluish-grey and forming a continuous band; tarsi pink to pinkish-brown.

Distribution and habitat In China, occurs in west Xinjiang. A rare vagrant to Beijing, found in scrub and shrubby areas in urban parks.

Behaviour and ecology Secretive, often hopping on ground beneath bushes or perched low in branches. Uses more prominent perch when singing. Song is variable and melodious, often with high-pitched metallic whistles and trills.

IUCN 红色名录 无危 (LC) 2018.8.9

日本歌鸲	*Larvivora akahige*	Japanese Robin	鹟科 Muscicapidae

雌鸟 Female

雄鸟 Male

IUCN 红色名录 无危 (LC) 2016.10.1

形态特征 小型鸣禽，体长 14~15 厘米。雄鸟上体红棕色，前额及颊橙红色；下体色彩鲜艳分明，喉、颈侧及胸上部橙红色，具黑色胸带，胸下部蓝灰色具黑斑，腹中部及尾下覆羽白色，两胁灰色或褐灰色延伸至腹部；尾羽栗红色；跗跖粉色或粉棕色。雌鸟羽色似雄鸟，但上体偏棕色，头部的橙色红范围小且色彩暗淡，下体多棕色，无黑色胸带。

分布与生境 在中国东部沿海主要为冬候鸟。在北京为罕见迷鸟，见于较浓密的城市园林灌丛。

习性与生态 常在灌木中及其地面活动。鸣啭较单一，常由一个单音节的先导音符和紧随其后的一串颤音组成，似红尾歌鸲但每句较短。

Morphological characteristics A small passerine (14-15 cm). Male has reddish-brown upperparts, forehead and entire cheek orange-red; throat, side of neck and upper breast orange-red, breast-band black, lower breast blue-grey with black spots, central belly and undertail-coverts white; flanks grey to brownish-grey; tail chestnut-red. Female is similar to male but more brownish on upperparts; orange area on head smaller and duller; and underparts more brownish, without black breast-band. Tarsi pink to pinkish-brown.

Distribution and habitat Winters in eastern coastal regions of China. In Beijing, a rare vagrant found in urban parks with good cover.

Behaviour and ecology Often hops low in bushes and on the ground. Song often begins with a single introductory note, followed by a short trill, similar to Rufous-tailed Robin but strophes shorter.

红尾歌鸲 *Larvivora sibilans* Rufous-tailed Robin 鹟科 Muscicapidae

形态特征 小型鸣禽，体长 13~15 厘米。雌雄相似。上体橄榄棕色，多具黑色细纹，尾羽栗红色，翼上覆羽及飞羽红褐色；眉纹皮黄色，多只见于眼前段，颊棕色具皮黄色或浅褐色细纹。下体皮黄色或白色，颏及喉部密布暗褐色斑点，胸部及上腹部密布暗褐色鳞状斑纹，两胁褐色或灰褐色；跗跖粉色或粉棕色。

分布与生境 繁殖于西伯利亚东部，越冬于中国南部及东南亚地区。在北京为不常见旅鸟，见于城市开阔林地或园林灌丛。

习性与生态 常在林下地面或灌丛中活动。鸣啭较单一，常由一长串颤音组成，似日本歌鸲但每句较长且不具先导音符。

Morphological characteristics A small passerine (13-15 cm). Sexes alike. Upperparts olive brown, densely covered with thin black stripes; tail chestnut-red; wing-coverts and remiges reddish-brown; supercilium buffy, often only preocular; cheek brown with buffy or pale brown thin stripes; underparts buffy or white; chin and throat densely covered with dark brown spots; breast and upper belly covered with dark brown scaly spots; flanks brown or greyish-brown. Tarsi pink to pink-brown.

Distribution and habitat Breeds in eastern Siberia and winters in south China and Southeast Asia. In Beijing, an uncommon migrant found in open woodland or scrub in urban parks.

Behaviour and ecology Often hops on ground in undergrowth. Song often with a long trill, similar to Japanese Robin, but syllables longer and with no introductory note.

IUCN 红色名录 无危 (LC) 2016.10.1

蓝歌鸲	*Larvivora cyane*	Siberian Blue Robin	鹟科 Muscicapidae

雄鸟 Male

雌鸟 Female

幼鸟 Juvenile

IUCN 红色名录　无危 (LC) 2016.10.1

形态特征　小型鸣禽，体长 13~14.5 厘米。喙黑色，冬季下喙基粉色。成年雄鸟上体钴蓝，眼先、下颊、颈侧及胸侧黑色；下体纯白，两胁有灰蓝色晕染；尾短，尾羽黑褐，多沾蓝色；初级飞羽黑褐且羽缘沾蓝，内侧飞羽及翼上覆羽钴蓝；跗跖粉色。雌鸟上体橄榄褐色，腰部沾蓝色；下体皮黄，颊、喉侧及胸具褐色鳞纹，胁浅棕。年轻雄鸟似雌鸟，但上体沾蓝，下体棕色较少。

分布与生境　在中国繁殖于华北及东北，越冬于东南亚。在北京为区域性常见的夏候鸟及旅鸟，过境时常见于城市园林灌丛底部，繁殖于山区森林。

习性与生态　常隐匿于密林地面，在繁殖地通常只闻其声不见其影。鸣啭婉转且响亮，通常由一串短促前导音和多变的颤音组成；鸣叫为简单的“ta”。营巢于山区林下地面或草丛中；窝卵数 4~6 枚。

Morphological characteristics A small passerine (13-14.5 cm). Bill is black, with base of lower mandible pink in winter. Adult male has cobalt blue upperparts; black lores, lower cheek, neck-sides and breast-sides; underparts pure white, with flanks and lower thigh tinged greyish-blue; tail short, blackish-brown, tinged blue; primaries blackish-brown with blue edges, inner remiges and wing-coverts cobalt blue. Tarsi pink. Female is olive-brown on upperparts, with rump tinged blue; underparts buff; cheeks, side of throat and breast have brown scaling; flanks pale brown. Immature male similar to female but with blue on upperparts and less brown on underparts.

Distribution and habitat Breeds in north and northeast China, wintering in Southeast Asia. Locally common in Beijing as a summer breeder in forests of mountain areas, and as a migrant found in dense undergrowth in city parks.

Behaviour and ecology Often skulks on the ground in dense vegetation. Best detected by song during breeding season. Song is loud and melodious, consisting of a few introductory notes followed by a varied trill. Call is a simple "tuck". Nests on the ground or in grass in montane forest. Clutch size: 4-6.

红喉歌鸲	*Calliope calliope*	Siberian Rubythroat	鹟科 Muscicapidae

形态特征 小型鸣禽，体长 14~16 厘米。成年雄鸟上体橄榄褐，额及头顶棕，眉纹及下髭纹雪白，边缘有细小的黑色纹路，眼先深黑，覆羽及飞羽暗棕；颏及喉部鲜红，有时可见黑色边缘，胸灰色，两胁褐黄，腹部及尾下覆羽皮黄；跗跖粉至褐色。雌鸟似雄鸟，但整体暗淡，颏及喉部白色或皮黄，眉纹及下髭纹细小沾污，眼先黑褐色，胸部沙褐色。

分布与生境 在中国繁殖于东北及西北地区，越冬于中国南部及东南亚地区。在北京为区域性常见旅鸟，见于芦苇丛、树林及城市园林。

习性与生态 常在林下地面、芦苇丛或灌丛中活动。鸣啭悦耳多变，擅效鸣；乐句中常包含若干固定的哨音作为起始音节，衔接短促的震颤音和婉转乐音；鸣叫为干涩的咂嘴声“chak”。

Morphological characteristics A small passerine (14-16 cm). Adult male has olive-brown upperparts, brown forehead and crown, snowy white supercilium and moustache, bordered by thin black stripes; lores dark black; wing feathers dark brown; chin and throat ruby-red, sometimes with black border; breast grey; flanks brownish-yellow; belly and undertail-coverts buffy. Tarsi pink to brown. Female is similar to male, but overall duller with white or buffy chin and throat; supercilium and moustache thinner and dirtier; lores black-brown and breast sandy-brown.

Distribution and habitat Breeds in northeast and northwest China, and winters in southern China and Southeast Asia. In Beijing, a locally common migrant found in reeds, scrub, forests and urban parks.

Behaviour and ecology Often found on the ground under trees, in reeds or in bushes. Song is variable and musical, with mimicry, often starting with some whistles as introductory notes, followed by short rapid trills and musical notes. Call is a dry smacking "chak".

北京市二级保护野生动物 | 国家二级保护野生动物 | IUCN 红色名录 无危 (LC) 2016.10.1

雄鸟 Male

雌鸟 Female

雄鸟 Male

白腹短翅鸲 *Luscinia phaenicuroides* White-bellied Redstart

鹟科
Muscicapidae

雄鸟 Male

雄鸟 Male

雌鸟 Female

IUCN 红色名录　无危 (LC)　2016.10.1

形态特征　中型鸣禽，体长 18~19 厘米。成年雄鸟上体深灰蓝色；初级飞羽蓝黑色，翼上余部黑褐且具灰蓝色羽缘，小翼羽上有醒目的白色端斑；中央尾羽蓝黑色，两侧尾羽基部橙红、端部蓝黑；颏、喉部及胸部深灰蓝，腹中部白色，两胁蓝灰；尾下覆羽黑色，端部白；跗跖肉粉至暗褐色。雌鸟上体橄榄褐，下体多浅棕色。亚成雄鸟似雌鸟但整体色暗，按不同年龄沾染不同范围的蓝色。

分布与生境　分布于中国中部、西南及西北山区。在北京为不常见夏候鸟，繁殖于高海拔山区密林。

习性与生态　常在浓密的林下灌丛中活动，擅隐蔽。鸣啭婉转且响亮，通常由 1 至 3 声先导哨音和音调高且拖长的多音节组成；鸣叫多为重复次数不一且音低的咂嘴声。营巢于灌丛中；窝卵数 2~4 枚。

Morphological characteristics　A medium-sized passerine (18-19 cm). Adult male has deep greyish-blue upperparts; primaries blue-black; remainder of wings blackish-brown with greyish-blue edges; distinctive white spots on tips of alula; central rectrices blue-black, outer retrices have orange-red base and blue-black tip; chin, throat and breast deep greyish-blue; central belly white; flanks bluish-grey; undertail-coverts black with white tips. Tarsi fleshy-pink to dark brown. Females have olive-brown upperparts and underparts mostly pale brown. Juvenile male is similar to female but overall darker, with more areas tinged blue, growing with age.

Distribution and habitat　Occurs in mountainous areas of central, southwest and northwest China. In Beijing, an uncommon summer breeder found in dense forests of mountainous regions at high elevations.

Behaviour and ecology　Secretive, often skulks in dense undergrowth. Song loud and melodious, often consisting of one to three introductory notes, followed by multiple high-pitched and prolonged syllables. Call is a repeated low-pitched smacking. Nests in bushes. Clutch size: 2-4.

蓝喉歌鸲	*Luscinia svecica*	Bluethroat	鹟科 Muscicapidae

形态特征 小型鸣禽，体长 13~15.5 厘米。喙黑色。成年雄鸟上体沙褐，眉纹污白，侧冠纹黑，眼先褐黑，颊褐或红褐色；中央尾羽黑褐，具沙褐色羽缘，两侧尾羽基部栗红，端部黑色；颏、喉及胸部色彩鲜艳多变，常由蓝、橙红、黑、白和栗红色的醒目带状斑纹组成，腹部及尾下覆羽污白。雌鸟似雄鸟但整体暗淡，眉纹褐黄；颏、喉及胸部皮黄，多具黑褐色斑纹且形成喉侧纵纹及胸带，有时沾染蓝色或橙色斑点；跗跖肉褐色。

分布与生境 在中国繁殖于东北及西北地区，越冬于南部及西南地区。在北京为区域性常见旅鸟，见于芦苇丛、园林灌丛。

习性与生态 常在林下地面、灌丛或芦苇中活动。鸣啭婉转多变且较长，具有模仿其他鸟类叫声的能力。

Morphological characteristics A small passerine (13-15.5 cm). Adult male has black bill, sandy-brown upperparts, whitish supercilium, black lateral crown-stripe, brownish-black lores, brown or rufous cheeks; central rectrices blackish-brown with sandy-brown edges; bases of other rectrices chestnut-red with black tips; chin, throat and breast colourful and variable, often consisting of distinctive bands of blue, orange-red, blue, black, white and chestnut-red; belly and undertail-coverts whitish. Female is similar to male but overall duller with brownish-yellow supercilium, buffy chin, throat and breast with black streaks and spots forming breast-band and lateral throat stripes, sometimes tinged blue or orange. Tarsi fleshy-brown.

Distribution and habitat In China, breeds in northeast and northwest regions, and winters in south and southwest regions. In Beijing, a locally common migrant found in scrub of city parks or at fringes of wetlands.

Behaviour and ecology Often found on the ground in scrub or reedbeds. Song variable, melodious and lengthy, with much mimicry.

北京市二级保护野生动物 | 国家二级保护野生动物 | IUCN 红色名录 无危 (LC) 2018.11.1

雄鸟 Male

雌鸟 Female

雄鸟 Male

红胁蓝尾鸲 *Tarsiger cyanurus* Orange-flanked Bluetail 鹟科 Muscicapidae

雄鸟 Male

雌鸟 Female

雄鸟 Male

IUCN 红色名录 无危 (LC) 2016.10.1

形态特征 小型鸣禽，体长 13~15 厘米。成年雄鸟喙黑色；上体深蓝色带辉光，眉纹白沾染蓝色，眼先及颊黑，耳羽暗褐，多黑色细纹；头顶两侧、肩、腰、翼上覆羽及尾上覆羽亮蓝色；尾羽黑褐，具蓝色羽缘；飞羽暗褐色，三级飞羽及内侧次级飞羽沾蓝；下体皮黄，喉侧及胸侧暗蓝，两胁橙红色可延伸至胸侧。雌鸟及未成年雄鸟上体橄榄褐，喉侧及胸侧无蓝色。

分布与生境 在中国繁殖于东北地区，越冬于长江流域及以南地区。在北京为常见的旅鸟，见于园林树林、灌丛；亦为罕见的山区夏候鸟及不常见冬候鸟。

习性与生态 多活动于林下及地面，捕食昆虫等。站立时常上下摆尾。鸣啭简短而婉转；鸣叫常为单一且不断重复的高音“yi”，或似蛙类的低音“kr”。常营巢于土洞、树根中或树洞中；窝卵数 3~7 枚。

Morphological characteristics A small passerine (13-15 cm). Adult male has black bill; glossy deep blue upperparts; supercilium white, tinged blue; lores and cheeks black; ear-coverts dark brown with black streaks; side of crown, scapulars, upperwing-coverts, rump and uppertail-coverts bright blue; tail blackish-brown with blue edges; Remiges dark brown; tertials and inner secondaries tinged blue; underparts buffy; side of throat and side of breast dark blue; flanks orange-red sometimes spreading to side of breast. Female and sub-adult male have olive-brown upperparts with no blue on side of throat or breast.

Distribution and habitat Breeds in northeast China, and winters to Yangtze River and its south. In Beijing, a common migrant found in trees and bushes of city gardens, also a rare breeder in the mountains and uncommon winter visitor.

Behaviour and ecology Often found on low branches in trees or shrubs from where it makes forays to the ground to catch insects; characteristically flicks tail downwards when perched. Song short and simple but melodious. Calls include a simple but repeated high-pitched note "ee", and a low-pitched frog-like "kr". Usually nests in holes in the ground, among tree roots on the ground or in tree holes. Clutch size: 3-7.

白喉红尾鸲	*Phoenicuropsis schisticeps*	White-throated Redstart	鹟科 Muscicapidae

形态特征 小型鸣禽，体长 14~16 厘米。雄鸟头黑色，前额至枕部灰蓝；眼先、颈侧及背黑，肩、腰及尾上覆羽栗红；尾羽黑褐；飞羽深褐色，翼上具一道醒目长翼斑；下体多栗红，颏及喉两侧黑，喉中央及下腹白色；跗跖黑色。雌鸟上体深橄榄褐色，中央尾羽黑褐色，外侧尾羽黑褐色具栗红色羽缘，下体浅褐，喉部白斑及白翼斑似雄鸟，但白色范围更小。

分布与生境 分布于中国西北及喜马拉雅山区，喜高山林缘灌丛及草甸灌丛。在北京为罕见迷鸟，目前仅见于西部山区。

习性与生态 性活泼，常翻飞跳跃于灌丛间捕食昆虫等，冬季主要以浆果为食。鸣啭婉转，乐句快速且短促。鸣叫为单一重复的拖长金属音，或一长串短小但快速重复的金属音。

Morphological characteristics A small passerine (14-16 cm). Male has black head, greyish-blue on forehead to nape, black lores, side of neck and back; chestnut-red lower mantle, rump and uppertail-coverts; tail deep blackish-brown; remiges deep brown; upperwing lesser coverts, inner medium coverts, inner greater coverts and tertials edged white, forming a long distinctive wing-bar; underparts mostly chestnut-red; chin and side of throat black; central throat white and lower belly white. Tarsi black. Female has deep olive-brown upperparts; central rectrices blackish-brown, other rectrices blackish-brown with chestnut red edges; underparts pale brown; white throat spot and white wing-bar similar to male but smaller, especially the throat spot.

Distribution and habitat Occurs in northwest China and the Himalayas where it is found in alpine scrub at forest margins and meadows. In Beijing, a rare vagrant to western montane regions.

Behaviour and ecology Active; often found foraging among bushes, from where it makes forays to the ground to catch insects; in winter feeds mostly on berries. Song is melodious, rapid and short. Calls include a simple repeated prolonged high-pitched metallic note, or a long series of short but rapid metallic notes.

IUCN 红色名录 无危 (LC) 2016.10.1

雄鸟 Male
雌鸟 Female

贺兰山红尾鸲	*Phoenicurus alaschanicus*	Alashan Redstart	鹟科 Muscicapidae

雄鸟 Male
雌鸟 Female

国家二级保护野生动物　IUCN 红色名录　近危 (NT) 2016.10.1

形态特征　小型鸣禽，体长 15~16.5 厘米。雄鸟前额、头顶、枕及颈侧蓝灰，背及腰棕红色；颊、颏、喉及下体大部棕红色，下腹中央及尾下覆羽白色；中央尾羽深黑褐色，两侧尾羽棕红色；飞羽黑褐色，初级覆羽白色呈醒目白斑，翼上小覆羽、内侧中覆羽、内侧大覆羽及三级飞羽羽缘白色，形成一道醒目长翼斑；跗跖黑色。雌鸟上体灰褐色，腰及尾部似雄鸟，翼斑似雄鸟但白色范围更小；下体浅灰褐色，尾下覆羽色浅。

分布与生境　主要分布于中国西北部地区，喜高山林缘灌丛及草甸灌丛。在北京为罕见冬候鸟，见于西部山区。

习性与生态　在地面或灌丛中觅食昆虫，冬季主要以浆果为食。鸣声资料极少，鸣啭为一长串多变的高音调颤音。

Morphological characteristics　A small passerine (15-16.5 cm). Male has greyish-brown forehead, crown, nape and side neck; rufous-red on lower back and rump; cheek, chin, throat and most of the underparts rufous-red, with central lower belly and undertail-coverts white; central rectrices deep blackish-brown, other rectrices rufous-red; remiges blackish-brown; primary coverts white and forming distinctive white spot; upperwing lesser coverts, inner medium coverts, inner greater coverts and tertials edged white, forming a long distinctive wing bar. Tarsi black. Female has greyish-brown upperparts, rump and tail similar to male; wing-bars similar to male but less prominent; underparts pale greyish-brown; undertail-coverts paler.

Distribution and habitat　Occurs in northwest China; found in alpine bushes at forest margins, alpine scrub and meadows. In Beijing, a rare winter visitor to the western montane regions.

Behaviour and ecology　Forages on the ground or in bushes for insects; in winter feeds mostly on berries. Vocalisation not well-known. Song includes a long series of varying high-pitched trills.

赭红尾鸲	*Phoenicurus ochruros*	Black Redstart	鹟科 Muscicapidae

形态特征 小型鸣禽，体长 13~16 厘米。成年雄鸟（*rufiventris* 亚种）黑红两色；头顶至腰部深烟黑色，尾上覆羽深橘红；头颈至胸部纯黑，下体余部深橘红；中央尾羽黑，其余尾羽深橘红。两翼纯黑，无翼斑或翼带；跗跖黑色。成年雌鸟上体暗灰褐，下体沙褐，耳羽略沾棕；尾上覆羽及尾羽似雄鸟。未成年雄鸟黑色较浅，翼上偏黑褐而不纯黑。幼鸟上体灰褐，羽缘多皮黄色点斑，下体皮黄色杂褐色鳞斑。

分布与生境 常见于中国西部广大地区，越冬于云南及南亚地区。在北京为罕见迷鸟，见于西部山区，城区公园有零星记录。

习性与生态 在地面或灌丛中觅食。鸣啭为不断重复的短句，较北红尾鸲等显单调；鸣叫为上挑的尖细哨音"típ"等。

Morphological characteristics A small passerine (13-16 cm). Adult male of race *rufiventris* is a two-tone black and red bird; crown to rump deep smoky-black; uppertail-coverts deep orange-red; head, neck and breast pure black; remainder of underparts deep orange-red; central rectrices black; remainder of tail deep orange-red; wings black, with no wing-bars or bands. Tarsi black. Adult female has dar greyish-brown upperparts and sandy brown underparts; ear-coverts slightly tinged brown; uppertail-coverts and tail similar to male. Immature male is overall less blackish, and wings are more blackish-brown instead of pure black. Juvenile has greyish-brown upperparts, with buffy spots on edges; underparts buffy with brown scaly spots.

Distribution and habitat Widespread and commonly seen in west China, wintering to Yunnan and South Asia. In Beijing, a rare vagrant found in western montane regions, with sporadic records in urban parks.

Behaviour and ecology Forages on the ground or in low bushes. Song is a series of repeated short strophes. Calls include an ascending sharp whistle "teep".

IUCN 红色名录　无危 (LC) 2018.8.9

雄鸟 Male
雌鸟 Female

北红尾鸲	*Phoenicurus auroreus*	Daurian Redstart	鹟科 Muscicapidae

雄鸟 Male

雌鸟 Female

IUCN 红色名录 无危 (LC) 2016.10.1

形态特征 小型鸣禽，体长 13~15 厘米。成年雄鸟头顶至后颈灰白，上背及肩羽黑色，下背至尾上覆羽橘红；眼先、颊、颏、喉至上胸黑；中央尾羽暗褐，其余尾羽橘红，最外侧尾羽外翈具棕色边缘；两翼大部黑色，次级飞羽基部白，形成白色三角形翼斑。雌鸟头及上体大部棕褐而无黑色，下体橘黄；两翼棕褐，白色翼斑较雄鸟小甚至无翼斑。

分布与生境 中国广泛分布。在北京为平原常见的旅鸟及冬候鸟，也是中、低海拔山区及近山平原的常见夏候鸟。喜灌丛、阔叶林及人工绿地等。

习性与生态 常站在显眼的枝头、电线等处，有抖尾等行为。鸣啭为多变的婉转短句；鸣叫为干涩尖锐的单调金属哨音“zīp”，以及短促低哑的告警声“gāk”。营巢于岩隙或树洞中；窝卵数 3~6 枚。

Morphological characteristics A small passerine (13-15 cm). Adult male has greyish-white on the crown to hindneck, black on upper back and scapulars and orange-red on lower back to uppertail-coverts; lores, cheek, chin, throat to upper breast all black; central rectrices dark blackish-brown; remainder of tail orange-red; outer webs of outermost rectrices edged dark brown; wings mostly black; base of secondaries white, forming a triangular white wing patch. Female has mostly brownish head and upperparts with no black; underparts orange; wings brown with a smaller white wing spot compared to male (sometimes invisible).

Distribution and habitat Widespread in China. In Beijing, a common migrant and winter visitor to the lowlands, and a common summer breeder in low to medium elevation mountainous regions, preferring bushes, broadleaved forests and artificial woodlands.

Behaviour and ecology Often perches conspicuously on bare branches or on wires, flicking its tail. Song consists of a series of musical and varied short strophes. Calls include a dry and harsh metallic whistle "zeep", and a short coarse alarm call "gak". Nests in crevices or tree holes. Clutch size: 3-6.

红腹红尾鸲 | *Phoenicurus erythrogastrus* | White-winged Redstart | 鹟科 Muscicapidae

形态特征 中型鸣禽，体长 16~19 厘米。成年雄鸟头顶雪白，头余部至背、肩纯黑，腰、尾上覆羽及尾羽棕红；颈部及上胸亦纯黑，下体余部均匀的棕红；翼上大部纯黑，飞羽基部雪白，形成大面积白翼斑。成年雌鸟周身褐色，腰至尾羽橘红，无白翼斑。未成年雄鸟整体羽色似雌鸟，但已具有较明显翼斑及部分黑、红体羽。

分布与生境 在中国为西部地区留鸟及夏候鸟，于华北及西南部分山区越冬。在北京为西部中高海拔山区的不常见冬候鸟及平原地区罕见旅鸟，偏好开阔的草甸、灌丛等环境。

习性与生态 在北京越冬时，常集小群，在沙棘丛附近取食沙棘。鸣啭婉转，模式似北红尾鸲；鸣叫为尖锐的“zīk”、干涩的“tak”及颤音“drrr-drrr”等。

Morphological characteristics A medium-sized passerine (16-19 cm). Adult male has snowy white crown with remainder of head, back and scapulars pure black; rump, uppertail-coverts and tail rufous-red; neck and upper breast pure black; remainder of underparts uniformly rufous-red; upperwing mostly black, with snowy white on the base of remiges, forming a large white wing patch. Adult female is uniformly brown overall, with orange on rump and tail and lacking white wing patches. Immature male is similar to female but with distinctive wing patch and some black and red feathers.

Distribution and habitat Resident or summer visitor to west China, wintering in mountainous regions in north and southwest China. In Beijing, an uncommon winter visitor to western high-altitude montane regions and a rare migrant in the plains, preferring open meadows and scrub.

Behaviour and ecology During winter in Beijing, often feeds on sea buckthorn berries, in pairs or small flocks. Song is melodious, with a similar pattern to that of Daurian Redstart. Calls include a sharp "zeek", a dry "tak" and trills "trrr-trrr".

IUCN 红色名录 无危 (LC) 2016.10.1

雄鸟 Male
雌鸟 Female

红尾水鸲	*Rhyacornis fuliginosa*	Plumbeous Water Redstart	鹟科 Muscicapidae

IUCN 红色名录　无危 (LC) 2016.10.1

形态特征　小型鸣禽，体长 12~13 厘米，体型圆胖而紧凑，尾羽较短。雄鸟头部、背部、胸腹部、两翼多蓝灰色；初级飞羽、次级飞羽沾棕褐色；尾羽及尾下覆羽红色。雌鸟头顶至背部铅灰色；两翼飞羽及覆羽深褐色具两道白色翼斑；下体偏白色密布鳞状斑纹；尾羽深褐色，外侧尾羽白色。

分布与生境　分布于中国华北、华中及南方地区。北京区域性常见于郊区的山涧溪流环境，为留鸟。

习性与生态　常单独或成对在水边或水中石头堆活动，站立时尾羽时常上下摆动或张开。雄鸟领域性强，会驱赶进入其领地的其他小型鸟类。一般营巢于近水岩隙中；窝卵数 3~5 枚。

Morphological characteristics　A small and plump passerine (12-13 cm) with relatively short tail. Male is mostly blue-grey on the head, back, breast, belly and wings; primaries and secondaries washed with tan; tail and undertail-coverts red. Female is slate grey from crown to back; dark brown wings with two white wing-bars; scaly pattern on off-white underparts; dark brown tail with white outermost rectrices.

Distribution and habitat　North, central and southern China. In Beijing, locally common along mountain streams in rural areas. Residential status: resident.

Behaviour and ecology　Usually solitary or in pairs on shore or on rock piles in the water. Tail is often fanned and bobs up and down while standing. Male is territorial, and may chase away any small birds in its territory. Nests in crevices near water. Clutch size: 3-5.

白顶溪鸲	*Chaimarrornis leucocephalus*	White-capped Water Redstart	鹟科 Muscicapidae

形态特征 中型鸣禽，体长 18~19 厘米。雌雄同色。喙黑色。除头顶白色外，头两侧、背及两翼皆为黑色。腰、尾上覆羽和尾羽栗红色，尾羽具显著的黑色端斑。颏、喉部至上胸黑色，下胸、腹部及尾下覆羽栗红色；跗跖近黑色。

分布与生境 分布于中国华北、华中、西部及南方地区。北京不常见于房山区、门头沟区等地的山涧溪流环境，多为留鸟。

习性与生态 常单独站立在山区河流、山涧小溪中的岩石上，主要在水边捕食昆虫。停落时常有点头伴随上下抖尾的行为。于水边岩隙中或近水树洞中营巢；窝卵数 3~5 枚。

Morphological characteristics A medium-sized passerine (18-19 cm). Sexes alike. Adult has black bill, white crown, black head-sides, back and wings. Rump, uppertail-coverts and tail chestnut, with prominent black terminal band on rectrices. Chin, throat and upper breast black; lower breast, belly and undertail-coverts chestnut; tarsi blackish.

Distribution and habitat North, central, western and southern China. In Beijing, uncommon along mountain streams in Fangshan and Mentougou Districts. Residential status: resident.

Behaviour and ecology Usually found singly on rocks in river or stream in mountain area. Feeds on insects near water. Often bobs and quivers tail when perched. Nests in crevices or tree holes near water. Clutch size: 3-5.

IUCN 红色名录 无危 (LC) 2016.10.1

紫啸鸫	*Myophonus caeruleus*	Blue Whistling Thrush	鹟科 Muscicapidae

IUCN 红色名录　无危 (LC) 2016.10.1

形态特征　大型鸣禽，体长 29~35 厘米。雌雄同色。北京可见 *caeruleus* 亚种。喙黑色，整体大致为深蓝紫色，头部、颈部、背部、胸腹部蓝紫色具浅蓝色点状斑纹。两翼蓝紫色，翼上覆羽具浅色亮斑。尾羽及尾下覆羽蓝紫色。

分布与生境　广泛分布于中国东部、西部及南方地区。北京不常见于山涧溪流环境，迁徙时有时可在城区公园看到，为旅鸟及夏候鸟。

习性与生态　常单独在溪流环境活动，喜在水边捕食昆虫。时常发出尖锐刺耳的叫声。营巢于近水岩石缝隙中；窝卵数 3~5 枚。

Morphological characteristics　A large passerine (29-35 cm). Sexes alike. The *caeruleus* subspecies can be found in Beijing. Bill black; body largely blue-purple, spotted with light blue on head, neck, back, breast and belly; wing-coverts blue-purple with pale spangles; blue-purple tail and undertail-coverts.

Distribution and habitat　Eastern, western and southern China. In Beijing, uncommon close to mountain streams, rarely in city parks on migration. Residential status: migrant and summer breeder.

Behaviour and ecology　Usually solitary; feeds on insects near stream. Often utters high-pitched sharp calls. Nests in crevices near water. Clutch size: 3-5.

黑喉石䳭	*Saxicola maurus*	Siberian Stonechat	鹟科 Muscicapidae

形态特征 小型鸣禽，体长 12~14 厘米。雄鸟繁殖羽头部黑色，颈侧具一明显白斑；背部及两翼黑褐色，有一白色翼斑；尾羽黑色。雌鸟头部棕褐色，头顶具黑色细纹；背部深褐色，白色翼斑面积较雄鸟小。雄鸟非繁殖羽似雌鸟，但羽色较深，白色翼斑面积较大。

分布与生境 中国见于东部地区。北京迁徙季常见于开阔地带的灌丛、草地、农耕地等环境，为旅鸟。

习性与生态 常在开阔灌丛地带的枝头站立，喜单独活动，主要以昆虫为食，常飞至空中捕食昆虫。营巢于地面或近地面处的缝隙、洞穴或浓密的灌丛中；窝卵数 5~8 枚。

Morphological characteristics A small passerine (12-14 cm). Breeding male has black head with distinct white patch on sides of neck; black-brown back and wings with a white wing-bar; black tail. Female has greyish-brown head with fine black lines on crown; dark brown back; white wing-bar smaller than male. Non-breeding male is similar to female but with larger white wing-bar.

Distribution and habitat Eastern China. In Beijing, common in open scrub, grassland and farmland on migration. Residential status: migrant.

Behaviour and ecology Often perches on top of shrub or grass in open areas, from where it launches forays into the air or to the ground to catch insects. Nests in crevices or holes in the ground, or in dense bush. Clutch size: 5-8.

IUCN 红色名录 无危 (LC) 2020.8.12

雄鸟繁殖羽 Breeding male

雄鸟非繁殖羽 Non-breeding male

雌鸟 Female

灰林䳭	*Saxicola ferreus*	Grey Bushchat	鹟科 Muscicapidae

雄鸟 Male

雌鸟 Female

IUCN 红色名录　无危 (LC) 2016.10.1

形态特征　小型鸣禽，体长 14~15 厘米。雄鸟整体深灰色；眼周为明显黑色，眉纹白色，头顶深灰色具黑色细纹；背部及两翼为深灰色和黑色；喉部白色，胸腹部烟灰色。雌鸟整体褐色；眼周为棕褐色，眉纹黄褐色；背部及两翼为黄褐色；喉部偏白色，胸腹部褐色。

分布与生境　广泛分布于中国南方地区。北京地区为甚罕见迷鸟。喜在林缘、灌丛、农耕地等环境活动。

习性与生态　喜单独活动，常站立在林缘地带的植被顶部。主要以昆虫为食。

Morphological characteristics A small passerine (14-15 cm). Male is overall dark grey; black around eyes with white supercilium; dark grey crown with black streaks; back and wings dark grey and black; throat white; breast and belly ashy-grey. Female is overall brown; dark brown around eyes with tan supercilium; back and wings yellowish-brown; throat whitish; breast and belly brown.

Distribution and habitat Southern China. In Beijing, very rare vagrant in forest edge, scrub and farmland.

Behaviour and ecology Usually solitary; often perches on top of shrub or grass at the forest edge. Feeds mainly on insects.

沙䳭	*Oenanthe isabellina*	Isabelline Wheatear	鹟科 Muscicapidae

形态特征 小型鸣禽，体长 15~16.5 厘米。雄鸟整体棕褐色；头顶褐色，眉纹皮黄色，贯眼纹黑色，颊皮黄色；背部灰褐色，两翼深褐色；腰部及尾上覆羽白色；中央尾羽黑色，其余尾羽白色，具黑色端斑，飞行时黑色部分形成“凸”字形斑纹；下体大致为白色或污白色。雌鸟似雄鸟，但眼先深褐色，背部偏褐色而少灰色。

分布与生境 分布于中国北部地区。北京地区为罕见旅鸟，于延庆和昌平有少量记录。喜在平原地区的荒漠、半荒漠地带活动。

习性与生态 常单独或成对活动，领域性强。喜在地面快速奔走，尾不断上下摇摆。主要以昆虫为食。

Morphological characteristics A small passerine (15-16.5 cm). Male is pale brown overall; crown brown with black eye-stripe, buffy supercilium and cheek; back grey-brown; wings dark brown; rump and uppertail-coverts white; central rectrices black; outer rectrices white with black terminal band, forming a distinctive T-shaped pattern in flight. Underparts largely white or off-white. Female is similar to male, but with dark brown lores; back browner, less grey.

Distribution and habitat Northern China. Rare passage migrant in Beijing, with a few records in Yanqing and Changping Districts. Favours open grassland and semi-desert areas in lowlands.

Behaviour and ecology Usually solitary or in pairs; highly territorial. Forages on the ground, running and bobbing its tail. Feeds mainly on insects.

IUCN 红色名录 无危 (LC) 2016.10.1

雄鸟 Male

雌鸟 Female

穗䳭	*Oenanthe oenanthe*	Northern Wheatear	鹟科 Muscicapidae

雄鸟 Male

雌鸟 Female

IUCN 红色名录　无危 (LC) 2018.8.9

形态特征　小型鸣禽，体长 14~16.5 厘米。雄鸟繁殖羽头顶蓝灰色，眉纹白色，贯眼纹至耳羽黑色；背部蓝灰色，两翼近黑色；腰部及尾上覆羽白色，中央尾羽黑色，其余尾羽白色而端部黑色，飞行时黑色部分形成黑色“凸”字斑纹；颏、喉及上胸为淡黄白色或污白色，下体余部白色。雌鸟似雄鸟，但贯眼纹、耳羽色浅；背部偏黄褐色而少灰色。

分布与生境　分布于中国北部地区。北京地区为罕见旅鸟，于延庆、房山、昌平等地有少量记录。喜在平原地区的草地、多石荒地等环境活动。

习性与生态　常单独或成对活动。喜在地面快速奔走，尾不断上下摇摆。主要以昆虫为食。

Morphological characteristics A small passerine (14-16.5 cm). Breeding male has bluish-grey crown, white supercilium, black eye-stripe and ear-coverts; back bluish-grey; wings blackish; rump and uppertail-coverts white; central rectrices black; outer rectrices white with black tips, forming distinctive T-shaped tail pattern in flight. Chin, throat and upper breast creamy or off-white; remainder of underparts white. Female is similar to male but has paler eye-stripe and ear-coverts; balck more yellowish-brown, less blue-grey.

Distribution and habitat Northern China. Rare migrant in Beijing, with a few records in Yanqing, Fangshan and Changping Districts. Favours open grassland or stony areas such as dry riverbeds in lowland.

Behaviour and ecology Feeding solitary or in pairs, highly territorial. Actively running on ground, bob their tails. Mainly feeds on insects.

白顶䳭	*Oenanthe pleschanka*	Pied Wheatear	鹟科 Muscicapidae

形态特征 小型鸣禽，体长 14~16.5 厘米。雄鸟繁殖羽头顶至颈部偏白色，头两侧至喉部黑色；背部、两翼近黑色；腰部及尾上覆羽白色，尾羽大部分为白色，但尾羽末端及中央尾羽黑色，尾羽末端黑色面积狭窄，飞行时形成“T”字形斑纹；下体白色；跗跖黑色。雄鸟非繁殖羽上体为斑驳的黑色，下体皮黄色。雌鸟头部及上体深褐色，下体白色。

分布与生境 分布于中国北部地区。北京地区为罕见旅鸟和夏候鸟。喜在平原地区的草地、半荒漠环境活动。

习性与生态 常单独或成对活动，站立挺拔。喜在地面快速奔走。主要以昆虫为食。营巢于岩石缝隙或土洞中；窝卵数 4~6 枚。

Morphological characteristics A small passerine (14-16.5 cm). Breeding male is whitish from crown to neck with black head-sides and throat; blackish back and wings; rump and uppertail-coverts white; narrow black band at end of tail and black central tail-feathers form distinctive T-shaped pattern in flight. Underparts white; tarsi black. Non-breeding male has mottled black upperparts and buff underparts. Female has dark brown head and upperparts, and white underparts.

Distribution and habitat Northern China. Rare migrant and summer breeder in Beijing. Favours open grassland or semi-desert areas.

Behaviour and ecology Usually solitary or in pairs. Feeds on the ground, where it forages for food by running or hopping before standing upright. Feeds mainly on insects. Nests in crevices or holes in the ground. Clutch size: 4-6.

IUCN 红色名录 无危 (LC) 2016.10.1

雄鸟 Male

雌鸟 Female

漠䳭	*Oenanthe deserti*	Desert Wheatear	鹟科 Muscicapidae

IUCN 红色名录　无危 (LC) 2016.10.1

形态特征　小型鸣禽，体长 14.5~15.5 厘米。雄鸟繁殖羽头顶棕黄色，眉纹白色，颊至耳羽黑色；背部棕黄色，两翼近黑色，肩羽白色；腰部及尾上覆羽白色，尾羽黑色；颏、喉部黑色，下体淡皮黄色或淡黄白色。雄鸟非繁殖羽头侧为斑驳的黑色。雌鸟头部两侧及喉部为褐色，无黑色。

分布与生境　分布于中国北部地区。北京地区为罕见迷鸟，于延庆、昌平有少量记录。喜在平原地区的多石荒漠、半荒漠环境活动。

习性与生态　常单独或成对活动。喜在地面快速奔走。主要以昆虫为食。窝卵数 4~6 枚。

Morphological characteristics　A small passerine (14.5-15.5 cm). Breeding male has tawny crown, white supercilium, black cheek and ear-coverts; tawny back; blackish wings with white scapulars; white rump and uppertail-coverts and black tail. Chin and throat black; remainder of underparts pale buff or pale yellowish-white. Non-breeding male has mottled black head-sides. Female has brown, not black, head-sides and throat.

Distribution and habitat　Northern China. In Beijing, rare vagrant in Yanqing and Changping Districts. Favours open stony or semi-desert areas in lowlands.

Behaviour and ecology　Usually solitary or in pairs. Forages on the ground. Feeds mainly on insects. Clutch size: 4-6.

白背矶鸫	*Monticola saxatilis*	Common Rock Thrush	鹟科 Muscicapidae

形态特征 中型鸣禽，体长 17~20 厘米。雄鸟头部、颈部、枕部、喉部蓝灰色；胸腹部橘色；背部白色，两翼近黑色；尾羽棕红色。雌鸟整体灰褐色，头部灰褐色具白色斑纹；背部深灰色，两翼近黑色；胸腹部白色沾橘色，具深灰色鳞状斑纹；尾羽棕红色。

分布与生境 中国主要见于西北地区。活动于海拔 1000~3000 米的多岩石灌丛之地。北京为甚罕见迷鸟。

习性与生态 繁殖期成对活动，非繁殖期常单独或集小群活动。喜在多山石环境捕食昆虫。

Morphological characteristics A medium-sized passerine (17-20 cm). Male is blue-grey on head, neck, nape and throat; orange on breast and belly; white on back with blackish wings; reddish-brown tail. Female is overall greyish-brown with greyish-brown head with white markings; deep grey back, blackish wings; breast and belly are white with orange tinge and dark-grey scaling; rufous tail.

Distribution and habitat Northwest China. Favours rocky and bushy habitats in montane area (1,000-3,000 m). Very rare vagrant in Beijing.

Behaviour and ecology Usually found in pairs during breeding season; alone or in small groups during non-breeding season. Feeds on insects.

IUCN 红色名录 无危 (LC) 2016.10.1

雄鸟 Male

雌鸟 Female

蓝矶鸫	*Monticola solitarius*	Blue Rock Thrush	鹟科 Muscicapidae

雄鸟 Male

雌鸟 Female

IUCN 红色名录　无危 (LC) 2016.10.1

形态特征　中型鸣禽，体长 20~23 厘米。北京可见的 *philippensis* 亚种雄鸟上体及颏、喉部皆为蓝色，尾羽深蓝色；胸、腹部至尾下覆羽棕红色。雌鸟整体深灰褐色，胸、腹部皮黄色，具深色鳞状斑纹。

分布与生境　中国广泛分布。北京不常见于多裸岩且接近河流的山区或丘陵地带，为夏候鸟。

习性与生态　繁殖期常成对活动，喜在多山石环境捕食昆虫。营巢于岩缝中；窝卵数 3~6 枚。

Morphological characteristics　A medium-sized passerine (20-23 cm). Subspecies *philippensis* can be found in Beijing. Male *philippensis* has blue upperparts, chin and throat; dark blue tail; rufous breast, belly and undertail-coverts. Female is mostly deep greyish-brown; breast and belly buff with dark scaling.

Distribution and habitat　Widespread in China. In Beijing, uncommon in bare rocky areas near water in low hills or mountains. Residential status: summer breeder.

Behaviour and ecology　Feeding in pairs in breeding season. Favours catching insects in rocky habitat. Nests in crevices. Clutch size: 3-6.

白喉矶鸫	*Monticola gularis*	White-throated Rock Thrush	鹟科 Muscicapidae

形态特征 中型鸣禽，体长 17~19 厘米。雄鸟头顶蓝色，眼周黑色，脸颊棕红色；背部及两翼黑色，肩羽蓝色，有一白色翼斑；喉部白色，胸腹部及两胁棕红色。雌鸟整体褐色，背部多鳞状斑纹；胸腹部白色具深色鳞状斑纹。

分布与生境 中国主要见于东北、华北、华东、东南等地区。北京不常见于中低海拔山区或城市公园的林地环境，为旅鸟及夏候鸟。

习性与生态 多单独或成对活动。繁殖期栖息于多水潮湿的阴坡，营巢于近水的岩石缝隙、洞穴等隐蔽处；窝卵数 4~8 枚。

Morphological characteristics A medium-sized passerine (17-19 cm). Male has blue crown with black around eyes and rufous cheek; black back and wings with a white wing-bar; scapulars blue; white throat contrasts with rufous breast, belly and flanks. Female is mostly brown with scaling on the back; breast and belly are white with dark scaling.

Distribution and habitat Northeast, north, east and southeast China. In Beijing, uncommon migrant and summer breeder in middle or lower mountain areas, and in city parks (mainly on migration).

Behaviour and ecology Solitary or in pairs. nests on wet, shady wooded slopes. Nests in cracks, crevices or holes. Clutch size: 4-8.

IUCN 红色名录 无危 (LC) 2016.10.1

雄鸟 Male

雌鸟 Female

灰纹鹟	*Muscicapa griseisticta*	Grey-streaked Flycatcher	鹟科 Muscicapidae

北京市二级保护野生动物		IUCN 红色名录 无危 (LC) 2018.8.9

形态特征 小型鸣禽，体长 13~15 厘米。雌雄同色。喙较小，大致呈黑色，下喙基黄色不明显。头部灰色，眼先深色，颈部、胸腹部白色，具显著的灰褐色纵纹，半颈环通常不甚明显。背部、两翼、尾羽灰色，初级飞羽长，翼尖通常达尾的三分之二。

分布与生境 中国主要见于东部地区，迁徙时偶见于内陆地区。北京不常见于城市或郊野的开阔林地，为旅鸟。

习性与生态 喜单独活动，常立于树枝上，然后突然飞至空中捕食昆虫。

Morphological characteristics A small passerine (13-15 cm). Sexes alike. Bill relatively small, mostly black, with indistinct yellow base to lower mandible. Grey head with dark lores; neck, breast and belly whitish with greyish-brown streaks; indistinct half neck-ring; grey on back, wings and tail; long primaries with wing tips extending to two thirds of tail.

Distribution and habitat Mainly eastern China, rarely seen in inland areas on migration. In Beijing, uncommon in urban and rural open woodland. Residential status: migrant.

Behaviour and ecology Alone. Often perches on branches and darts out into the air to catch insects.

乌鹟	*Muscicapa sibirica*	Dark-sided Flycatcher	鹟科 Muscicapidae

形态特征 小型鸣禽，体长 12~14 厘米。雌雄同色。喙较小，近黑色，下喙基黄色不明显。头部灰色。眼先深色。颈部、胸腹部偏白色，具模糊的深色斑纹，白色半颈环明显。背部、两翼、尾羽灰色，初级飞羽较长，翼尖通常达尾的一半以上。幼鸟尾下覆羽常具褐色鳞状斑，于成鸟则不清晰。

分布与生境 中国见于东北、华北、华东、华南、东南、西南等地区。北京区域性常见于城市公园或郊野的开阔林地环境，为旅鸟。

习性与生态 单独活动。喜在林中层或冠层活动，常立于树木横枝上，发现昆虫追至空中捕捉。在树侧枝上营杯状编织巢，偶尔营巢于树洞中；窝卵数 4~5 枚。

Morphological characteristics A small passerine (12-14 cm). Sexes alike. Adult has relatively small, blackish bill, with indistinct yellow base to the lower mandible. Grey head with dark lores; neck, breast and belly whitish with dark smudgy streaks; distinctive half neck-ring. Grey on back, wings and tail; rather long primaries with wing tips extending beyond half the length of the tail. Juvenile usually has brown scales on undertail-coverts (indistinct in adults).

Distribution and habitat Northeast, north, east, south, southeast and southwest China. In Beijing, locally common passage migrant in city parks or open forest.

Behaviour and ecology Solitary. Favours middle or top level of canopy. Often darts out from branch to catch insects. Builds cup-shaped nests on tree branches; occasionally nests in tree holes. Clutch size: 4-5.

北京市二级保护野生动物		IUCN 红色名录　无危 (LC) 2016.10.1

北灰鹟	*Muscicapa dauurica*	Asian Brown Flycatcher	鹟科 Muscicapidae

北京市二级保护野生动物

IUCN 红色名录　无危 (LC) 2016.10.1

形态特征　小型鸣禽，体长 12~14 厘米。雌雄同色。喙较大，下喙基黄色明显。头部灰色，眼先为浅色。颈部、胸腹部偏白色而少斑纹，白色半颈环不甚明显。背部、两翼、尾羽灰色，初级飞羽较短，翼尖通常达尾的一半。

分布与生境　中国见于东北、华北、华南、华东、西南、东南等地区。北京区域性常见于城市公园或郊野的开阔林地环境，为旅鸟。

习性与生态　似乌鹟。

Morphological characteristics A small passerine (12-14 cm). Sexes alike. Relatively large bill with obvious yellow base to the lower mandible. Grey head with pale lores; few streaks on whitish neck, breast and belly; indistinct half neck-ring. Grey on back, wings and tail; relatively short primaries, with wing tips usually reaching halfway to tip of the tail.

Distribution and habitat Northeast, north, east and southern China. In Beijing, locally common migrant in city parks and suburban open forest.

Behaviour and ecology Similar to Dark-sided Flycatcher.

褐胸鹟	*Muscicapa muttui*	Brown-breasted Flycatcher	鹟科 Muscicapidae

形态特征 小型鸣禽，体长 12~14 厘米。雌雄同色。喙宽大，上喙色深，下喙黄褐色或肉色。头部灰褐色，眼先皮黄色。颈部、胸腹部偏皮黄色，皮黄色半颈环不明显。背部灰褐色，两翼深褐色，尾羽及尾上覆羽红褐色。

分布与生境 中国见于南部及华东等地区。北京罕见于门头沟中海拔山区林地环境，居留型尚不明确。

习性与生态 常单独活动于中低海拔的阔叶林、混交林环境，性胆怯，常在较为茂密的林中层活动，常立于树木横枝上，少至地面活动。

Morphological characteristics A small passerine (12-14 cm). Sexes alike. Broad bill with dark upper mandible and yellowish-brown lower mandible; greyish-brown head with buff lores; buff neck, breast and belly with indistinct buff half neck-ring; back greyish brown, wings dark brown, tail and uppertail-coverts rufous.

Distribution and habitat Southern China and east China. In Beijing, rare in mountain forest at mid-elevations in Mentougou District. Residential status: unclear.

Behaviour and ecology Usually solitary in broadleaved and mixed forest at middle and low elevations. Timid; often perches on branches in dense woods, rarely on ground.

IUCN 红色名录 无危 (LC) 2016.10.1

白眉姬鹟	*Ficedula zanthopygia*	Yellow-rumped Flycatcher	鹟科 Muscicapidae

雄鸟 Male

雌鸟 Female

北京市二级保护野生动物　　IUCN 红色名录　无危 (LC) 2016.10.1

形态特征　小型鸣禽，体长 12~14 厘米。雄鸟头顶、头两侧、颈部、枕部黑色，具一明显的白色眉纹；上背黑色，下背及腰黄色；两翼黑色具一道长条形白色翼斑；尾羽黑色；喉部、胸部、腹部为艳黄色；尾下覆羽白色。雌鸟头至上体橄榄绿色，无眉纹，下体污白色。

分布与生境　中国除西部外广泛分布。北京区域性常见于城市公园或山区的落叶阔叶林和混交林等环境，为旅鸟和夏候鸟。

习性与生态　常单独或成对活动，喜在中低海拔的林中层或冠层活动，性活跃，常在树丛中跳跃或觅食，时常至空中捕捉食物，较少至地面活动。营巢于树洞中；窝卵数 4~7 枚。

Morphological characteristics A small passerine (12-14 cm). Male has black head with bold white supercilium; upper back black; lower back and rump yellow; black wings with a long white patch on the wing; tail black; chin, throat, breast and belly bright yellow; white undertail-coverts. Female has olive-green head and upperparts; lacks supercilium; off-white below.

Distribution and habitat Widespread in China except for western China. In Beijing, locally common in broadleaf deciduous forests and mixed forests in mountains and also in urban parks. Residential status: migrant and summer breeder.

Behaviour and ecology Usually solitary or in pairs. Active in the middle or crown level of canopy in woodland at medium to low elevations. Catches insects in the air; rarely on the ground. Nests in tree holes. Clutch size: 4-7.

黄眉姬鹟	*Ficedula narcissina*	Narcissus Flycatcher	鹟科 Muscicapidae

形态特征 小型鸣禽，体长 13~14 厘米。雄鸟头至上背黑色，具显著的黄色眉纹；下背及腰黄色；尾羽黑色；两翼黑色，具一道长圆形白色翼斑；颏及喉部橙黄色，胸部、腹部为艳黄色，臀至尾下覆羽白色。雌鸟上体橄榄褐色，喉至胸淡皮黄色，下体余部污白色。

分布与生境 中国主要见于东部地区，迁徙时偶见于内陆地区。北京罕见于城市公园或郊野的开阔林地环境，为旅鸟。

习性与生态 常单独或成对活动，喜在中低海拔的林中层或冠层活动，性活跃，常在树丛中跳跃或觅食，时常至空中捕捉食物，较少至地面活动。营巢于树洞中；窝卵数 3~5 枚。

Morphological characteristics A small passerine (13-14 cm). Male has black head with obvious yellow supercilium; upperparts largely black with yellow lower back and rump; black wings with a distinctive white wing-patch; tail black; chin and throat orange-yellow; breast and belly bright yellow; vent and undertail-coverts white. Female has olive-brown upperparts; throat and breast pale buff. Remainder of underparts off-white.

Distribution and habitat Mainly eastern China; accidental inland during migration. In Beijing, rare in open woodlands of parks and suburbs. Residential status: migrant.

Behaviour and ecology Usually solitary or in pairs; active in middle or crown level of forest at mid to low elevations. Fly-catches for insects, rarely on the ground. Nests in tree holes. Clutch size: 3-5.

北京市一级保护野生动物 | | IUCN 红色名录 无危 (LC) 2016.10.1

雄鸟 Male
雌鸟 Female

绿背姬鹟	*Ficedula elisae*	Green-backed Flycatcher	鹟科 Muscicapidae

雄鸟 Male

雌鸟 Female

北京市一级保护野生动物 *

IUCN 红色名录 无危 (LC) 2016.10.1

形态特征 小型鸣禽，体长 12~14 厘米。雄鸟头及上体大致为橄榄绿色，具一较短的黄色眉纹；腰黄色；两翼黑褐色，具一道较窄而长的白色翼斑；尾羽黑褐色；下体柠檬黄色。雌鸟似雄鸟，但羽色较暗淡，无眉纹，翼斑极不明显。

分布与生境 中国见于华北、华东、东南等地区。北京区域性常见于山区落叶阔叶林（海拔 800~1400 米），迁徙过境时亦见于城市公园，为旅鸟和夏候鸟。

习性与生态 似黄眉姬鹟。繁殖期擅鸣，典型的鸣啭常由 2~5 个婉转乐音起始，旋即衔接以细而高频的金属颤音，再转为一串轻柔的悦耳哨音；鸣叫包含轻柔的“hǜ”等，常重复。

● * 注：《北京市一级保护野生动物名录》未收录（更新 / 接受）绿背姬鹟和黄眉姬鹟的物种划分，故此处沿用黄眉姬鹟的数据。

Morphological characteristics A small passerine (12-14 cm). Male has olive-green head with relatively short yellow supercilium; upperparts largely olive-green; rump yellow; wings blackish-brown with a narrow, long wing-patch; tail blackish-brown; underparts lemon-yellow. Female similar to male but duller with faint wing-bars.

Distribution and habitat Eastern China. In Beijing, locally common in montane areas with deciduous forest (800-1,400 m). Also occurs in urban parks on migration. Residential status: migrant and summer breeder.

Behaviour and ecology Similar to Narcissus Flycatcher. Highly vocal in breeding season. Typical song usually starts with two to five warbling notes, followed by a thin and high-pitched metallic trilling, then turn to a series of soft whistling. Calls include a soft and repetitive "huet".

鸲姬鹟	*Ficedula mugimaki*	Mugimaki Flycatcher	鹟科 Muscicapidae

形态特征 小型鸣禽，体长 12~14 厘米。雄鸟头部黑色，于眼后有一短粗的白色眉纹；背部及两翼黑色，翼上具一显著的宽阔白色翼斑；尾羽黑色；颏、喉部、胸部橙色，腹部偏白色；尾下覆羽白色。雌鸟头部及上体褐色，翼斑甚窄；喉部、胸部淡橙色。

分布与生境 中国见于东部及北部地区。北京不常见于城市公园的混交林或林缘环境，为旅鸟。

习性与生态 常单独或成对活动，喜在林冠层活动。性活跃，主要以昆虫为食。在树枝间营碗状编织巢；窝卵数 4~8 枚。

Morphological characteristics A small passerine (12-14 cm). Male has black head with short, white supercilium behind eyes; back and wings black with a prominent broad, white wing-patch; tail black; orange chin, throat and breast; whitish belly and undertail-coverts. Female has brown head and upperparts; wing-bars narrow; pale orange throat and breast.

Distribution and habitat Northern and eastern China. In Beijing, uncommon on migration in mixed forest and forest edge in city parks. Residential status: migrant.

Behaviour and ecology Usually solitary or in pairs; active in the mid- to upper canopy, feeds on insects. Builds bowl-shaped nests on tree branches. Clutch size: 4-8.

北京市二级保护野生动物 | IUCN 红色名录 无危 (LC) 2016.10.1

雄鸟 Male
雌鸟 Female

锈胸蓝姬鹟 | *Ficedula sordida* | Slaty-breasted Flycatcher | 鹟科 Muscicapidae

雄鸟 Male

雌鸟 Female

IUCN 红色名录　无危 (LC)　2016.10.1

形态特征　小型鸣禽，体长 12~14 厘米。雄鸟头部、上体蓝灰色；颏、喉部、胸腹部橙色，下腹部白色；尾羽蓝灰色，外侧尾下基部白色，尾下覆羽白色。雌鸟整体灰褐色，胸腹部深灰色。

分布与生境　中国见于华北、华南、西南等地区。北京罕见于中高海拔的山地针阔混交林，为夏候鸟。

习性与生态　常单独或成对活动。性活跃，主要以昆虫为食。常于浓密的灌丛中筑杯状巢；窝卵数 3~4 枚。

Morphological characteristics　A small passerine (12-14 cm). Male has bluish-grey head and upperparts; orange chin, throat, breast and belly; white lower belly; bluish-grey tail with white undertail-coverts. Female is overall greyish-brown with dark grey breast and belly.

Distribution and habitat　North, south and southwest China. Rare summer breeder in Beijing. Occurs in broadleaf and mixed forest at high elevations in the mountains.

Behaviour and ecology　Usually solitary or in pairs. Active; feeds on insects. Usually builds cup-shaped nests in dense scrub. Clutch size: 3-4.

橙胸姬鹟	*Ficedula strophiata*	Rufous-gorgeted Flycatcher	鹟科 Muscicapidae

形态特征 小型鸣禽，体长 13~15 厘米。雄鸟头部深蓝灰色，于眼先有一明显的较短的白色眉纹；头顶至背部、两翼均为深褐色；胸部蓝灰色具一橙色月牙形斑纹，腹部灰白色；尾羽近黑色，外侧尾羽基部白色。雌鸟似雄鸟，但整体色浅，胸部橙色斑纹不明显。

分布与生境 中国见于南部地区。通常在分布区见于海拔 1000~3000 米的山地常绿阔叶林、混交林、林缘灌丛等环境，非繁殖期见于较低海拔。北京地区为甚罕见迷鸟。

习性与生态 常单独活动，喜在林中层或灌丛活动。性活跃，主要以昆虫为食。

Morphological characteristics A small passerine (13-15 cm). Male is dark brown from crown, back to wings; distinctive short white supercilium in front of eye; orangey gorget on bluish-grey breast; pale belly; blackish tail with white at base of outer rectrices. Female is similar to male but overall duller; orange gorget inconspicuous.

Distribution and habitat Southern China. Within its range, it occurs in broadleaved evergreen forest, mixed forest and forest edges in the mountains (1,000-3,000 m). Descends in non-breeding season. Very rare vagrant in Beijing.

Behaviour and ecology Usually solitary in mid-level forest or scrub. Active; feeds mainly on insects.

IUCN 红色名录 无危 (LC) 2016.10.1

红胸姬鹟	*Ficedula parva*	Red-breasted Flycatcher	鹟科 Muscicapidae

IUCN 红色名录 无危 (LC) 2018.8.9

形态特征 小型鸣禽，体长 12~14 厘米。上喙铅灰色，下喙偏肉色，白色眼圈明显。雄鸟颊灰色，头顶至背部深褐色，两翼近黑色；尾羽近黑色，外侧尾羽基部白色；喉部橙红色下延至胸部，面积较红喉姬鹟明显大，胸部偏灰色，腹部灰白色。雌鸟似雄鸟，但整体色浅，喉部无橙红色。

分布与生境 中国零星见于华北、华东、西北等地区。北京于秋冬季节罕见于低海拔的开阔林地和灌丛环境，为迷鸟。

习性与生态 常单独活动，喜在林中层或灌丛活动。性活跃，尾常上下抖动，主要以昆虫为食。

Morphological characteristics A small passerine (12-14 cm). Male has greyish cheek and conspicuous white eye-ring; bill bi-coloured with lead-grey upper mandible and flesh-coloured lower mandible; dark brown from crown to back; blackish wings; blackish tail, with white at base of outer rectrices; orange-red throat patch is similar to Taiga Flycatcher but larger, extending to breast; greyish lower breast; pale belly. Female similar to male but duller with no orange on throat.

Distribution and habitat Scarce in north, northwest and east China. In Beijing, rare vagrant in late autumn and winter to parks, open woodland and bushes at low elevations.

Behaviour and ecology Usually solitary; active in mid- to low levels of trees or scrub. Active; flicks tail frequently; feeds on insects.

红喉姬鹟	*Ficedula albicilla*	Taiga Flycatcher	鹟科 Muscicapidae

形态特征 小型鸣禽，体长 11~13 厘米。雄鸟喙铅灰色，白色眼圈明显；颊偏灰色，头顶至背部深褐色，两翼近黑色；尾羽近黑色，外侧尾羽基部白色；喉部橙红色，胸部偏灰色，腹部灰白色。雌鸟似雄鸟，但整体色浅，喉部无橙红色。

分布与生境 中国除西部地区外广泛分布。北京甚常见于低海拔的阔叶林、混交林、林缘灌丛等环境，为旅鸟。

习性与生态 常单独活动，迁徙时亦常集成松散的小群，喜在林中层或灌丛活动。性活跃，尾常上下抖动，主要以昆虫为食。鸣叫声为单调、短促的“ta”声。营巢于树洞中；窝卵数 4~7 枚。

Morphological characteristics A small passerine (11-13 cm). Male has lead-grey bill and clear white eye-ring; cheeks greyish; crown and back dark brown, blackish wings; tail largely blackish; outer rectrices white at base; orange-red throat, greyish breast, buff belly. Female is similar to male but duller and lacks orange on throat.

Distribution and habitat Widespread in China except for the west. In Beijing, abundant on migration to almost all wooded habitats, including broadleaved and mixed forest and scrub. Residential status: migrant.

Behaviour and ecology Usually solitary but sometimes in small loose groups on migration; Feeds at mid- to low levels of trees or shrubs. Abundant on migration. Active, flicks tail; feeds on insects. Call is a short, dry "ta". Nests in tree holes. Clutch size: 4-7.

北京市二级保护野生动物		IUCN 红色名录 无危 (LC) 2016.10.1

雄鸟 Male

雌鸟 Female

灰蓝姬鹟	*Ficedula tricolor*	Slaty-blue Flycatcher	鹟科 Muscicapidae

雄鸟 Male

雌鸟 Female

IUCN 红色名录　无危 (LC) 2016.10.1

形态特征　小型鸣禽，体长 10~13 厘米。喙黑色。雄鸟头部深蓝灰色，头顶至背部蓝灰色，两翼近黑色；尾羽蓝灰色，尾羽外侧白色明显；喉偏白色，胸腹部灰褐色。雌鸟整体褐色，下体较上体色浅，尾羽棕红色。

分布与生境　中国见于华北、华南、西南等地区。繁殖期见于海拔 1500~4000 米的阔叶林、混交林、林缘灌丛等环境，非繁殖期向较低海拔移动。北京地区为甚罕见迷鸟。

习性与生态　常单独活动，喜在林中层或灌丛活动。性活跃，较惧人，停落时尾常上下抖动，主要以昆虫为食。

Morphological characteristics A small passerine (10-13 cm). Bill black. Male has dark blue-grey head; bluish-grey from crown to back; blackish wings; bluish-grey tail with distinct white on sides; buff throat; greyish-brown breast and belly. Female is overall brown with paler underparts and rufous tail.

Distribution and habitat North, south and southwest China. Occurs in montane broadleaved, mixed forest and scrub at forest edge (1,500-4,000 m). Seen also in lowlands in non-breeding season. Very rare vagrant in Beijing.

Behaviour and ecology Usually solitary in the mid-level of forest. Active but timid, flicking tail when perched. Feeds on insects.

白腹蓝鹟	*Cyanoptila cyanomelana*	Blue-and-white Flycatcher	鹟科 Muscicapidae

形态特征 中型鸣禽，体长 16~17 厘米。雄鸟喙黑色，颊蓝黑色，头顶蓝色具金属光泽，上体、两翼和尾羽蓝色；喉部、胸部近黑色，腹部白色。雌鸟整体灰褐色，尾羽偏深棕红色。第一年雄鸟除两翼蓝色外，其余体色似雌鸟。

分布与生境 广泛分布于中国东部地区。北京罕见于中低海拔的山地阔叶林、混交林、林缘灌丛等环境，为旅鸟和夏候鸟。

习性与生态 常单独活动，喜在林中层活动。常站立于树枝横枝上休息或寻找食物，主要以昆虫为食。营巢于树洞或树根间；窝卵数 3~7 枚。

Morphological characteristics A medium-sized passerine (16-17 cm). Male has bluish-black cheek with black bill; metallic blue crown; bluish upperparts, wings and tail; blackish throat and breast; white belly. Female is overall greyish-brown with dark brown-red tail. First-year male is similar to female except wings are blue.

Distribution and habitat Widespread in eastern China. In Beijing, rare in broadleaved and mixed forests, and forest edge scrub in the mountains. Residential status: migrant and summer breeder.

Behaviour and ecology Usually solitary. Often perches on horizontal branches while resting or foraging. Feeds on insects. Nests in tree holes or among roots of trees. Clutch size: 3-7.

北京市二级保护野生动物 | IUCN 红色名录 无危 (LC) 2016.10.1

雄鸟 Male

雌鸟 Female

白腹暗蓝鹟	*Cyanoptila cumatilis*	Zappey's Flycatcher	鹟科 Muscicapidae

雄鸟 Male

雌鸟 Female

北京市二级保护野生动物 * | 国家二级保护野生动物 | IUCN 红色名录 近危 (NT) 2016.10.1

形态特征 中型鸣禽，体长 15~17 厘米。雄鸟眼先至颊蓝黑色；头、上体、两翼、颏、喉、胸皆为湖蓝色，仅眼先为深蓝色，腹部及尾下覆羽白色。雌鸟整体灰褐色，尾羽偏深棕红色。似白腹蓝鹟，但颏、喉、胸羽色较淡。

分布与生境 中国见于华北、华南、西南等地区。北京不常见于山区阔叶林、混交林和林缘灌丛，为旅鸟和夏候鸟。

习性与生态 习性似白腹蓝鹟，但二者鸣声存在差异：白腹暗蓝鹟和白腹蓝鹟的鸣啭通常为 5~12 个音节组成的悦耳短句，其典型乐句的区别体现在前者有更多重复音节（如单音节重复 5~10 次）、节奏稍慢、音调低（最高频约 5.5 kHz）；后者乐句更多变（常含转折音、快速颤音等）且音调高（最高频常超过 6 kHz）。

* 注：《北京市二级保护野生动物名录》未收录（更新 / 接受）白腹暗蓝鹟和白腹蓝鹟的物种划分，故此处沿用白腹蓝鹟的数据。

Morphological characteristics A medium-sized passerine (15-17 cm). Male has blackish-blue lores and cheek; turquoise-blue head, upperparts, wings, chin, throat and breast; belly and undertail-coverts white. Female is overall greyish-brown with dark brown-red tail. Very similar to Blue-and-white Flycatcher but has turquoise-blue throat and breast.

Distribution and habitat North, south and southwest China. In Beijing, uncommon in broadleaved and mixed forest, as well as forest edge scrub in montane areas. Residential status: migrant and summer breeder.

Behaviour and ecology Similar to Blue-and-white Flycatcher but has different volcalisation. Songs of Zappey's and Blue-and-white Flycatcher usually consist of melody short strophes with 5-12 notes. The typical song of the former is more repetitive (e.g. five to ten repetitions of one single note), slightly slower-paced, and lower-pitched (with the highest frequency ~ 5.5 kHz), whereas the latter is more variable (including twists and rapid trills) and higher-pitched (with the highest frequency often > 6 kHz).

铜蓝鹟	*Eumyias thalassinus*	Verditer Flycatcher	鹟科 Muscicapidae

形态特征 中型鸣禽，体长 15~17 厘米。雄鸟整体铜蓝色；喙黑色，眼先黑色；尾下覆羽具白色鳞状斑纹。雌鸟似雄鸟，但整体色浅，眼先灰色。

分布与生境 中国见于南方地区，于华北有零星记载。北京甚罕见于低海拔的阔叶林、混交林、林缘灌丛等环境，推测为迷鸟。

习性与生态 常单独活动，喜在林中层活动。常站立于树枝横枝上休息或寻找食物，主要以昆虫为食。一般营巢于树洞、土洞或岩隙中；窝卵数 3~5 枚。

Morphological characteristics A medium-sized passerine (15-17 cm). Male is generally copper-blue with black bill and black lores; undertail-coverts have white scales. Female is similar to male but duller with grey lores.

Distribution and habitat Mainly southern China, scarce in north China. In Beijing, very rare in broadleaved and mixed forest, and forest edge scrubs at low elevations. Residential status: vagrant.

Behaviour and ecology Usually solitary in mid-level of forest. Often perches on transverse branches when resting or foraging. Feeds on insects. Nests in holes in the ground or in tress, also in rock crevices. Clutch size: 3~5.

IUCN 红色名录 无危 (LC) 2016.10.1

雄鸟 Male

雌鸟 Female

白喉林鹟	*Cyornis brunneatus*	Brown-chested Jungle Flycatcher	鹟科 Muscicapidae

	国家二级保护野生动物	IUCN 红色名录　易危 (VU) 2016.10.1

形态特征　小型鸣禽，体长 13~16 厘米。雌雄同色。喙较大，上喙黑色，下喙黄色。整体棕褐色。头部棕褐色，背部、两翼棕褐色。尾羽、腰部及尾上覆羽略偏红褐色。喉部偏白色，胸腹部棕褐色。

分布与生境　中国主要见于华南、东南地区。北京罕见于低海拔的林地、竹林、林缘灌丛等环境，为迷鸟。

习性与生态　常单独活动，喜在林中层活动。常站立于树枝横枝上休息或寻找食物，有时亦至地面觅食。

Morphological characteristics　A small passerine (13-16 cm). Sexes alike. Relatively large bill with black upper mandible and yellow lower mandible. Overall brown plumage; brownish head, back and wings; tail, rump and uppertail-coverts reddish-brown; whitish throat; brownish breast and belly.

Distribution and habitat　South and southeast China. In Beijing, rare in woods and scrub near forest edge at low elevations. Residential status: vagrant.

Behaviour and ecology　Usually solitary and found in mid-level forest. Often perches on horizontal branches when resting or foraging; occasionally feeds on the ground.

戴菊	*Regulus regulus*	Goldcrest	戴菊科 Regulidae

北京市二级保护野生动物		IUCN 红色名录　无危 (LC) 2018.8.9

形态特征　小型鸣禽，体长 9~10 厘米。雄鸟头部灰色，具一甚为明显的白色眼圈，头顶黄色，具一不显著的橙色条纹；胸腹部、背部淡绿色，两翼绿色，具两道白色翼斑；尾暗绿色。雌鸟甚似雄鸟，于野外不易识别，头顶黄色条纹中无橙色。

分布与生境　中国广泛分布。北京不常见于中低海拔的针叶林、混交林环境，为冬候鸟及旅鸟。

习性与生态　冬季常成小群活动，喜在茂密的针叶林中层活动。性活跃，主要以昆虫为食。于针叶树上营碗状巢；窝卵数 7~12 枚。

Morphological characteristics A small passerine (9-10 cm). Male has distinctive white eye-ring on grey head; yellow crown with an orange central streak; breast, belly and back are pale green; green wings with two whitish wing-bars; tail dark green. Female is similar to male, but yellow crown lacks orange central streak.

Distribution and habitat Widespread in China. In Beijing, uncommon in coniferous and mixed forest in lowlands. Residential status: winter visitor and migrant.

Behaviour and ecology Usually found in small groups in dense coniferous forest or amongst evergreen trees in parks. Very active. Feeds on insects. Builds bowl-shaped nests in coniferous trees. Clutch size: 7-12.

太平鸟	*Bombycilla garrulus*	Bohemian Waxwing	太平鸟科 Bombycillidae

形态特征 中型鸣禽，体长 19~23 厘米。雄鸟头部棕褐色，贯眼纹黑色，头顶具甚显著的冠羽；两翼棕褐色，翼尖黑色具两道白色翼斑；次级飞羽红色突出明显，初级飞羽外缘黄色；尾灰褐色，尾端黄色；颏、喉黑色，胸部淡粉褐色，尾下覆羽栗棕色。雌鸟似雄鸟，但次级飞羽红色突出部分及初级飞羽外缘黄色不明显。

分布与生境 中国见于西北、东北、华北、华南、华东等地区。北京不常见于针叶林和混交林，为冬候鸟及旅鸟。

习性与生态 冬季常成群活动，喜在树上层活动。时常与小太平鸟等混群，主要以植物果实为食。于近水的针叶树上营巢；窝卵数 4~7 枚。

Morphological characteristics A medium-sized passerine (19-23 cm). Male has dark brown head with a conspicuous crest on crown and black eye-stripe; wings with two white wing patches and black wing tip; distinctive red patches to secondaries and yellow waxy fringes to primaries; greyish-brown tail with yellow tip; black throat; pale pinkish-buff breast and belly; undertail-coverts chestnut. Female is similar to male, but tend to have smaller black chin-patch and fewer red or yellow waxy fringes on the wing.

Distribution and habitat Northwest, northeast, north, east and south China. In Beijing, uncommon in coniferous forest and mixed forest, including urban parks. Residential status: winter visitor and migrant.

Behaviour and ecology Usually found in flocks feeding on berries in winter. Often forms mixed flocks with Japanese Waxwings. Nests in coniferous trees near water. Clutch size: 4-7.

北京市二级保护野生动物 | IUCN 红色名录 无危 (LC) 2016.10.1

小太平鸟	*Bombycilla japonica*	Japanese Waxwing	太平鸟科 Bombycillidae

北京市二级保护野生动物		IUCN 红色名录　近危 (NT) 2018.8.9

形态特征　中型鸣禽，体长 18~20 厘米。雄鸟头部棕褐色，贯眼纹黑色，头顶具显著的冠羽；两翼棕褐色，肩羽红色，翼尖黑色，初级飞羽外缘红色；尾灰褐色，尾端红色；喉部黑色，胸腹部棕褐色，腹部中央具一淡黄白色斑块，尾下覆羽栗棕色。雌鸟似雄鸟，喉部黑色边缘较模糊，初级飞羽外翈白色。

分布与生境　中国见于东北、华北、华南、华东、东南等地区。北京不常见于针叶林、混交林环境，为冬候鸟及旅鸟。

习性与生态　冬季常成群在林冠层及灌丛觅食。时常与太平鸟混群，春、夏季主食昆虫，秋、冬季则主要以植物果实为食。

Morphological characteristics　A medium-sized passerine (18-20 cm). Male has dark brown head with a conspicuous crest on crown and black eye-stripe; greyish-brown wings have black wingtip; red fringes on primaries; scapular red; greyish-brown tail with red tip; black throat, greyish-brown breast with a pale creamy-white belly-patch; undertail-coverts chestnut-brown. Female is similar to male, but the black border on throat is diffuse and primaries have white fringes with little or no red.

Distribution and habitat　Northeast, north, south, east and southeast China. In Beijing, uncommon in coniferous forest and mixed forest, including urban parks. Residential status: winter visitor and migrant.

Behaviour and ecology　Usually found feeding in canopy and bush in flocks in winter, sometimes mixed with Bohemian Waxwings. Feeds on berries in autumn and winter, and insects in spring and summer.

领岩鹨	*Prunella collaris*	Alpine Accentor	岩鹨科 Prunellidae

形态特征 中型鸣禽，体长 15~18 厘米。成鸟喙灰黑色，下喙基黄色。头深灰色，背部棕褐色具深色纵纹，两翼棕色具两道白色翼斑。腰及尾上覆羽棕红色。尾深褐色。喉部沾白色，胸部深灰色，腹部及两胁皮黄色具橘色三角形斑纹，尾下覆羽无白色。

分布与生境 广泛分布于中国长江以北地区。北京不常见于海拔 900 米以上的裸岩和草地环境，为冬候鸟及旅鸟。

习性与生态 冬季常成小群活动，喜在岩石或草地上站立。性活跃，不甚惧人，主要以昆虫、草籽为食。多在石缝和石洞中营巢；窝卵数 3~5 枚。

Morphological characteristics A medium-sized passerine (15-18 cm). Adult has blackish-grey bill with yellow base to the lower mandible; dark grey head; brown wings with two white wing-bars; rump and uppertail-coverts rufous; tail dark brown; greyish-brown back with dark streaks; throat washed with white; breast dark grey, buff belly and flanks with orange triangular marks; vent and undertail-coverts off-white.

Distribution and habitat Widespread in the north of Yangtze River in China. In Beijing, uncommon in rocky habitats at mid- to upper elevations (above 900 m). Residential status: winter visitor and migrant.

Behaviour and ecology Often feeds in small groups, occasionally large flocks in winter. Forages on ground and sometimes perches on rocks or grassy tussocks. Energetic, feeds on insects and seeds. Nests in rock crevices and holes. Clutch size: 3-5.

IUCN 红色名录　无危 (LC) 2016.10.1

棕眉山岩鹨	*Prunella montanella*	Siberian Accentor	岩鹨科 Prunellidae

IUCN 红色名录　无危 (LC) 2016.10.1

形态特征　小型鸣禽，体长 13~16 厘米。成鸟头顶及头两侧近黑色，眉纹棕黄色，较为宽阔。背部棕褐色具深色纵纹，两翼棕色具一道不显著白色翼斑。尾近黑色。喉部、胸腹部棕黄色，两胁具深棕色纵纹。

分布与生境　中国主要见于东北和华北，亦偶见于西北、中部和华东地区。北京区域性常见于山地、平原地区的灌丛、草地环境，为冬候鸟及旅鸟。

习性与生态　常成对或单独活动，有时和鹀类混群。主要以昆虫、草籽为食。常营巢于灌丛中或灌丛旁的地面；窝卵数 4~6 枚。

Morphological characteristics　A small passerine (13-16 cm). Adult has blackish crown and head-sides; supercilium broad, buff; dark-streaked brown back; brown wings with indistinct white wing-bar; blackish tail; buff throat, breast and belly with dark brown streaks on flank;

Distribution and habitat　Mainly north and northeast China, also scarce in northwest, east and central China. In Beijing, locally common in scrubby habitats in both lowlands and medium-high elevations. Residential status: winter visitor and migrant.

Behaviour and ecology　Usually in small groups in winter, sometimes in loose mixed flocks with buntings. Feeds on insects and seeds. Nests in bushes or on the ground, shelterd by bushes. Clutch size: 4-6.

褐岩鹨	*Prunella fulvescens*	Brown Accentor	岩鹨科 Prunellidae

形态特征 小型鸣禽，体长 15~16 厘米。成鸟头顶黑褐色，头两侧黑色，眉纹白色。背部黄褐色具深色纵纹，两翼棕色具一道不显著皮黄色翼斑。尾近黑色。喉部皮黄色，胸腹部皮黄色，两胁无纵纹。

分布与生境 中国见于西北、华北、西南等地区。北京甚罕见于山区地带的灌丛、草地环境，为冬候鸟。

习性与生态 常成对或单独活动，喜在多灌丛、碎石山坡觅食。主要以昆虫、草籽为食。一般在岩隙间或灌丛中营杯状巢；窝卵数 4~5 枚。

Morphological characteristics A small passerine (15-16 cm). Adult has blackish-brown crown, white supercilium and black head-sides; tawny back with dark streaks, brown wings with a vague buffy wing- bar; tail blackish; buff throat, breast and belly; unstreaked flanks.

Distribution and habitat Northwest, north and southwest China. In Beijing, very rare in rocky scrub in mountainous habitats. Residential status: winter visitor.

Behaviour and ecology Usually solitary or in pairs; prefers to forage in scrubby and rocky hillsides. Feeds on insects and seeds. Usually builds cup-shaped nest in rock crevices or in bushes. Clutch size: 4-5.

IUCN 红色名录　无危 (LC) 2018.8.9

家麻雀	*Passer domesticus*	House Sparrow	雀科 Passeridae

雌鸟 Female

雄鸟 Male

雌鸟 Female

IUCN 红色名录 无危 (LC) 2016.10.1

形态特征 小型鸣禽，体长 14~16 厘米。喙铅灰色。雄鸟头顶灰色，颊白色，枕至后颈棕红色，背部棕褐色具深色纵纹；尾黑褐色；喉部黑色，下体白色。雌鸟总体棕黄色，头部具一皮黄色眉纹，上体棕褐色具深色纵纹，下体淡皮黄色，无斑纹。

分布与生境 中国主要见于西北和东北，于其他地区亦有零星记载。于分布区内常在城市乡村有人类居住的环境活动。甚罕见于北京，仅在朝阳公园有一笔迷鸟记录。

习性与生态 性活跃，不甚惧人。常成小群活动，喜在农田、村落附近觅食。有时与麻雀混群。窝卵数 5~7 枚。

Morphological characteristics A small passerine (14-16 cm). Bill lead-grey. Male has grey crown, black eye-stripe and whitish cheek; nape and hindneck chestnut; brown upperparts with black streaks; blackish-brown tail; black throat, off-white below. Female is overall tawny with a buffy supercilium; brownish upperparts with black streaks; unmarked pale buff underparts.

Distribution and habitat Mainly northwest and northeast China, scarce in other areas in China. Common in populated cities and villages within its range. Very rare in Beijing, with only one record of a vagrant at Chaoyang Park.

Behaviour and ecology Active and bold. Usually found in small groups in cultivated areas and villages, mostly near human habitation. Sometimes in mixed flock with Eurasian Tree Sparrows. Feeds mainly on insects and seeds. Clutch size: 5-7.

山麻雀	*Passer cinnamomeus*	Russet Sparrow	雀科 Passeridae

形态特征 小型鸣禽，体长 11~14 厘米。喙较厚实，铅灰色。雄鸟头顶栗红色，颊白色，贯眼纹黑色，较短；背部栗红色具黑色点斑，两翼多黑色，羽缘白色，具一道白色翼斑；尾近黑色；喉部黑色，下体余部白色。雌鸟头部褐色，具一淡皮黄色眉纹；上体及两翼棕褐色，具深色短纵纹，两翼具一显著的近白色翼斑；下体皮黄色几无斑纹。

分布与生境 中国见于华北、中部及南方等地区。北京不常见于山区和丘陵地带的林缘、草地、农田等环境，为夏候鸟及旅鸟。

习性与生态 常成对活动。不甚惧人，主要以昆虫、草籽为食。营巢于树洞或岩石缝隙中；窝卵数 4~6 枚。

Morphological characteristics A small passerine (11-14 cm). Bill deep, lead-grey. Male has rufous-chestnut crown, white cheek with short, black eye-stripe; russet back with black spots; wings blackish with white fringes and a white wing-bar; tail blackish; black throat; white below. Female has brown head with pale buff supercilium; short, blackish streaks on brown upperparts and wings with a prominent whitish wing-bar; unmarked underparts.

Distribution and habitat North, central and southern China. In Beijing, uncommon in forest edge, villages and cultivated areas in mountains and hills. Residential status: summer breeder and migrant.

Behaviour and ecology Usually found in pairs. Bold. Feeds mainly on insects and seeds. Nests in tree holes or rock crevices. Clutch size: 4-6.

IUCN 红色名录 无危 (LC) 2016.10.1

雄鸟 Male

雌鸟 Female

雄鸟 Male

麻雀	*Passer montanus*	Eurasian Tree Sparrow	雀科 Passeridae

IUCN 红色名录 无危（LC）2016.10.1

形态特征 小型鸣禽，体长 12~15 厘米。雌雄相似。具厚实的深铅灰色喙。成鸟头顶棕红色，脸颊白色具一黑色斑，贯眼纹黑色，颈部具一近白色颈环；背部棕色，具黑色纵纹；两翼近黑色，羽缘棕色，大覆羽和中覆羽羽端近白色。尾近黑色。颏及喉部黑色，下体大致为污白色，两胁皮黄色。

分布与生境 广泛分布于中国海拔 4500 米以下的各种生境。北京甚常见于有人居住的城镇、乡村，以及农田、荒野等各种环境，是北京地区最常见的鸟类之一；为留鸟。

习性与生态 常在繁殖期集小群，非繁殖季集大群活动。不甚惧人，主要以昆虫、草籽为食。一般营巢于人类建筑、石缝或树洞中，亦会占据家燕和金腰燕的巢；窝卵数 5~6 枚。

Morphological characteristics A small passerine (12-15 cm). Sexes alike. Bill deep, lead-grey. Adult has red-brown crown, white cheek with a black patch; short, black eye-stripe; a whitish neck ring; back brown, streaked black; wings largely blackish, fringed brown; greater and median coverts tipped whitish; tail blackish; chin and throat black; whitish below with buff flanks.

Distribution and habitat Widespread in China (below 4,500 m). In Beijing, abundant human populated areas in both urban and rural, as well as open country and farmland. It is one of the most common bird species in Beijing. Residential status: resident.

Behaviour and ecology Often found in small groups in breeding season and large flocks in non-breeding season. Bold. Feeds mainly on insects in summer and seeds in winter. Nests in human buildings, rock crevices or tree holes; also occupies nests of Barn Swallows and Red-rumped Swallows. Clutch size: 5-6.

石雀	*Petronia petronia*	Rock Sparrow	雀科 Passeridae

形态特征 中型鸣禽，体长 15~17 厘米。喙较厚，喙峰灰色，其余黄色。雌雄相似。成鸟整体棕黄色，头顶深褐色，眉纹皮黄色，颊褐色。背部深褐色，具两道皮黄色翼斑。尾深褐色。胸部具一黄色斑块，胸腹部灰褐色具棕褐色纵纹。

分布与生境 中国见于西北、东北、华北、西南等地区。北京为甚罕见迷鸟。主要栖息于荒山、岩石荒坡、稀疏灌丛等环境。

习性与生态 非繁殖季常成群活动，喜在地面觅食。不甚惧人，主要以昆虫、草籽为食。

Morphological characteristics A medium-sized passerine (15-17 cm). Sexes alike. Bill deep; culmen grey; rest yellow. Adult is overall brownish-yellow with dark brown crown, brown cheek and a buff supercilium; dark brown back with two buff wing-bars; dark brown tail. yellow patch between throat and breast; breast and belly greyish-brown with brown streaks.

Distribution and habitat Northwest, northeast, north and southwest China. Occurs on bare hills, rocky barren slopes and areas with sparse shrubs. Very rare vagrant in Beijing.

Behaviour and ecology Found in flocks in non-breeding season. Feeds mainly on insects and seeds on the ground.

IUCN 红色名录 无危 (LC) 2016.10.1

山鹡鸰	*Dendronanthus indicus*	Forest Wagtail	鹡鸰科 Motacillidae

IUCN 红色名录 无危 (LC) 2016.10.1

形态特征 中型鸣禽，体长 16~18 厘米。上喙近黑色，下喙粉色。成鸟头顶橄榄色，眉纹皮黄色或白色。背部橄榄色，两翼近黑色具两道白色翼斑。尾羽较长，两枚中央尾羽橄榄褐色，两枚最外侧尾羽白色，其余尾羽黑褐色。下体大致为白色，上胸部具一道显著的黑色横带，下胸具一断开的深褐色横带；跗跖浅褐色。

分布与生境 中国除西部地区外广泛分布。主要活动于阔叶林和混交林及林缘，亦见于城市林地。北京为不常见旅鸟和罕见夏候鸟。

习性与生态 常单独活动。树栖型，喜在树枝上行走或休息。与其他鹡鸰不同，山鹡鸰尾常左右摇摆。营巢于乔木侧枝上；窝卵数 4~5 枚。

Morphological characteristics A medium-sized passerine (16-18 cm). Sexes alike. It has blackish upper mandible and pinkish lower mandible. Adult has olive crown with buff or white supercilium; olive back; two white wing-bars on blackish wings. Tail long; central rectrices olive-brown; two outermost rectrices white; rest of rectrices blackish-brown. Underparts largely white; a prominent black band on upper breast; a broken dark brown band on lower breast. Tarsi pale brown.

Distribution and habitat Widespread in China except for the west. Usually occurs in broadleaved forest, mixed forest (including forest edge) and urban woodland. Uncommon migrant and rare summer breeder in Beijing.

Behaviour and ecology Usually solitary. Often walking or perching on low branches. Forest Wagtail wags its tail horizontally, not vertically as in other wagtails. Nests in trees. Clutch size: 4-5.

西黄鹡鸰 | *Motacilla flava* | Western Yellow Wagtail | 鹡鸰科 Motacillidae

形态特征 中型鸣禽，体长 15~18 厘米。亚种较多，*beema* 亚种雄鸟头顶灰色，脸颊灰色，耳羽具白色点斑。胸腹部黄色，背部橄榄绿色；尾近黑色，外侧尾羽白色。*feldeggi* 亚种雄鸟头部黑色，背部橄榄绿色，下体全为柠檬黄色；跗跖黑色。各亚种雌鸟下体均较淡（通常为淡皮黄色）。

分布与生境 中国主要见于西北地区。主要栖息于淡水沼泽湿地、河流、农田等环境。北京为甚罕见迷鸟，仅 *feldeggi* 亚种于 2013 年 5 月在昌平沙河水库有过一次记载。

习性与生态 习性似黄鹡鸰，非繁殖期常成小群活动。在水边行走时尾常不停上下摆动。

Morphological characteristics A medium-sized passerine (15-18 cm) with many subspecies. Male *beema* has grey crown and cheek with white spots on ear-coverts; yellow breast and belly, olive back and blackish tail with white outermost rectrices. Subspecies *feldeggi* has black head; olive back; underparts lemon-yellow. Tarsi black. Female of each subspecies has paler underparts (usually pale buff).

Distribution and habitat Northwest China. Occurs in freshwater wetlands, including rivers, marshes and irrigated agricultural land. Very rare vagrant in Beijing with only one record (subspecies *feldeggi*) at Shahe Reservoir, Changping District (May 2013).

Behaviour and ecology Behaviours similar to Eastern Yellow Wagtail. Found in small groups in non-breeding season. Constantly pumps tail while walking near water.

IUCN 红色名录 无危 (LC) 2018.8.9

黄鹡鸰	*Motacilla tschutschensis*	Eastern Yellow Wagtail	鹡鸰科 Motacillidae

IUCN 红色名录　无危 (LC) 2019.8.15

形态特征　中型鸣禽，体长 16~18 厘米。亚种较多，其中 *tschutschensis*、*macronyx* 和 *taivana* 亚种在北京有记录。*macronyx* 亚种雄鸟头部灰色。*taivana* 亚种雄鸟头顶橄榄绿色，眉纹黄色，耳羽深橄榄色。*tschutschensis* 亚种雄鸟头部灰色，眉纹白色；跗跖黑色。雌鸟头部灰色或灰褐色，上体橄榄褐色，下体淡皮黄色。

分布与生境　广泛分布于中国各地。北京为区域性常见旅鸟。主要栖息于较低海拔的淡水沼泽湿地、河流、农田等环境。

习性与生态　非繁殖期常成群活动，有时可见百只以上的大群。喜在近水湿地行走，尾常不停上下摆动。主要以昆虫为食。常在近河流或溪流的草丛中筑巢；窝卵数 5~6 枚。

Morphological characteristics A medium-sized passerine (16-18 cm) with many subspecies. Subspecies *tschutschensis*, *macronyx* and *taivana* have been recorded in Beijing. Male *macronyx* has grey head. Male *taivana* has olive crown with yellow supercilium and dark olive ear-coverts. Male *tschutschensis* has grey head with white supercilium. Tarsi black. Female has greyish-brown crown, olive-brown upperparts and pale buff underparts.

Distribution and habitat Widespread in China. In Beijing, locally common in lowland freshwater wetlands including along rivers, on irrigated agricultural land and along the edges of lakes and reservoirs. Residential status: passage migrant.

Behaviour and ecology Found in flocks in non-breeding season and on migration, sometimes numbering hundreds of individuals. Constantly pumps tail while walking near water. Feeds mainly on insects. Usually nests in grass near rivers or streams. Clutch size: 5-6.

黄头鹡鸰	*Motacilla citreola*	Citrine Wagtail	鹡鸰科 Motacillidae

形态特征 中型鸣禽，体长 16~20 厘米。雄鸟头、胸至腹部皆为柠檬黄色；后颈黑色；背部黑色或深灰色，两翼黑色具白色斑纹；尾羽近黑色，最外侧尾羽白色；尾下覆羽白色；跗跖黑色。雌鸟整体色较雄鸟淡，头灰褐色，具一较宽的淡黄色眉纹向后延伸并环绕耳羽外周；背部灰褐色或橄榄褐色。

分布与生境 广泛分布于中国各地。北京为区域性常见旅鸟。主要栖息于沼泽湿地、河流、农田等环境。

习性与生态 非繁殖期常成小群活动。喜在近水湿地行走，尾羽不停摆动。于地面或草丛中营巢；窝卵数 4~5 枚。

Morphological characteristics A medium-sized passerine (16-20 cm). Male has lemon-yellow head, breast and belly; hindneck black; back blackish or dark grey; black wings with double white wing-bars; tail black with white outermost rectrices; undertail-coverts white; tarsi black. Female has paler plumage; head greyish-brown; supercilium pale yellow, extending around ear-coverts. Upperparts greyish-brown or olive-brown.

Distribution and habitat Widespread in China. In Beijing, locally common at wetlands, including along rivers and at the edges of lakes and reservoirs. Residential status: passage migrant.

Behaviour and ecology Found in small groups in non-breeding season. Constantly pumps tail while walking. Feeds mainly on insects. Nests on the ground or in grasses. Clutch size: 4-5.

IUCN 红色名录　无危 (LC) 2019.8.15

灰鹡鸰	*Motacilla cinerea*	Grey Wagtail	鹡鸰科 Motacillidae

IUCN 红色名录　无危 (LC) 2016.10.1

形态特征　中型鸣禽，体长 16~18 厘米。雄鸟繁殖羽头部灰色，具长且宽的白色眉纹，颊纹白色；上体大致为灰色，两翼近黑色；尾上覆羽黄色，尾羽黑褐色，外侧尾羽白色；颏、喉部黑色，其余下体柠檬黄色；跗跖粉色。雌鸟和非繁殖期雄鸟喉部为白色，有时具斑驳的黑色区域。

分布与生境　广泛分布于中国各地。北京为常见旅鸟、夏候鸟，冬季偶有记录。主要栖息于山涧河流、淡水沼泽湿地环境。

习性与生态　常单独或成对活动，迁徙过境时亦形成松散的小群。喜在近水湿地行走，尾常不停上下摆动。有时涉水觅食，主要以昆虫为食。营巢于山区溪流和河流附近；窝卵数 4~6 枚。

Morphological characteristics　A medium-sized passerine (16-18 cm). Male in breeding plumage has grey head with long, broad, white supercilium and white moustachial stripe; upperparts largely grey; wings blackish; uppertail-coverts yellow; tail blackish-brown; outermost rectrices white; black chin and throat; lemon-yellow below; pink tarsi. Female and non-breeding male has white throat, sometimes mottled black.

Distribution and habitat　Widespread in China. Common migrant, summer breeder and scarce winter visitor in Beijing. Occurs in mountain creeks and lowland wetlands.

Behaviour and ecology　Usually solitary or in pairs, often in small loose groups on migration. Constantly pumps tail up while walking near water. Sometimes wades when foraging. Feeds mainly on insects. Nests near streams and rivers in montane areas. Clutch size: 4-6.

白鹡鸰	*Motacilla alba*	White Wagtail	鹡鸰科 Motacillidae

形态特征 中型鸣禽，体长 16~19 厘米。亚种众多，北京可见中国有分布的 6 个亚种（*leucopsis, ocularis, baicalensis, lugens, alba* 及 *personata*），整体为黑白两色或黑白灰三色。雄鸟通常前额白色，头顶黑色，头两侧白色（部分亚种，如 *ocularis* 和 *lugens* 具黑色贯眼纹），尾羽黑色，外侧尾羽白色，胸黑色，其余下体白色。雌鸟似雄鸟但整体色浅。幼鸟头两侧沾黄色。

分布与生境 广泛分布于中国各地。北京为甚常见旅鸟、夏候鸟，亦有零散的冬候鸟记录。主要栖息于淡水沼泽湿地、农田等环境。

习性与生态 喜在近水湿地行走，尾常不停上下摆动。主要以昆虫为食，飞行时呈明显的波浪形。营巢于接近水域的各种生境，巢位甚为隐蔽；窝卵数 5~6 枚。

Morphological characteristics A medium-sized passerine (16-19 cm). Six subspecies (*leucopsis, ocularis, baicalensis, lugens, alba* and *personata*) have been recorded in Beijing. Both sexes have a combination of black and white or black, white and grey plumage, depending on subspecies. Male generally has white forehead; black crown; white head-sides (sometimes a black eye-stripe, such as *ocularis* and *lugens*); black tail with white outermost rectrices; black breast and white below. Female generally paler than male and more greyish. Juvenile has yellowish head-sides.

Distribution and habitat Widespread in China. Abundant passage migrant, summer breeder and scarce winter visitor in Beijing. Found in freshwater wetlands, including rivers, lakes and reservoirs.

Behaviour and ecology Constantly pumps tail while walking near water. Flight undulating. Feeds on insects. Nests in a variety of habitats near water. The nest is carefully hidden from viewing. Clutch size: 5-6.

IUCN 红色名录 无危 (LC) 2018.10.30

田鹨	*Anthus richardi*	Richard's Pipit	鹡鸰科 Motacillidae

IUCN 红色名录 无危 (LC) 2016.10.1

形态特征 中型鸣禽，体长 17~18 厘米。雌雄相似。喙较长且粗壮，上喙深灰色，下喙粉色。成鸟整体黄褐色。头顶黄褐色具黑色细纹，眉纹皮黄色，背部黄褐色具较模糊的黑色斑纹，中覆羽黑斑较尖。尾羽黄褐色，最外侧尾羽白色。喉部偏白色，胸腹部黄褐色，胸侧具黑色纵纹；跗跖和后爪较长。

分布与生境 除青藏高原外广泛分布于中国各地。北京为区域性常见旅鸟和夏候鸟。主要栖息于平原地区的旷野、草地、农田等环境。

习性与生态 常单独或成对活动，迁徙时可见小群。站姿较为挺拔，飞行时轨迹呈波浪形。主要以昆虫为食。营巢于地面凹坑中，多隐蔽在草丛中；窝卵数 4~6 枚。

Morphological characteristics A medium-sized passerine (17-18 cm). Sexes alike. Bill is long and stout; upper mandible dark grey; lower mandible pink. Adult is overall yellow-brown; yellow-brown crown with fine black streaks; buff supercilium; irregular dark streaks on yellow-brown mantle, pointed black centres to the median coverts; rectrices yellow-brown; outermost rectrices white; whitish throat and brownish breast and belly; black streaks on breast-sides; long tarsi and hindclaw.

Distribution and habitat Widespread in China except for the Tibetan Plateau. In Beijing, locally common in open grassland, meadows and agricultural land. Residential status: passage migrant and summer breeder.

Behaviour and ecology Usually encountered in small groups on migration. Stands upright; flight undulating. Feeds on insects. Nests in a depression in the ground, usually well-hidden in grass. Clutch size: 4-6.

布氏鹨	*Anthus godlewskii*	Blyth's Pipit	鹡鸰科 Motacillidae

形态特征 中型鸣禽，体长 15~17 厘米。雌雄相似。喙较田鹨短小。成鸟整体黄褐色。头顶黄褐色具黑色细纹，眉纹皮黄色。背部黄褐色具较明显的黑色斑纹，中覆羽黑斑较钝，具淡皮黄色或淡黄白色羽缘。尾羽黄褐色。胸腹部黄褐色，胸侧部具黑色纵纹。似田鹨但跗跖和后爪较田鹨短，且中覆羽和大覆羽的浅色羽缘更为显著。

分布与生境 中国见于东北、华北、华南、西南等地。北京为不常见旅鸟。主要栖息于平原或山区的旷野、农田、林缘等环境。

习性与生态 常集成松散的小群或与田鹨混群迁徙过境。与田鹨相比，站姿较为低平。营巢于有草丛、灌丛、岩石等遮蔽的地面凹坑内；窝卵数 3~5 枚。

Morphological characteristics A medium-sized passerine (15-17 cm). Sexes alike. Bill shorter than Richard's Pipit. Adult overall yellow-brown; yellow-brown crown with fine black streaks; buff supercilium; yellow-brown mantle with black streaks; blunt black centres to the median coverts with pale buff or pale yellowish-white fringes; tail yellow-brown. Yellow-brown breast and belly with black streaks on breast-sides. Resembles Richard's Pipit but has shorter bill, tarsi and hindclaw, and has more prominent pale fringes on median and greater coverts.

Distribution and habitat Northeast, north, south and southwest China. In Beijing, uncommon in open habitats such as grassland, vegetated dry riverbeds and agricultural land. Residential status: passage migrant.

Behaviour and ecology Usually encountered in small and loose groups, often with Richard's Pipit on migration. Posture more typical of a small pipit than the upright stance of Richard's Pipit. Nests in a depression in the ground, sheltered by grass, bush or rocks. Clutch size: 3-5.

IUCN 红色名录 无危 (LC) 2016.10.1

草地鹨	*Anthus pratensis*	Meadow Pipit	鹡鸰科 Motacillidae

IUCN 红色名录 近危（NT） 2018.8.9

形态特征 小型鸣禽，体长 14~15.5 厘米。雌雄相似。成鸟整体大致为黄褐色。头顶褐色具黑色清晰点斑，几无眉纹。髭纹几不可见。上体褐色，具较明显的黑色纵纹。大覆羽和中覆羽中部黑色，羽缘淡皮黄色。颏、喉淡黄白色，下体余部淡皮黄色，胸部和两胁具黑色斑纹。跗跖黄褐色。

分布与生境 中国主要见于西北地区。主要栖息于平原地区多草地的环境。北京为甚罕见迷鸟。

习性与生态 非繁殖期常集小群活动，但北京的记载多为单只。主要以昆虫为食，飞行轨迹呈波浪形。

Morphological characteristics A small passerine (14-15.5 cm). Sexes alike. Adult is largely earthy-brown; black streaks on brown crown; face relatively unmarked; upperparts brown, streaked black; greater and median coverts centred black, fringed yellowish-buff; chin and throat creamy white; rest of underparts yellowish-buff; breast and flanks streaked black; tarsi yellowish-brown.

Distribution and habitat Mostly found in northwest China. Found in open grassy habitats in lowland, such as grassland and meadows. Very rare vagrant in Beijing.

Behaviour and ecology Usually in small flocks outside the breeding season but in Beijing most records are of singles. Insectivorous; flight undulating.

林鹨	*Anthus trivialis*	Tree Pipit	鹡鸰科 Motacillidae

形态特征 小型鸣禽，体长 14~16 厘米。雌雄相似。成鸟整体大致为橄榄褐色。头顶橄榄褐色，眉纹皮黄色，耳羽无黑斑，脸颊具一不甚明显的皮黄色点斑。上体橄榄褐色，具清晰的黑色斑纹，胸腹部黄褐色，黑色斑纹较重；跗跖近粉色。

分布与生境 中国主要见于西北地区。于分布区内主要栖息于针叶林、混交林、林缘等环境。北京为甚罕见迷鸟，仅于 2013 年 5 月在朝阳区有一笔记载。

习性与生态 常单独或成对活动，非繁殖期亦集小群。尾常不停上下摆动，飞行时呈现波浪形轨迹。主要以昆虫为食。

Morphological characteristics A small passerine (14-16 cm). Sexes alike. Adult has overall olive-brown plumage; olive-brown crown has dark streaks; buff supercilium, no obvious black mark on ear-coverts; faint buff spot on cheek; upperparts streaked dark brown; whitish below with pale buff wash; heavy dark streaking on breast, finer on belly and flanks; pinkish tarsi.

Distribution and habitat Mostly found in northwest China. Occurs in coniferous, mixed forest and forest edge in its range. Very rare vagrant in Beijing with only one record in Chaoyang District in May 2013.

Behaviour and ecology Usually solitary or in pairs, sometimes in small groups in non-breeding season. Often pumps tail; flight undulating. Feeds mainly on insects.

IUCN 红色名录 无危 (LC) 2018.8.9

树鹨	*Anthus hodgsoni*	Olive-backed Pipit	鹡鸰科 Motacillidae

IUCN 红色名录　无危 (LC) 2016.10.1

形态特征　中型鸣禽，体长 15~17 厘米。雌雄相似。成鸟头顶及上体橄榄褐色，眉纹皮黄色或白色，耳羽具清晰黑色新月形斑和白色点斑。胸腹部黄褐色，黑色斑纹较重。背部橄榄绿色具不甚清晰的黑褐色斑纹。跗跖粉色或淡黄褐色。

分布与生境　中国除西部地区外广泛分布。北京为常见旅鸟，冬季偶有记录。主要栖息于平原、丘陵、山地的林缘、草地、农田耕地等环境。

习性与生态　非繁殖期常成小群活动。尾常不停上下摆动。主要以昆虫为食。营巢于林间有草地、灌丛或岩石遮蔽的地面凹坑内；窝卵数 4~6 枚。

Morphological characteristics A medium-sized passerine (15-17 cm). Sexes alike. Adult has olive-brown head and upperparts; buff or white supercilium; distinctive black crescent and white spot on ear-coverts; olive-green back has blackish streaks; breast and belly are pale yellow-brown with heavy dark streaks; tarsi pink or pale yellowish-brown.

Distribution and habitat Widespread in China except for the west. In Beijing, common migrant and scarce winter visitor in farmland, grassland and forest edge in lowland and hills.

Behaviour and ecology Often in small groups in non-breeding season. Constantly pumps tail. Feeds mainly on insects. Nests in a depression in the ground in grassland, bush or beneath rocks in forests. Clutch size: 4-6.

北鹨	*Anthus gustavi*	Pechora Pipit	鹡鸰科 Motacillidae

形态特征 小型鸣禽，体长 14~15 厘米。雌雄相似。成鸟喙较短，整体黄褐色。头顶黄褐色具黑色斑纹，眉纹较短，呈皮黄色。上体棕黄色，背部具较粗的黑色纵纹和两道清晰的白色纵纹。颏、喉、腹部乳白色，胸及两胁黄褐色，具清晰的黑色纵纹；跗跖近粉色。

分布与生境 中国主要见于东部地区。北京为罕见旅鸟。主要栖息于平原、丘陵地带的湿润多草地环境。

习性与生态 常单独活动，性谨慎，行为隐秘。尾常不停上下摆动。主要以昆虫为食。营巢于草丛中；窝卵数 4~6 枚。

Morphological characteristics A small passerine (14-15 cm). Sexes alike. Adult has relatively short bill and overall brown, streaky plumage; black streaks on pale brown crown; short, supercilium buff; upperparts warm brown, with broad black streaks and two distinct white "braces" on back; whitish chin, throat and belly; pale brownish breast and flanks with distinctive black streaks; pinkish tarsi.

Distribution and habitat Mostly found in eastern China. In Beijing, rare at margins of wetlands and grassland in lowlands and low hills. Residential status: migrant.

Behaviour and ecology Usually solitary and shy. Constantly pumps tail. Feeds mainly on insects. Nests in grass. Clutch size: 4-6.

IUCN 红色名录 无危 (LC) 2018.8.9

粉红胸鹨	*Anthus roseatus*	Rosy Pipit	鹡鸰科 Motacillidae

非繁殖羽 Non-breeding

繁殖羽 Breeding

IUCN 红色名录 无危 (LC) 2019.6.13

形态特征 小型鸣禽，体长 15~16.5 厘米。雌雄相似。成鸟繁殖羽眉纹粉红色，上体橄榄褐色，具黑色纵纹；翼上大覆羽和中覆羽黑色，羽缘褐色，端部皮黄色，形成两道皮黄色翼斑；尾羽深褐色；颏、喉、胸至上腹部粉红色，胸侧及两胁具黑色纵纹，下体余部皮黄色；跗跖粉色。非繁殖羽眉纹及下体皆为皮黄色而无粉色，胸部纵纹更为浓重。

分布与生境 中国除东北地区外广泛分布。北京为不常见旅鸟和夏候鸟。主要繁殖于较高海拔的高山草甸及多草、多石地带，迁徙及越冬时亦见于平原、丘陵地带草地、农田环境，通常距离湿地较近。

习性与生态 常单独活动或集成小群。尾常不停上下摆动。主要以昆虫为食。一般营巢于草丛中或石缝内；窝卵数 3~5 枚。

Morphological characteristics A small passerine (15-16.5 cm). Sexes alike. Breeding adult has rosy-pink supercilium; olive-brown upperparts with dark streaks; median and great coverts black, fringed brown and tipped buff, forming two buff wingbars; tail dark brown; chin, throat, breast and upper belly rosy-pink; remainder of underparts buff; tarsi pink. Non-breeding adult has buff supercilium and underparts, and more extensive streaks on breast.

Distribution and habitat Widespread in China except for the northeast. Uncommon migrant and summer breeder in Beijing. Breeds in high altitude alpine meadows and rocky, grassy habitats; on migration also found in lowland grassland, usually near water.

Behaviour and ecology Usually solitary or in small groups. Constantly pumps tail. Feeds mainly on insects. Nests in grass or rock crevices. Clutch size: 3-5.

红喉鹨	*Anthus cervinus*	Red-throated Pipit	鹡鸰科 Motacillidae

形态特征 小型鸣禽，体长 14~16 厘米。繁殖羽头、颏、喉及胸红色或棕红色（雄鸟羽色较雌鸟鲜亮）；头顶具黑色纵纹，背部具黑白相间之纵纹；腹部皮黄色或污白色，两胁具显著的黑色纵纹；跗跖偏肉色。非繁殖羽头部、胸部不见红色或仅残留些微红色；头为褐色，具一皮黄色眉纹；颏、喉、胸亦为皮黄色。

分布与生境 中国主要见于东北、华北及南方地区。北京为区域性常见旅鸟。主要栖息于平原湿地、农田、水田等环境。

习性与生态 喜单独或成小群活动，尾常不停上下摆动。主要以昆虫为食。营巢于苔原近水处；窝卵数 4~6 枚。

Morphological characteristics A small passerine (14-16 cm). Sexes alike. Breeding adult has rufous or red head, throat and breast (brighter in male); crown streaked black; back streaked black and white; belly off-white or buff; flanks streaked black; tarsi flesh-coloured. Non-breeding adult has little or no red on head and breast. Head brown with buff supercilium; chin, throat and breast buff.

Distribution and habitat Northeast, north and southern China. In Beijing, locally common migrant in wetlands, grassland and open areas, usually near water.

Behaviour and ecology Usually found singly or in small groups. Constantly pumps tail. Feeds mainly on insects. Nests on tundra near water. Clutch size: 4-6.

IUCN 红色名录 无危 (LC) 2018.10.30

繁殖羽 Breeding

非繁殖羽 Non-breeding

黄腹鹨	*Anthus rubescens*	Buff-bellied Pipit	鹡鸰科 Motacillidae

繁殖羽 Breeding

非繁殖羽 Non-breeding

IUCN 红色名录　无危 (LC) 2018.8.9

形态特征　中型鸣禽，体长 15~17 厘米。雌雄相似。成鸟繁殖羽整体偏灰色；头顶灰色具模糊点斑，眉纹于眼后模糊；三角形髭纹较重，背部灰色，具模糊的深色纵纹；下体皮黄色，胸部和两胁具深色纵纹；跗跖黄褐色。成鸟非繁殖羽上体偏褐，下体纵纹显著，特别是喉侧及颈侧的黑色纵纹更为浓重，在每侧各形成一近似三角形的黑色区域。

分布与生境　广泛分布于中国各地。北京为区域性常见冬候鸟、旅鸟。主要栖息于河流、湿地等环境。

习性与生态　非繁殖期常集小群活动，迁徙过境时亦常集成较大群体。喜在近水湿地行走，尾常不停上下摆动。营巢于草丛或岩隙中；窝卵数 3~7 枚。

Morphological characteristics　A medium-sized passerine (15-17 cm). Sexes alike. Breeding adult is greyish overall; grey crown with light streaking; supercilium fades behind eye; dark triangular malar; relatively plain grey back with only faint dark markings; underparts orange-buff or yellow-buff; bold black streaks on breast and flanks; tarsi yellowish-brown. Non-breeding adult has browner upperparts and more extensively streaked underparts. Especially, the black streaks on throat-sides and neck-sides are prominent, forming a triangular black patch on each side.

Distribution and habitat　Widespread in China. In Beijing, locally common in wetland habitats. Residential status: passage migrant and winter visitor.

Behaviour and ecology　Usually in groups and sometimes large flocks on migration. Forages around edges of wetland or in grassland, constantly pumping tail. Nests in grass or crevices. Clutch size: 3-7.

水鹨	*Anthus spinoletta*	Water Pipit	鹡鸰科 Motacillidae

形态特征 中型鸣禽，体长 15~18 厘米。雌雄相似。成鸟繁殖羽头部偏灰色，胸腹部近粉色。非繁殖羽头顶偏黄褐色具清晰点斑，白色眉纹清晰，髭纹不明显。背部黄褐色，具不显著的深褐色纵纹。下体淡黄白色，具较细的灰黑色斑纹；跗跖黑褐色。成鸟非繁殖羽胸部及两胁的纵纹较为显著。甚似黄腹鹨，但喉侧、颈侧及胸部的纵纹不如其浓重。

分布与生境 广泛分布于中国各地。北京为区域性常见冬候鸟、旅鸟。主要栖息于河流、湿地等环境，冬季亦见于城区公园湿地。

习性与生态 常单独活动或集小群。喜在近水湿地行走，尾常不停上下摆动。主要以昆虫为食。营巢于草丛中或有植被遮挡的岩缝内；窝卵数 4~6 枚。

Morphological characteristics A medium-sized passerine (15-18 cm). Breeding adult has greyish head, pinkish breast and belly. Non-breeding adult has clearly streaked yellowish-brown crown, distinct white supercilium; obscure malar; relatively plain greyish-brown back with weak dark brown streaks; creamy underparts with fine greyish-black streaks; tarsi blackish-brown. Non-breeding adult has more extensive streaks on breast and flanks. Closely resembles Buff-bellied Pipit but has weaker streaks on throat-sides, neck-sides and breast.

Distribution and habitat Widespread in China. In Beijing, locally common in wetland habitats, including in parks in winter. Residential status: winter visitor and migrant.

Behaviour and ecology Usually solitary or in small groups. Forages at wetland edge, pumping tail. Feeds mainly on insects. Nests in grass or rock crevices shelter by plants. Clutch size: 4-6.

IUCN 红色名录 无危 (LC) 2018.8.9

非繁殖羽 Non-breeding

繁殖羽 Breeding

苍头燕雀	*Fringilla coelebs*	Common Chaffinch	燕雀科 Fringillidae

形态特征 小型鸣禽，体长 14~16 厘米。喙锥状。雄鸟头顶至后颈蓝灰色，背褐色，腰黄绿色，肩羽、翼斑白色；尾羽蓝黑色，最外侧尾羽大部分为白色；脸颊、喉、胸、两胁淡栗红色，腹至尾下覆羽白色。雌鸟头顶至背暗褐色，各部羽色暗。

分布与生境 分布于中国北方地区。见于北京郊区各类林地及城市公园，为罕见冬候鸟。

习性与生态 在北京多单独活动或与燕雀混群。常在树上、灌丛或地面活动，以植物果实和种子为食，兼食昆虫。

Morphological characteristics A small passerine (14-16 cm). Triangular bill. Male has blue-grey crown and nape; brown back; rump yellowish-green; mantle and wing-patch white; tail blackish-blue; outermost rectrices mostly white; cheek, throat, breast and flanks reddish-brown; white belly and undertail-coverts. Female has pale brown crown and back; overall duller than male.

Distribution and habitat Northern China. In Beijing, rare in rural woodlands and also urban parks. Residential status: winter visitor.

Behaviour and ecology Usually solitary or joins flocks of Bramblings in Beijing. Feeds on fruits, seeds and insects in trees, bush or on the ground.

IUCN 红色名录 无危 (LC) 2018.8.9

雄鸟 Male

雌鸟 Female

燕雀	*Fringilla montifringilla*	Brambling	燕雀科 Fringillidae

雄鸟 Male

雌鸟 Female

北京市二级保护野生动物

IUCN 红色名录　无危 (LC) 2016.10.1

形态特征　小型鸣禽，体长 14~16 厘米。雄鸟繁殖羽喙黑色，头、后颈至上背黑色，下背、腰、尾上覆羽白色；具白色肩羽，棕色翼斑，尾羽主要为黑色，呈浅叉形；颏、喉、胸橙色，两胁具黑色斑点，下体余部白色。雄鸟非繁殖羽喙黄色，喙端黑色；头部黑褐色，背部各羽具褐色羽缘。雌鸟头灰褐色，其余各部羽色亦较雄鸟暗淡。

分布与生境　分布于中国西北、东部及南部地区。为北京城郊各类林地及公园中的甚常见冬候鸟和旅鸟。

习性与生态　非繁殖期常集大群活动。以植物嫩芽、果实和种子为食，兼食昆虫。于乔木侧枝处营杯状巢；窝卵数 4~7 枚。

Morphological characteristics A small passerine (14-16 cm). Breeding male has black bill; black head, hindneck and upper back; white lower back, rump and upper rectrices; white scapulars and brown wing-patch; slightly forked tail mostly black; chin, throat and breast orange; flanks spotted black; rest of underparts white. Non-breeding male has yellowish bill with black tip; head brownish-black; back fringed brown. Female is duller overall and has greyish-brown head.

Distribution and habitat Northwest, eastern and southern China. In Beijing, abundant in both urban and rural woodlands. Residential status: winter visitor and migrant.

Behaviour and ecology Often in large flocks in non-breeding season. Feeds on tender shoots, fruits and seeds, also insects. Builds cup-shaped nests on tree branches. Clutch size: 4-7.

锡嘴雀	*Coccothraustes coccothraustes*	Hawfinch	燕雀科 Fringillidae

形态特征 中型鸣禽，体长 16~20 厘米。喙较大，甚为粗厚，呈铅蓝灰色。雄鸟头顶至枕部、头侧及颊部均为棕黄色，背羽棕褐色，腰浅棕黄色；飞羽黑色带有蓝紫色金属光泽，具白色翼斑；尾羽基部黑色，端部白色；下体浅棕带有肉桂色。雌鸟与雄鸟相似，头部多浅棕灰色，飞羽缺少金属光泽。

分布与生境 除青海高原、海南外，广布于中国各地。迁徙期区域性常见于北京山地林区、平原林地、城市公园。在北京主要为旅鸟，少量冬候鸟。

习性与生态 常单独或成对活动。喜食松、柏的种子、葵花籽以及其他植物的种子、果实和少量昆虫。在阔叶树的侧枝上营巢；窝卵数 3~7 枚。

Morphological characteristics A medium-sized passerine (16-20 cm). Bill massive, blue-grey. Male has buff-brown crown, nape, head-side and cheeks; rufous-brown back; pale buff-brown rump; wings glossy bluish-purple with white wing-bar; tail base black with white tip; underparts pale brown, tinged cinnamon. Female similar to male but has pale greyish-brown head and remiges lack gloss.

Distribution and habitat Widespread in China except for the Tibetan Plateau and Hainan Province. In Beijing, locally common in woodlands in mountain and lowland, including parks. Residential status: migrant and winter visitor.

Behaviour and ecology Solitary or in pairs. Feeds on seeds of pine and cypress, sunflower and other seeds, also insects in summer. Nests on branches of broadleaved trees. Clutch size: 3-7.

北京市二级保护野生动物 | IUCN 红色名录 无危 (LC) 2016.10.1

黑尾蜡嘴雀	*Eophona migratoria*	Chinese Grosbeak	燕雀科 Fringillidae

北京市二级保护野生动物 | IUCN 红色名录 无危 (LC) 2016.10.1

形态特征 中型鸣禽，体长 17~21 厘米。喙为粗厚的锥状，呈黄色。雄鸟头黑色，具金属光泽，背部、腰部到尾上覆羽，由灰褐色转为灰色。两翼和尾羽大致为黑色；初级覆羽端部和初级飞羽端部白色；下体淡灰褐色，两胁锈红色。雌鸟的头部和背部均为灰褐色，初级覆羽及初级飞羽端部的白斑较窄，胁部锈红色较淡。

分布与生境 除西部地区、海南外，广布于中国各地，是各地的留鸟或候鸟。常年见于北京低地林区、平原林地、城市公园，为常见留鸟。

习性与生态 喜食松、柏及白蜡树的种子及果实，以及葵花籽和其他植物的种子、果实、嫩芽等，繁殖期捕食昆虫。在阔叶树的侧枝树叉上营碗状或杯状巢；窝卵数 4~5 枚。雌雄鸟共同参与育雏。

Morphological characteristics A medium-sized passerine (17-21 cm). Powerful, triangular yellow bill. Male has glossy black head; greyish-brown to grey from back and rump to uppertail-coverts; wings and rectrices black; primary coverts and primaries white-tipped; underparts pale greyish-brown; flanks rusty-red. Female has greyish-brown head and back, narrower white tips on primary coverts and primaries, and paler rusty-red flanks.

Distribution and habitat Widespread in China except for western China and Hainan Province. In Beijing, common in lowland woodland and parks. Residential status: resident.

Behaviour and ecology Feeds on seeds and fruits of pine, cypress and ash tree, sunflower and other seeds, and tender shoots; also foraged for insects during breeding season. Builds cup-shaped or bowl-shaped nests on branches of broadleaved trees. Clutch size: 4-5. Both female and male share breeding duties.

黑头蜡嘴雀 | *Eophona personata* | Japanese Grosbeak | 燕雀科 Fringillidae

形态特征 中型鸣禽，体长 21~24 厘米。雌雄相似。喙为粗厚的锥状，呈黄色。头黑色带金属光泽，头部黑色区域明显小于黑尾蜡嘴雀雄鸟。上体及下体皆以灰色为主。两翼和尾羽黑色，具白色翼斑。

分布与生境 中国见于东部及南部广大地区。北京见于低地林区、平原林地、城市公园，为不常见旅鸟。

习性与生态 喜食松树种子、葵花籽和其他植物的种子、果实、嫩芽等，繁殖期喜食昆虫。在阔叶树或针叶树的枝叉上营巢；窝卵数 3~4 枚。

Morphological characteristics A medium-sized passerine (21-24 cm). Sexes alike. Heavy, triangular yellow bill; glossy black head, similar to Chinese Grosbeak but with restricted black; grey upperparts and underparts; black wings and rectrices, with white wing-patch.

Distribution and habitat Widespread in eastern and southern China. In Beijing, uncommon in lowland woodland and parks. Residential status: migrant.

Behaviour and ecology Feeds on seeds, fruits and tender shoots; also foraged for insects during breeding season. Nests on branches of broadleaved or coniferous trees. Clutch size: 3-4.

北京市二级保护野生动物 | IUCN 红色名录 无危 (LC) 2018.8.9

红腹灰雀	*Pyrrhula pyrrhula*	Eurasian Bullfinch	燕雀科 Fringillidae

IUCN 红色名录 无危 (LC) 2018.8.9

形态特征 中型鸣禽，体长 15~17 厘米。喙短粗，铅灰色。雄鸟头顶、眼周、眼先及颏均为黑色，背部灰色，腰部白色，两翼和尾羽黑色，翼上覆羽多为灰色，具一白色或灰白色翼斑；颊、喉、前颈、胸、上腹部粉红色或橙红色；下腹部及尾下覆羽白色。雌鸟与雄鸟相似，只是在雄鸟具红色的羽区都是灰白色。

分布与生境 分布于中国东北及西北地区，冬季亦偶见于华北。罕见于北京低山、平原林地，为冬候鸟。

习性与生态 喜食草籽、落叶松种子、长白松种子等。繁殖于西伯利亚泰加林。巢常筑在针叶树侧枝末端的枝杈处；窝卵数 4~6 枚。

Morphological characteristics A medium-sized passerine (15-17 cm). Short but deep bill lead grey. Male has black crown, eye-ring, lores and chin, grey back, white rump, black wings and tail; grey upperwing-coverts with broad, off-white wing-bar; cheeks, throat, breast and upper belly pinkish-red or orange-red; white lower belly and undertail-coverts. Female has greyish-white replacing corresponding red parts of male.

Distribution and habitat Northwest and northeast China, scarce in north China in winter. In Beijing, rare in open woodlands in lowlands. Residential status: winter visitor.

Behaviour and ecology Feeds on seeds of grass, larch and pine. Breeds in Siberian taiga forest. Nests on coniferous trees. Clutch size: 4-6.

蒙古沙雀	*Bucanetes mongolicus*	Mongolian Finch	燕雀科 Fringillidae

形态特征 小型鸣禽，体长 11~14 厘米。喙短呈锥状，黄色。雄鸟繁殖羽上体沙褐色或黄褐色，具深褐色纵纹；眉纹粉红色；腰部及尾上覆羽淡粉色；尾羽黑褐色，呈浅叉状；翼上有两块边缘不清的粉白色翼斑；喉、前颈、胸、上腹部及两胁淡粉红色或粉白色；下腹部及尾下覆羽白色。雌鸟和雄鸟非繁殖羽上体羽色较暗淡，下体粉色更浅淡。

分布与生境 分布于中国西北至东北地区，生境为半荒漠地区的稀疏灌丛、草甸及沙石滩。在北京为甚罕见迷鸟。

习性与生态 主要以植物种子为食，也吃嫩叶、芽及麦种。繁殖期 5~6 月，常在土崖或石头与灌丛旁的地面用嘴掘洞筑巢；窝卵数 3~5 枚。

Morphological characteristics A small passerine (11-14 cm). Short, triangular yellow bill. Breeding male has yellowish-brown or buff-brown upperparts, streaked dark brown; pink supercilium; pale pink rump and uppertail-coverts; dark brown rectrices slightly forked; two pinkish-white wing-patches; pinkish-red or pinkish-white throat, breast, upper belly and flanks; lower belly and undertail-coverts white. Non-breeding male and female have duller upperparts and paler underparts, lacking pink.

Distribution and habitat Resident in northwest and northeast China. Occurs in areas of sparse shrubs, meadows and sandy beaches in semi-desert areas. Very rare vagrant to Beijing.

Behaviour and ecology Feeds on plant seeds, tender shoots and sprouts, and wheat seeds. Breeds in May and June. Often nests in holes in earth cliffs or on ground next to rocks and shrubs. Clutch size: 3-5.

IUCN 红色名录 无危 (LC) 2018.8.9

粉红腹岭雀	*Leucosticte arctoa*	Asian Rosy Finch	燕雀科 Fringillidae

IUCN 红色名录　无危 (LC) 2018.8.9

形态特征　中型鸣禽，体长 15~18 厘米。喙锥状，黄色。雄鸟上体羽呈棕褐色，背部具黑色纵纹；腰和尾上覆羽具粉红色鳞状斑；翼上覆羽末端形成粉红色斑；下体大致为近黑色，胸部具白色点斑，腹部密布淡粉色纵纹，尾下覆羽粉白色。雌鸟与雄鸟相似，体羽中的粉红色部分较浅淡。

分布与生境　分布于中国东北、华北及西北地区，主要栖息于高山草甸等高海拔环境，为冬候鸟，仅有少数个体在东北满洲里、大兴安岭及新疆的阿勒泰有繁殖记录。见于北京较高海拔山坡多岩石的生境，为不常见冬候鸟。

习性与生态　非繁殖期集群活动。喜食植物种子及灌木果实，也食部分昆虫等。繁殖期 6~7 月，巢常筑在岩石缝隙间，喜集群繁殖；窝卵数 3~4 枚。

Morphological characteristics A medium-sized passerine (15-18 cm). Triangular bill yellow. Male has brown back, streaked black, pinkish-red scaled rump and uppertail-coverts; pinkish-red tipped upperwing-coverts; underparts largely blackish, with white spotted breast, pale pinkish-red streaked belly and pinkish-white undertail-coverts. Female is similar to male but duller pinkish-red.

Distribution and habitat Northeast, north and northwest China, mainly inhabits high-altitude alpine meadows and rocky slopes as winter visitor; a few breeding records in Manzhouli and Daxinganling in northeast China, and Altay in Xinjiang. In Beijing, uncommon on rocky mountain slopes. Residential status: winter visitor.

Behaviour and ecology In flocks in non-breeding season. Feeds on seeds, fruits and insects. Breeds in June and July. Often nests colonially in rock crevices. Clutch size: 3-4.

普通朱雀	*Carpodacus erythrinus*	Common Rosefinch	燕雀科 Fringillidae

形态特征 中型鸣禽，体长 15~17 厘米。喙短粗呈锥状，铅灰色。雄鸟头红色，贯眼纹深褐色；背部和肩橄榄褐色，两翼和尾黑褐色，具淡红色羽缘；腰、喉、胸为红色，下体近白色。雌鸟上体橄榄褐色，上体及下体皆具暗褐色纵纹，两翼和尾黑褐色；下体灰白色，有时沾粉色。

分布与生境 中国分布广泛。见于北京低山、平原疏林，为区域性常见旅鸟。

习性与生态 单只、成对或集小群活动。喜食植物的果实、种子、花序、嫩芽等。巢常筑在多刺灌丛或小树枝杈上；窝卵数 4~5 枚；由雌鸟孵卵，雄鸟在雌鸟孵卵期常寻食饲喂雌鸟，雌雄鸟共同育雏。

Morphological characteristics A medium-sized passerine (15-17 cm). It has short but deep lead-grey triangular bill. Male has red head; eye-stripe dark brown, sometimes indistinct; olive brown back and scapulars; dark brown wings and tail with pale red fringes; red rump, throat and breast; underparts whitish. Female has olive-brown upperparts, dark brown streaked above and below, dark brown wings, underparts greyish-white, sometimes washed pink.

Distribution and habitat Widespread in China. In Beijing, locally common in lightly wooded areas in the hills and lowlands. Residential status: migrant.

Behaviour and ecology Solitary, in pairs or in small flocks. Feeds on fruits, seeds, tender shoots and insects. Nests on thorny shrubs or branches of young trees. Clutch size: 4-5. Male feeds female when female is incubating; female and male share feeding of offspring.

IUCN 红色名录 无危 (LC) 2016.10.1

雄鸟 Male

雌鸟 Female

中华朱雀	*Carpodacus davidianus*	Chinese Beautiful Rosefinch	燕雀科 Fringillidae

雄鸟 Male

雌鸟 Female

北京市二级保护野生动物 | | IUCN 红色名录 无危 (LC) 2018.8.9*

形态特征 小型鸣禽，体长 13~15 厘米。喙呈锥状。雄鸟前额、眉纹及颊粉红色，贯眼纹深褐色，头顶和其余上体灰褐色并具黑褐色纵纹；腰部粉红色；两翼和尾羽黑褐色，翼上具两道粉红色翼斑，下体粉红色。雌鸟不具粉色或红色，上体灰褐色或土黄色，下体淡黄色或灰褐色。

分布与生境 分布于中国华北至中部地区。见于北京 1200~2000 米的山间林地，是北京山地生境中区域性常见的留鸟。

习性与生态 繁殖期单独或成对活动，非繁殖期集群。喜食草籽、浆果、嫩芽及作物种子等。繁殖于山地灌丛。巢常筑在带刺灌丛或小树上；窝卵数 3~6 枚。

● *注：《北京市二级保护野生动物名录》和 IUCN 红色名录均未收录（更新 / 接受）中华朱雀和红眉朱雀的物种划分，故此处沿用红眉朱雀的数据。

Morphological characteristics A small passerine (13-15 cm). Triangular bill. Male has pinkish-red forehead, supercilium and cheeks; eye-stripe dark brown; crown and remainder of upperparts greyish-brown with dark brown streaks; rump pinkish-red; dark brown wings and tail; rosy-red wing-patches; underparts pinkish-red. Female lacks pink or red, instead greyish-brown or buff-brown upperparts, pale yellow or greyish-brown underparts.

Distribution and habitat North and central China. In Beijing, locally common in montane woodlands at 1,200-2,000 m. Residential status: resident.

Behaviour and ecology Solitary or in pairs in breeding season; in flocks in non-breeding season. Feeds on seeds, berries, shoots and insects. Breeds in montane shrubs. Nests on thorny shrubs and small trees. Clutch size: 3-6.

长尾雀	*Carpodacus sibiricus*	Long-tailed Rosefinch	燕雀科 Fringillidae

形态特征 小型鸣禽，体长 13~18 厘米。喙较短而厚实。雄鸟繁殖羽全身大致为粉红色，头部、腰部和下体的红色更显著，背部具暗色纵纹；两翼黑褐色，具两道显著的白色翼斑；尾羽长，外侧尾羽白色。非繁殖羽雄鸟上体沙褐色具暗色纵纹，腰部、下体的粉红色浅淡。雌鸟全身为沙褐色，腰部、尾上覆羽棕黄色。

分布与生境 分布于中国东北、西北、华北山区和丘陵。在北京为不常见冬候鸟，一般见于灌丛或开阔林地。

习性与生态 繁殖期单独或成对活动，非繁殖期集小群。喜食草籽等植物种子、浆果、果实、嫩叶及少量昆虫。巢常筑在灌丛和小树上；窝卵数 4~6 枚；由雌雄鸟轮流孵卵，共同育雏。

Morphological characteristics A small passerine (13-18 cm). Deep, short bill. Male has pinkish-red plumage with bright red head, rump and underparts; dark streaked back; blackish-brown wings with two distinctive white wing-patches; long, blackish tail with white outer rectrices. Non-breeding male has buff-brown upperparts with dark streaks, and pale pinkish-red rump and underparts. Female has yellowish-brown plumage, buff-brown rump and uppertail-coverts.

Distribution and habitat Northeast, northwest and north China. In Beijing, uncommon on scrubby hills, and open woodlands. Residential status: winter visitor.

Behaviour and ecology Solitary or in pairs in breeding season; forms small flocks in non-breeding season. Feeds on seeds, berries, fruits, shoots and insects. Nests in shrubs and small trees. Clutch size: 4-6. Female and male takes turns to incubate and share feeding of offspring.

IUCN 红色名录 无危 (LC) 2018.8.9

雄鸟 Male
雌鸟 Female

北朱雀	*Carpodacus roseus*	Pallas's Rosefinch	燕雀科 Fringillidae

北京市二级保护野生动物 | 国家二级保护野生动物 | IUCN 红色名录 无危 (LC) 2018.8.9

形态特征 中型鸣禽，体长 15~17 厘米。喙较短而粗厚，呈灰色。雄鸟额及头顶前部珍珠白色，并具粉红色羽缘，形成显著的白色鳞状斑，头顶后部、头侧、枕部及后颈粉红色；背部灰褐色，具粉红色羽缘，腰和尾上覆羽鲜亮的粉红色；两翼黑褐色，翼上具一道或两道白色沾粉红色翼斑；尾羽黑褐色；颊、颏、喉具银白色点斑，下体粉红色。雌鸟上体沙褐色，腰和尾上覆羽粉红色，尾黑褐色，翼斑不及雄鸟显著；颏灰白色，喉、胸和两胁粉红色；下体其余部分白色或淡灰褐色。

分布与生境 分布于中国东北、西北、华北地区。冬季区域性常见于北京较高海拔的山区林地，为冬候鸟。

习性与生态 喜集 10 只以内的小群活动。喜食种子、灌木芽和昆虫等。繁殖于西伯利亚东部和中部，中国没有繁殖记录。

Morphological characteristics A medium-sized passerine (15-17 cm). It has short, deep, grey bill. Male has pearly-white forehead, tipped pinkish-red and with prominent white scales; hindcrown, head-sides, nape and hindneck pinkish-red; greyish-brown back tipped pinkish-red; rump and uppertail-coverts pink; wings blackish-brown with one or two pinkish-white wingbars; tail greyish-brown; silver-spotted cheeks, chin and throat, and pinkish-red underparts. Female has yellowish-brown upperparts, pinkish-red rump and uppertail-coverts, dark brown tail and greyish-white chin; pinkish-red throat, breast and flanks; remainder of underparts white or light grey-brown.

Distribution and habitat Northeast, northwest and north China. In Beijing, locally common in wooded hills, usually at higher elevations. Residential status: winter visitor.

Behaviour and ecology Often in small flocks (less than 10 individuals). Feeds on seeds, buds and insects. Breeds in eastern and central Siberia. No breeding records in China.

金翅雀	*Chloris sinica*	Grey-capped Greenfinch	燕雀科 Fringillidae

形态特征 小型鸣禽，体长 12~14 厘米。喙粗壮，呈粉色。雄鸟头顶深灰色，背部栗褐色，腰部黄色或黄绿色，初级飞羽基部黄色，形成显著的黄色翼斑，飞行时格外醒目；颏、喉黄绿色，胸、腹部褐色，尾下覆羽黄色。雌鸟与雄鸟相似，但头顶偏褐色，金黄色羽区稍小且不如雄鸟鲜艳。

分布与生境 广泛分布于中国南北各地，为留鸟。常见于北京平原、中低山的林地和城市公园，为留鸟。

习性与生态 冬季有数百只以上的集群现象，可能有短距离的迁移行为。主要食植物果实、种子等，也食少量昆虫。鸣叫声为轻柔而似鼻音的“嘀 嘀”声。巢常筑在枝叶茂密的乔木侧枝上；每年可繁殖 2~3 窝，窝卵数 4~5 枚；由雌鸟孵卵，雌雄鸟共同育雏。

Morphological characteristics A small passerine (12-14 cm). Bill strong, pink. Male has dark greyish crown, brown back and yellow or greenish-yellow rump. Primaries blackish with yellow base, forming distinctive yellow wing-patches. Chin and throat greenish-yellow; breast and belly brown; undertail-coverts yellow. Female similar but has browner crown and golden parts are smaller and paler.

Distribution and habitat Widespread in China. In Beijing, common in woodlands in lowlands and low hills. Residential status: resident.

Behaviour and ecology In winter, forms flocks of up to several hundred; some may do short distance movements. Feeds on fruit, seeds and insects. Calls soft, nasal "di di". Often nests on branches of trees. Female incubates eggs; male and female share feeding duties. Clutch size: 4-5, with 2-3 broods each year.

北京市二级保护野生动物

IUCN 红色名录　无危 (LC) 2018.8.7

白腰朱顶雀	*Acanthis flammea*	Common Redpoll	燕雀科 Fringillidae

雄鸟 Male

雌鸟 Female

北京市二级保护野生动物

IUCN 红色名录　无危 (LC) 2018.8.9

形态特征　小型鸣禽，体长 12~14 厘米。喙较短而粗厚，呈黄色。雄鸟前额、眼先、颏黑色，头顶朱红色，上体皮黄褐色具黑褐色纵纹，腰部白色沾粉红色且具暗色纵纹，喉、胸粉红色，两胁具深褐色纵纹，腹部白色。雌鸟的喉、胸缺少粉红色，其余部分与雄鸟相似。

分布与生境　分布于中国北方及华东地区。见于北京低山、平原林地及草地等，为不常见旅鸟、冬候鸟。

习性与生态　常集群活动。喜食草籽，也食作物种子。营巢在树木的低枝上或灌丛中；窝卵数 4~6 枚。

Morphological characteristics A small passerine (12-14 cm). Bill short, deep, yellow. Male has black forehead, lores and chin; scarlet crown, buff-brown upperparts with black streaks; white rump tinged pinkish-red and with black streaks; pinkish-red throat and breast; flanks streaked dark brown; remainder of belly white. Female similar to male but lacks pinkish-red on throat and breast.

Distribution and habitat Northern and east China. In Beijing, uncommon in any wooded areas and grassland, including in the hills and lowlands. Residential status: passage migrant and winter visitor.

Behaviour and ecology Usually in flocks. Feeds on seeds. Nests in shrubs or on low branches of trees. Clutch size: 4-6.

极北朱顶雀	*Acanthis hornemanni*	Arctic Redpoll	燕雀科 Fringillidae

形态特征 小型鸣禽，体长 12~14 厘米。喙黄色。头淡褐色，具深色细纵纹，眼先和颏黑色。上体浅灰白色，具黑白相间的纵纹；腰白色。下体大致为白色，胸部淡粉色（雌鸟无粉色），两胁具较细的深褐色纵纹。甚似白腰朱顶雀，然本种喙较小，头顶红色斑块较小，且腰部几无纵纹。

分布与生境 中国有限分布于西北及东北，偶至北方其他地区。繁殖期栖息于北极苔原开阔林地和灌丛，非繁殖期栖息于低山及山脚平原的林地和农田。在北京为甚罕见冬候鸟。

习性与生态 非繁殖期常集群或与白腰朱顶雀混群。喜食草籽和桦树种子，也食植物嫩芽、果实等。营巢于桦树、灌丛或岩缝中；窝卵数 4~6 枚。

Morphological characteristics A small passerine (12-14 cm). Bill yellow. Adult has pale brown head with fine dark streaks; lores and chin black; upperparts largely pale greyish-white with black and white streaks; rump white; underparts mostly white; breast pale pink (female lacks pink); flanks streaked dark brown. In field easily confused with Common Redpoll but lacks dark streaks on white rump and has less red on crown and a smaller bill.

Distribution and habitat Northwest and northeast China, scarce in other areas in northern China. Breeds in low, open forests and bushes on Arctic tundra; in non-breeding season, found in open woodland or farmland in lowlands and low hills. Very rare winter visitor in Beijing.

Behaviour and ecology Forms flocks or joins flocks of Common Redpolls in non-breeding season. Feeds on grass and birch seeds, shoots and fruits. Nests in birch trees, bushes or rock crevices. Clutch size: 4-6.

北京市二级保护野生动物 | IUCN 红色名录 无危 (LC) 2018.8.9

红交嘴雀	*Loxia curvirostra*	Red Crossbill	燕雀科 Fringillidae

雄鸟 Male

雄鸟 Male

雌鸟 Female

北京市二级保护野生动物　国家二级保护野生动物　IUCN 红色名录　无危 (LC) 2016.10.1

形态特征　中型鸣禽，体长 15~17 厘米。喙粗厚，深灰色，上下喙在喙端交叉。雄鸟通体朱红色或绛红色，两翼和尾羽黑褐色；尾下覆羽灰白色；跗跖黑色。雌鸟上体灰褐色，头顶、腰、下体皆为黄绿色。相似种白翅交嘴雀的翼上有两道明显的白色翼斑，而本种不具翼斑。

分布与生境　分布于中国北方地区及华东和西南。多见于针叶林或以针叶树为主的针阔混交林。秋冬季见于北京低山、平原林地，为不常见冬候鸟。

习性与生态　喜食松柏树的种子，擅以交叉的喙剥开球果的种鳞取食里面的种子，也食嫩叶、浆果及少量昆虫。营巢在松树的侧枝上；窝卵数 3~5 枚。

Morphological characteristics A medium-sized passerine (15-17 cm). Bill deep, dark grey; upper and lower mandibles cross over at tip. Male has vermilion or crimson plumage with dark brown wings and tail. Undertail-coverts greyish-white; tarsi black. Female has greyish-brown upperparts, yellowish-green crown, rump and underparts. Similar to White-winged Crossbill but lacks white wing-patch.

Distribution and habitat Northern, east and southwest China. Breeds in coniferous forest and also mixed forest dominated by conifer trees. In Beijing, uncommon in forests in lowlands and low hills. Residential status: winter visitor.

Behaviour and ecology Feeds on pine and cypress seeds; often accesses nuts inside cones by using crossed bill; also feeds on tender shoots, berries and insects. Nests on branches of pine trees. Clutch size: 3-5.

白翅交嘴雀	*Loxia leucoptera*	White-winged Crossbill	燕雀科 Fringillidae

形态特征 中型鸣禽，体长 15~17 厘米。灰色的喙较红交嘴雀细长，上下喙在喙端交叉。雄鸟通体朱红色或绛红色，眼先近黑色；两翼和尾羽黑褐色，具两道较宽的白色翼斑；腹部及尾下覆羽近白色；跗跖黑色。雌鸟上体灰褐色，头顶和腰部沾黄绿色；下体淡黄绿色。

分布与生境 中国主要见于东北至华北。冬季见于北京低山、平原林地，为罕见冬候鸟。

习性与生态 非繁殖期常集小群活动。食性似红交嘴雀，喜食松、柏树的种子。繁殖期选择泰加林或落叶松林栖息。巢大多建在落叶松的侧枝上；窝卵数 3~5 枚。

Morphological characteristics A medium-sized passerine (15-17 cm). Bill grey, more slender than Red Crossbill. Upper mandible and lower mandible cross over at tip. Male has vermilion or crimson plumage; lores blackish; wings and tail dark brown. Median and greater coverts have broad white tips, forming two prominent white wing-bars. Belly and undertail-coverts whitish; tarsi black. Female has greyish-brown upperparts, yellowish-green tinged crown and rump, and pale greenish-yellow underparts.

Distribution and habitat Northeast and north China. In Beijing, rare in forests in lowlands and low hills. Residential status: winter visitor.

Behaviour and ecology Usually in small flocks in non-breeding season. Diet similar to Red Crossbill: feeds mainly on pine and cypress seeds. Breeds in taiga forest or larch forest. Nests on branches of larch trees. Clutch size: 3-5.

IUCN 红色名录 无危 (LC) 2016.10.1

雄鸟 Male

雌鸟 Female

黄雀	*Spinus spinus*	Eurasian Siskin	燕雀科 Fringillidae

雄鸟 Male

雌鸟 Female

北京市二级保护野生动物

IUCN 红色名录　无危 (LC) 2016.10.1

形态特征　小型鸣禽，体长 11~12 厘米。喙较细而尖锐，呈粉色。雄鸟从额至头顶及颏部黑色，上体黄色或黄绿色，两翼和尾羽黑褐色，但翼斑和尾羽基部鲜黄色；喉、胸至上腹黄色，下腹及尾下覆羽白色，两胁具黑色纵纹；跗跖黑色。雌鸟体羽黄色部分较雄鸟暗淡，头顶和颏部无黑色，余部似雄鸟。

分布与生境　中国除青藏高原和西南地区外广泛分布。春秋季见于北京低山、平原林地及城市公园，为不常见旅鸟，少量为冬候鸟。

习性与生态　非繁殖期常集群活动。喜食树木果实、草籽，也食作物种子及少量昆虫。营巢在松树树冠的侧枝上；窝卵数 5~6 枚。

Morphological characteristics　A small passerine (11-12 cm). Bill relatively fine, pink. Male has black forehead, crown and chin, yellow or yellowish-green upperparts; wings and tail dark brown; bright yellow wing markings and bases of rectrices; throat, breast and upper belly yellow; lower belly and undertail-coverts white; flanks streaked black; tarsi black. Female similar to male but with duller yellow plumage and lacking black on crown and chin.

Distribution and habitat　Widespread in China except for the Tibetan Plateau and southwest China. Uncommon passage migrant and scarce winter visitor in Beijing. Occurs in the hills and lowland forests, also in urban parks.

Behaviour and ecology　Usually forms flocks in non-breeding season. Feeds on fruit, seeds and insects. Nests on branches in canopy of pine tree. Clutch size: 5-6.

铁爪鹀	*Calcarius lapponicus*	Lapland Longspur	铁爪鹀科 Calcariidae

繁殖羽 Breeding

非繁殖羽 Non-breeding

IUCN 红色名录　无危 (LC) 2016.10.1

形态特征　小型鸣禽，体长 14~17 厘米。喙较短，呈黄色，喙端深色。雄鸟繁殖羽头部和胸黑色，白色眉纹向后延伸与颈侧白色条带相连，后颈棕红色；背部灰褐色，具深色条纹；尾羽黑褐色，最外侧尾羽白色；腹部白色，跗跖黑色。雄鸟非繁殖羽头、胸为棕褐色，眉纹、颊、颈侧为皮黄色；下体大致为污白色，喉、胸及两胁具深褐色纵纹或斑点。

分布与生境　主要分布于中国北部、中部及华东地区。繁殖于北极苔原带，不在中国繁殖。在北京活动于郊区平原草地，为不常见冬候鸟。

习性与生态　非繁殖期常集群活动，可见数百只至上千只的大群。喜食植物种子、果实及少量昆虫。鸣叫轻柔而短促，似“嘟－嘟”声。喜在苔原地上有灌丛隐蔽的土丘边或凹坑内造巢；窝卵数 4~6 枚。

Morphological characteristics　A small passerine (14-17 cm). It has short, yellow bill with dark tip. Breeding male has black head, chin, throat and breast; broad, white supercilium behind eye reaching down to white band on neck-sides; hindneck red-brown; dark streaked greyish-brown back; tail blackish-brown, with white outermost rectrices; belly white; tarsi black. Female and non-breeding male has brown head and breast; buff supercilium, cheek and neck-sides. Underparts largely off-white; throat, breast and flanks streaked or spotted dark brown.

Distribution and habitat　Northern, central and east China. Breeds in Arctic tundra. In Beijing, uncommon on lowland grasslands in rural areas. Residential status: winter visitor.

Behaviour and ecology　In non-breeding season, often in large flocks of hundreds, and sometimes thousands. Feeds on seeds, fruits and insects. Calls soft "tur tur". Nests on concealed shrubs on mound edge or in hollows on tundra. Clutch size: 4-6.

雪鹀	*Plectrophenax nivalis*	Snow Bunting	铁爪鹀科 Calcariidae

形态特征 中型鸣禽，体长 15~18 厘米。喙较短，呈黑色（繁殖期）或黄色（非繁殖期）。雄鸟繁殖羽背部、初级飞羽、三级飞羽为黑色，其余部分皆以白色为主；中央尾羽黑色，外侧尾羽白色，具黑色端斑；跗跖黑色。雌鸟繁殖羽似雄鸟繁殖羽，但头顶及耳羽为黑灰色，背部黑褐色。成鸟非繁殖羽头顶及耳羽橙棕色，背部褐色，具近黑色纵纹；胸侧褐色，其余下体白色。

分布与生境 分布于中国西北及东北地区，偶见于华北。活动于平原和低山的草地或其他开阔区域。在北京是甚罕见冬候鸟，仅于 2017 年 1 月在密云水库有过一次记载。

习性与生态 非繁殖期集群活动。喜食植物种子、果实，繁殖期亦食昆虫。营巢于岩壁缝隙、岩洞中、岩石间；窝卵数 4~6 枚。

Morphological characteristics A medium-sized passerine (15-18 cm). Bill short, black (breeding season) or yellow (non-breeding season). Breeding male is largely white with black back, primaries and tertials. Central rectrices black; outer rectrices white with black terminal band. Tarsi black. Breeding female similar to breeding male but has greyish-black crown and ear-coverts, and blackish-brown back. Non-breeding adults has orange-brown crown and ear-coverts; back brown with blackish streaks. Breast-sides brown; remainder of underparts white.

Distribution and habitat Mainly northwest and northeast China, scarce in north China. Occurs on grasslands or other open ground in lowlands and hills. Very rare winter visitor in Beijing, recorded only once at Miyun Reservoir in Miyun District (January 2017).

Behaviour and ecology In flocks in non-breeding season. Feeds on seeds, fruits and, in breeding season, insects. Nests in crevices in rocks or caves. Clutch size: 4-6.

IUCN 红色名录　无危 (LC) 2016.10.1

蓝鹀	*Emberiza siemsseni*	Slaty Bunting	鹀科 Emberizidae

雄鸟 Male

雌鸟 Female

	国家二级保护野生动物	IUCN 红色名录　无危 (LC) 2016.10.1

形态特征　小型鸣禽，体长 12~14 厘米。喙呈灰色。雄鸟通体深灰蓝色；两翼和尾羽黑褐色，具灰蓝色羽缘；最外侧尾羽具楔形白斑；腹部至尾下覆羽白色；跗跖黄褐色。雌鸟体羽以棕黄色为主，腰部和尾上覆羽蓝灰色；两翼和尾羽灰褐色。

分布与生境　本种是中国特有种，主要分布于中国华中、华南、东南及西南地区。栖息于 2000 米以下的山间林地，非繁殖期至较低海拔的林缘、灌丛及竹林中觅食。在北京为甚罕见迷鸟，目前仅于 2016 年 5 月在颐和园有过一次记载。

习性与生态　常单独、成对或集小群活动。鸣叫声为甚尖锐的"嗞嗞"声。尚缺少繁殖研究。

Morphological characteristics A small passerine (12-14 cm). Bill grey. Male overall dark slaty-blue; wings and tail blackish-brown, fringed slaty-blue; white wedges on outermost rectrices; white belly and undertail-coverts; tarsi yellowish-brown. Female largely buff-brown with slate-grey rump and uppertail-covert; wings and tail greyish-brown.

Distribution and habitat Endemic to China, found mainly south, southeast, southwest and central China. Inhabits montane forest below 2,000 m; forages in lowland scrub, bamboo forest and forest edge in non-breeding season. Very rare vagrant in Beijing, with only one record at the Summer Palace (May 2016).

Behaviour and ecology Usually solitary, in pairs or in small flocks. Calls very sharp "zi zi". Breeding biology poorly known.

黄鹀	*Emberiza citrinella*	Yellowhammer	鹀科 Emberizidae

形态特征 中型鸣禽，体长 16~18 厘米。喙呈锥状，灰色。雄鸟繁殖羽头部黄色，头顶和耳羽具深褐色纵纹，颊纹褐色；上体为斑驳的棕褐色，具黑色纵纹，腰部及尾上覆羽棕红色；尾羽较长，大致为黑褐色，外侧尾羽具白斑；颏、喉及胸部柠檬黄色，胸至两胁具显著的褐色纵纹，下体余部为淡黄色或柠檬黄色。雌鸟及雄鸟非繁殖羽黄色部分较淡，下体深色纵纹更为浓密。

分布与生境 分布于中国西北、东北及华北。为北京地区的罕见迷鸟，活动于林地、林缘及附近有稀疏树木和灌丛的农田、草地等生境。

习性与生态 喜食植物种子、果实、嫩芽、昆虫等。营巢于地面有低矮植被遮蔽的草丛、浅坑中；窝卵数 4~5 枚。

Morphological characteristics A medium-sized passerine (16-18 cm). Triangular, grey bill. Breeding male has prominent yellow head; crown and ear-coverts streaked dark brown; moustachial stripe brown; upperparts brown, steaked black; rump and uppertail-coverts rufous-brown; tail long, mostly blackish-brown; outermost rectrices have white patch. Chin, throat and breast yellow with distinctive brown streaks from breast to flanks; remainder of underparts pale yellow or lemon-yellow; tarsi brown. Female and non-breeding male have paler yellow and more extensive streaks on underparts.

Distribution and habitat Northwest, northeast and north China. Rare vagrant in Beijing. Occurs in forest and forest edge, seen also on grassland or agricultural land with a few trees or shrubs.

Behaviour and ecology Feeds on seeds, fruits, shoots and insects. Nests in grass or a depression in the ground, sheltered by grass or bush. Clutch size: 4-5.

IUCN 红色名录 无危 (LC) 2016.10.1

白头鹀	*Emberiza leucocephalos*	Pine Bunting	鹀科 Emberizidae

IUCN 红色名录 无危 (LC) 2016.10.1

形态特征 中型鸣禽，体长 16~18 厘米。喙灰色。雄鸟前额至侧冠纹近黑色（非繁殖期黑色较为斑驳）；顶冠纹白色；颊白色，其外周具黑色轮廓；眉纹、颏、喉为栗红色，背褐色并具深色纵纹，两翼黑褐色，各羽具棕红色羽缘；尾羽较长，呈黑褐色，最外侧尾羽基部白色；胸侧至两胁具浓密的栗棕色纵纹，下体余部白色。雌鸟和雄鸟相似，但颊为淡褐色，颏、喉白色，具较细的黑色纵纹。

分布与生境 广布于中国北部地区，在南方有零散记录。区域性常见于北京低山、平原林间空地、农田草地等，为冬候鸟。

习性与生态 非繁殖期常集小群活动。喜食植物种子和昆虫。营巢于地上幼树或灌木根旁的枯草丛中；窝卵数 4~5 枚。

Morphological characteristics A medium-sized passerine (16-18 cm). Bill grey. Male has blackish forehead and lateral crown-stripe (mottled black in non-breeding season), white median crown-stripe; white cheeks bordered black; chestnut supercilium, chin and throat; brown back with dark streaks; wings dark brown, fringed rufous-brown; tail long, blackish-brown; outermost rectrices based white. Breast-sides and flanks have extensive rufous streaks; remainder of underparts white. Female similar to male but has pale brownish cheek; chin and throat white with fine, black streaks.

Distribution and habitat Widespread in northern China; scarce in southern China. In Beijing, locally common in the hills and lowland farmland and grassland with scattered bushes and trees. Residential status: winter visitor.

Behaviour and ecology Often in small flocks in non-breeding season. Feeds on seeds and insects. Nests on the ground in withered grass, sheltered by low trees or shrubs. Clutch size: 4-5.

灰眉岩鹀	*Emberiza godlewskii*	Godlewski's Bunting	鹀科 Emberizidae

形态特征 中型鸣禽，体长 15~17 厘米。喙蓝灰色。雄鸟顶冠纹、眉纹、颊灰色，侧冠纹、贯眼纹深栗棕色，颊纹近黑色；上体红褐色或棕色，具黑色纵纹；两翼黑褐色，各羽具棕色羽缘，翼上有两条白色或淡皮黄色翼斑；尾羽深棕褐色，外侧尾羽具楔形白斑；颏、喉、胸灰色，下体余部棕褐色。雌鸟与雄鸟相似，但头部及下体羽色稍淡。

分布与生境 主要分布于中国北方、中部及西南等地区。见于北京海拔 500~2000 米的山区、丘陵的农田草地，冬季移动至较低海拔，为常见留鸟。

习性与生态 繁殖期成对活动，非繁殖期集小群，或与其他鹀类混群。喜食植物种子、果实、昆虫等。鸣啭为尖细而有节奏感的重复短句；鸣叫为鹀类典型的尖细"嗞"声。营巢在灌丛下部或草丛中；窝卵数 3~5 枚。

Morphological characteristics A medium-sized passerine (15-17 cm). Bill blue-grey. Male has grey median crown-stripe, supercilium and cheek; dark chestnut lateral crown-stripe and eye-stripe; blackish moustachial stripe; reddish-brown or brown upperparts with black streaks; wings blackish-brown with brown fringes and two white or creamy-buff wingbars; dark brown tail-feathers with white wedges on outermost rectrices; chin, throat and breast grey; remainder of underparts rufous-brown; Female similar to male but has paler head and underparts.

Distribution and habitat Northern, central and southwest China. In Beijing, common in open scrubby areas in mountains and hills (500-2,000 m); descends in winter. Residential status: resident.

Behaviour and ecology In pairs in breeding season; in small flocks or join mixed flocks of other buntings in non-breeding season. Feeds on seeds, fruits and insects. Songs high-pitched, repetitive phrases. Gives high-pitched call "zi" which is typical in buntings. Nests in low bush or in grass. Clutch size: 3-5.

IUCN 红色名录 无危 (LC) 2018.8.9

三道眉草鹀	*Emberiza cioides*	Meadow Bunting	鹀科 Emberizidae

雄鸟 Male

雌鸟 Female

北京市二级保护野生动物

IUCN 红色名录　无危 (LC) 2016.10.1

形态特征　中型鸣禽，体长 15~18 厘米。喙灰色。雄鸟头顶及耳羽栗棕色；眉纹白色，甚宽阔；眼先黑色；颊白色，颊纹黑色；背部棕色，具黑色纵纹；两翼黑褐色，具栗棕色羽缘；腰及尾上覆羽栗棕色；尾羽较长，中央尾羽棕红色，最外侧尾羽具楔形白斑，其余尾羽近黑色；颏、喉白色，胸及两胁栗棕色，下体余部白色。雌鸟似雄鸟，但头部及下体的羽色较淡。

分布与生境　除青藏高原和西南地区外，在中国广泛分布。见于北京中低山、丘陵及农田、草地，冬季亦在较低海拔及城市公园活动，为常见留鸟。

习性与生态　繁殖期成对，非繁殖期集群或与其他鹀类混群活动。鸣啭为一连串清脆的颤音；鸣叫为连续 2~4 声重复的尖细“嗞”声。营巢在低矮的树枝上，或在灌丛、草丛等靠近地面的隐蔽处；窝卵数 4~5 枚。

Morphological characteristics　A medium-sized passerine (15-18 cm). Bill grey. Male has chestnut-brown crown and ear-coverts; broad, white supercilium and cheek; black moustachial stripe; back brown with black streaks; wings blackish-brown with rufous-brown fringes; rump and uppertail-coverts rufous-brown; tail long; central rectrices rufous-brown; outermost rectrices have white patches; other rectrices blackish; chin and throat white; breast and flanks rufous-brown; remainder of underparts white. Female similar to male but duller and paler on head and underparts.

Distribution and habitat　Widespread in China except for the Tibetan Plateau and southwest China. In Beijing, common in grassland, farmland and scrub, mostly at middle and low elevation hills and mountains. Occurs in lower elevation and also in urban parks in winter. Residential status: resident.

Behaviour and ecology　In pairs in breeding season; in flocks or joins mixed flocks with other buntings in non-breeding season. Songs a series of trills; calls two to four notes high-pitched "zi". Nests in low branches of trees, in bushes or in grass. Clutch size: 4-5.

栗斑腹鹀	*Emberiza jankowskii*	Jankowski's Bunting	鹀科 Emberizidae

形态特征 中型鸣禽，体长 15~17 厘米。喙深灰色。雄鸟头顶及后颈栗褐色；眉纹白色，甚长且宽；眼先黑色，耳羽灰色；颊白色，颚纹黑褐色；上体棕红色，背部具黑色纵纹；尾羽黑褐色，外侧尾羽具白色区域；下体近白色，胸侧及两胁沾棕红色，腹部中央具一较大深栗色块斑。雌鸟似雄鸟，但头顶及下体羽色较淡，腹部栗色斑较小或缺失。似三道眉草鹀和灰眉岩鹀，但本种头顶、耳羽和下体羽色较淡，且雄鸟腹部具显著的栗色块斑。

分布与生境 中国有限分布于东北及华北。一般活动于较低海拔具稀疏灌丛的开阔草地。在北京为罕见冬候鸟。

习性与生态 繁殖期成对、非繁殖期集小群活动。营巢在地面草丛中或在山杏、小榆树及小黑松下部的枝杈上；窝卵数 4~6 枚。

Morphological characteristics A medium-sized passerine (15-17 cm). Bill dark grey. Male has rufous-brown crown, nape and hindneck; white supercilium long, broad; lores black; ear-coverts grey; cheek white; malar blackish-brown; upperparts rufous-brown with black streaks on back; tail mostly blackish-brown, with white patches on outermost rectrices; underparts largely off-white; breast-sides and flanks pale rufous-brown. It has dark chestnut patched belly-centre. Female like male, but duller and has smaller or lacks chestnut belly-patch. Godlewski's Bunting, Meadow Bunting and Jankowski's Bunting are similar, first two lack chestnut belly patch, while Jankowski does.

Distribution and habitat Restricted areas in northeast and north China. Occurs open grassland with scattered bushes in lowlands or low hills. Rare winter visitor in Beijing.

Behaviour and ecology In pairs in breeding season; in small flocks in non-breeding season. Nests on ground in long grass or on lower branches of wild apricot, young elm and black pine. Clutch size: 4-6.

	国家一级保护野生动物	IUCN 红色名录 濒危 (EN) 2018.8.9

雄鸟 Male

雌鸟 Female

白眉鹀	*Emberiza tristrami*	Tristram's Bunting	鹀科 Emberizidae

IUCN 红色名录 无危 (LC) 2016.10.1

形态特征 小型鸣禽，体长 12~15 厘米。喙灰色，下喙基部粉色。雄鸟头黑色，具显著的白色顶冠纹、眉纹和颚纹；上体大致为栗色，背部具黑色纵纹；尾羽黑褐色，外侧尾羽白色；颏、喉部黑色；胸及两胁具栗褐色纵纹，其余下体白色。雌鸟与雄鸟相似，但头顶和头侧的黑色区域为棕褐色。

分布与生境 广泛分布于中国东部和南部地区。迁徙过境时更偏好沿海林地和灌丛生境。春秋两季见于北京低山和平原的林地、林缘及灌丛区域，为不常见的旅鸟。

习性与生态 非繁殖期多集群活动。叫声为单调尖细的"嗞"声。繁殖期营巢在林间溪流边植被丰富的地面；窝卵数 5~6 枚。

Morphological characteristics A small passerine (12-15 cm). Bill grey, base of lower mandible pink. Male has black head; white median crown-stripe, supercilium and malar; upperparts largely chestnut-brown with black streaks on back; tail blackish-brown; white on outermost rectrices; chin and throat black; breast and flanks streaked rufous-brown; remainder of underparts white. Female similar to male, but with dark brown, not black, crown and head-sides.

Distribution and habitat Eastern and southern China. More common in coastal woodlands or bushes on migration. In Beijing, uncommon in wooded and scrubby areas in the hills and lowlands in spring and autumn. Residential status: migrant.

Behaviour and ecology Forms flocks in non-breeding season. Gives high-pitched calls "zi". Nests on ground in well-vegetated habitats, usually near streams in forests. Clutch size: 5-6.

栗耳鹀	*Emberiza fucata*	Chestnut-eared Bunting	鹀科 Emberizidae

形态特征 小型鸣禽，体长 14~16 厘米。喙灰色，下喙基部粉色或淡黄色。雄鸟头顶至后颈呈灰色，颊部至耳羽栗色，颊纹淡黄色或污白色，背部栗棕色，具黑褐色纵纹；尾羽深褐色；下体大致为白色，黑色的颚纹向下延伸至胸形成黑色胸带，两胁栗棕色。雌鸟与雄鸟相似，但头顶及上体偏褐色，胸带为栗棕色而非黑色，两胁为淡皮黄色。

分布与生境 广泛分布于中国东部、中部至南部地区。见于北京低山、平原的灌丛、草地及苇丛中，是不常见旅鸟。

习性与生态 非繁殖期常集 10 只以内的小群活动。性谨慎。叫声为短促而尖细的“嗞”声。营巢在林缘有灌丛且近水的草地中；窝卵数 4~6 枚。

Morphological characteristics A small passerine (14-16 cm). Grey bill; pink at base of lower mandible. Male is grey from crown to nape and hindneck, chestnut from cheek to ear-coverts; moustachial stripe white; back rufous-brown with blackish-brown streaks; tail dark brown; chin, throat and breast white or cream; underparts mostly white; black malar extends to breast, forming a black breast-band; flanks rufous-brown. Female similar to male but has brownish crown, upperparts, and a paler rufous-chestnut breast-band and pale buff flanks.

Distribution and habitat Widespread in eastern, central and southern China. In Beijing, uncommon in meadows, grassland and reedbeds in low hills and lowlands. Residential status: passage migrant.

Behaviour and ecology Forms small flocks (usually less than 10 individuals) in non-breeding season. Shy. Calls short, high-pitched "zi". Nests in scrubby wet meadows at forest edge; nest hidden on ground or in shrub close to ground. Clutch size: 4-6.

IUCN 红色名录　无危 (LC) 2016.10.1

小鹀	*Emberiza pusilla*	Little Bunting	鹀科 Emberizidae

IUCN 红色名录　无危 (LC) 2016.10.1

形态特征　小型鸣禽，体长 12~14 厘米。喙较短，呈灰色。雌雄相似。成鸟繁殖羽顶冠纹栗色，侧冠纹黑色，宽阔的眉纹呈栗色（眉纹羽色于眼后变淡），颊至耳羽亦为栗色，边缘具黑色轮廓，颚纹黑色；背部褐色，具黑色与污白色相间的纵纹；尾羽黑褐色，最外侧尾羽具显著白色部分；颏、喉栗棕色，下体余部白色，胸至两胁具清晰的黑色纵纹；跗跖黄褐色。成鸟非繁殖羽与繁殖羽大致相似，但头部及背部羽色较淡，颏、喉为淡黄白色或污白色。

分布与生境　中国除青藏高原外广泛分布。在北京是甚常见旅鸟，少数为冬候鸟，见于各种草地、农田、灌丛、苇丛和开阔林地等生境。

习性与生态　于非繁殖期集群活动或与其他鹀类混群。鸣叫声为清脆的“唧”声。繁殖在西伯利亚苔原和泰加林区域，营巢在地面的灌木、草丛或低矮的树丛中；窝卵数 4~6 枚。

Morphological characteristics A small passerine (12-14 cm). Male and female are similar. Bill short, grey. Breeding adult has chestnut median crown-stripe, black lateral crown-stripe and broad, chestnut supercilium (paler behind eyes); cheek and ear-coverts chestnut, bordered black; malar black; back brown, streaked black and off-white; tail blackish-brown; outermost rectrices partly white; chin and throat rufous-brown; breast and flanks streaked black; remainder of underparts white; tarsi yellowish-brown. Non-breeding adult is similar but has paler head and back, and creamy or off-white chin and throat.

Distribution and habitat Widespread in China except for the Tibetan Plateau. Abundant migrant with a few overwintering. Occurs on grassland and farmland, also in reedbeds, scrub and open woodland, including parks.

Behaviour and ecology In flocks or in mixed flocks with other buntings in non-breeding season. Gives clipped calls "chi". Breeds on the Siberian tundra and in taiga forest. Nests on ground in scrub, grass or low down in trees. Clutch size: 4-6.

黄眉鹀 *Emberiza chrysophrys* Yellow-browed Bunting

鹀科
Emberizidae

形态特征 小型鸣禽，体长 14~16 厘米。上喙灰色，下喙粉色或淡黄色。雄鸟顶冠纹白色，侧冠纹黑色，眉纹前段黄色，后端转为白色；眼先、颊至耳羽黑色，颊纹白色，耳羽后方具一白斑，颚纹黑色；背部褐色具深色纵纹，腰部和尾上覆羽棕红色，尾羽黑褐色，最外侧两对尾羽具白斑；下体大致为白色，胸部和两胁具黑色纵纹。雌鸟与雄鸟相似，但头部的条纹及斑块为棕褐色而非黑色。

分布与生境 分布于中国东部至华南地区。见于北京平原及低山林地、湿地、灌丛、草地和农田，是不常见旅鸟。

习性与生态 非繁殖期集群活动。鸣叫声为鹀类典型的短促“嗞”声。繁殖在西伯利亚的泰加林，营巢于树上较低处；窝卵数 4 枚。

Morphological characteristics A small passerine (14-16 cm). Upper mandible grey, lower mandible pale yellow or pink. Male has white median-crown stripe and black lateral crown-stripe; yellow fore supercilium and white rear supercilium; black lores, cheeks and ear-coverts; white moustachial stripe; white patch behind ear-coverts, black malar, brown back with dark streaks, reddish-brown rump and uppertail-coverts, dark brown tail with white on outermost tail-feathers; underparts largely white; breast and flanks streaked black. Female similar to male, but with brown, not black, crown and head-sides.

Distribution and habitat Eastern and south China. In Beijing, uncommon in woodlands, wetlands, scrub, grassland and farmland in lowland and low hills. Residential status: migrant.

Behaviour and ecology Forms flocks in non-breeding season. Gives short call "zi" which is typical in buntings. Breeds in Siberian taiga forest. Nests low down in trees. Clutch size: 4.

IUCN 红色名录 无危 (LC) 2016.10.1

雄鸟 Male

雌鸟 Female

田鹀	*Emberiza rustica*	Rustic Bunting	鹀科 Emberizidae

IUCN 红色名录 易危 (VU) 2016.10.1

形态特征 小型鸣禽，体长 14~16 厘米。喙较粗厚，呈粉色，喙端深灰色。雄鸟繁殖羽头黑色，具冠羽，眉纹及颊纹白色，耳羽后方具一小块白斑；上体棕红色，背羽具近黑色纵纹；两翼黑褐色，各羽具棕色羽缘，大覆羽和中覆羽端部白色，形成两条翼斑；尾羽黑褐色，最外侧尾羽部分白色；下体白色，胸部及两胁具浓密的栗色纵纹。雌鸟及雄鸟非繁殖羽头顶及颊为黄褐色而非黑色，整体羽色亦不及雄鸟繁殖羽鲜亮。

分布与生境 分布于中国东部、华中及华南地区，亦偶见于西北。在北京是区域性常见冬候鸟，活动于低山、丘陵的开阔林地及农田草地，亦见于城市公园。

习性与生态 非繁殖期集群活动，常与其他鹀类混群。鸣叫为短促的“嗞”声。通常营巢于枯草丛中；窝卵数 4~5 枚。

Morphological characteristics A small passerine (14-16 cm). Bill deep, pink, with dark grey tip. Breeding male has black head with a crest; white supercilium and moustachial stripe; a white patch behind ear-coverts; rufous-brown upperparts with blackish streaks on back; wings blackish-brown, fringed rufous; greater and median coverts have white tips, forming two distinctive wingbars; tail mostly blackish-brown with white on the outermost tail-feathers; underparts largely white, with extensive chestnut streaks on breast and flanks. Female and non-breeding male has overall duller plumage and buff-brown, not black, crown and cheek.

Distribution and habitat Eastern, central and south China; scarce in northwest. Locally common winter visitor in Beijing. Occurs in open woodland, farmland, grassland and urban parks.

Behaviour and ecology Forms flocks or in mixed flocks with other buntings in bon-breeding season. Calls short "zi". Nests in withered grass. Clutch size: 4-5.

黄喉鹀	*Emberiza elegans*	Yellow-throated Bunting	鹀科 Emberizidae

形态特征 小型鸣禽，体长 15~16 厘米。喙深灰色。雄鸟头顶黑色，眉纹至枕部为柠檬黄色，具显著的黑黄相间之冠羽；眼先、颊及耳羽黑色；背部棕色，具深褐色纵纹；尾羽深褐色，最外侧尾羽具白斑；颏、喉黄色，胸部具三角形黑斑，两胁具褐色纵纹，其余下体白色。雌鸟似雄鸟，但头部为棕褐色而非黑色，胸部黑斑不明显或缺失。

分布与生境 广布于中国东部、中部及南部地区。为北京常见的旅鸟、冬候鸟和夏候鸟，见于山区至平原的林地、灌丛、农田和草地（但繁殖地仅限于山区）。

习性与生态 繁殖期成对活动，非繁殖期集群或与其他鹀类混群。鸣啭为清脆的一连串颤音；鸣叫声为短促的“嗞”声。营巢于林地和林缘地面的草丛中，或灌丛及低矮的小树上；窝卵数 4~6 枚。

Morphological characteristics A small passerine (15-16 cm). Bill dark grey. Male has black crown, yellow supercilium and nape, forming black and yellow crest; black lores, cheeks and ear-coverts; brown back with dark brown streaks; tail dark-brown with white wedge on outermost rectrices; yellow chin and throat; triangular black breast-band; flanks streaked brown; remainder of underparts white. Female similar but with brown head instead of black and lacking black breast-band.

Distribution and habitat Widespread in eastern, central and southern China. In Beijing, common in woodland and scrub, including parks in both mountains and lowlands. Residential status: passage migrant, winter visitor and summer breeder (breeds only in mountains).

Behaviour and ecology In pairs in breeding season; forms flocks or joins mixed flocks with other buntings in non-breeding season. Songs a series of clear trills; calls short "zi". Nests in grass, shrubs or low trees at forest edge or in forest. Clutch size: 4-6.

北京市二级保护野生动物 | | IUCN 红色名录 无危 (LC) 2018.8.9

雄鸟 Male

雌鸟 Female

黄胸鹀 *Emberiza aureola* Yellow-breasted Bunting

鹀科 Emberizidae

雄鸟 Male

雄鸟 Male

雌鸟 Female

国家一级保护野生动物　IUCN 红色名录　极危 (CR) 2017.10.1

形态特征　小型鸣禽，体长 14~15 厘米。上喙灰色，下喙淡黄色。雄鸟繁殖羽头部及颏为黑色，头顶后部、背、腰及尾上覆羽皆为栗棕色；两翼黑褐色，各羽羽缘栗棕色，翼上具一道或两道白色翼斑；尾羽黑褐色，最外侧尾羽部分白色；下体黄色，胸部具一显著的栗色横带。雄鸟非繁殖羽头部为斑驳的褐色，背部具黑色纵纹。雌鸟上体黄褐色，眉纹淡黄色，耳羽淡褐色，下体淡黄色，胸无横带。

分布与生境　中国除青藏高原外广泛分布。见于北京中低山及平原的农田和湿地。近年在北京是不常见旅鸟。

习性与生态　近年来全球种群数量锐减。非繁殖季常集群活动。营巢在湿地边的灌丛、草丛遮挡的地面；窝卵数 4~5 枚。

Morphological characteristics A small passerine (14-15 cm). Upper mandible grey, lower mandible pale yellow. Breeding male has black head and chin; chestnut-brown from hindcrown to back, rump and uppertail-coverts; wings blackish-brown with chestnut-brown fringes and one or two white wing-patches; tail blackish-brown, with white on the outermost rectrices; yellow underparts with a chestnut breast-band. Non-breeding male has mottled brown head and black-streaked back. Female has buff-brown upperparts, buffy-yellow supercilium, pale brown ear-coverts and pale buff underparts, lacking breast-band.

Distribution and habitat Widespread in China except for the Tibetan Plateau. In Beijing, uncommon in farmlands and wetlands in lowlands and low hills. Residential status: migrant.

Behaviour and ecology The global population have decreased dramatically in recent years. Often in flocks in non-breeding season. Nest concealed on ground in grass or at base of bush at wetlands. Clutch size: 4-5.

栗鹀 | *Emberiza rutila* | Chestnut Bunting | 鹀科 Emberizidae

形态特征 小型鸣禽，体长 14~15 厘米。喙灰色，部分个体下喙基部较淡。雄鸟繁殖羽头、上体、颏、喉、胸皆为栗色，尾羽黑褐色；下体黄色，跗跖黄褐色。雄鸟非繁殖羽的头部和上体较雄鸟繁殖羽暗淡。雌鸟具不清晰的皮黄色眉纹，头顶及上体灰褐色，具深褐色纵纹；下体淡黄色，两胁具深褐色纵纹。

分布与生境 广泛分布于中国东部及南部各省。春秋两季见于北京中低山、平原的林地、灌丛及农田，为不常见旅鸟。

习性与生态 非繁殖季常集小群活动。鸣叫声为鹀类典型的短促“嗞”声。营巢在地面的枯草丛中；窝卵数 4~5 枚。

Morphological characteristics A small passerine (14-15 cm). Bill grey; sometimes paler at based of lower mandible. Breeding male has chestnut-red head, upperparts, neck, throat and upper breast; dark brown wings and tail, yellow underparts; tarsi yellowish-brown. Non-breeding male has duller head and upperparts. Female has indistinct buff supercilium; greyish-brown crown and upperparts with dark brown streaks; underparts pale yellow; flanks have dark brown streaks.

Distribution and habitat Widespread in eastern and southern China. In Beijing, uncommon in woodland, scrub and farmland in lowlands and low hills. Residential status: passage migrant.

Behaviour and ecology Often in small flocks in non-breeding season. Gives short call "zi", which is a typical call in buntings. Nests on ground in grass. Clutch size: 4-5.

IUCN 红色名录 无危 (LC) 2016.10.1

雌鸟 Female

雄鸟繁殖羽 Breeding male

雄鸟非繁殖羽 Non-breeding male

褐头鹀	*Emberiza bruniceps*	Red-headed Bunting	鹀科 Emberizidae

雄鸟 Male

雌鸟 Female

IUCN 红色名录　无危 (LC) 2016.10.1

形态特征　中型鸣禽，体长 16~18 厘米。喙灰色，呈锥状。雄鸟繁殖羽头、颏、喉至上胸红褐色，背部橄榄绿色，具黑褐色纵纹，腰黄色；两翼黑褐色，翼上各羽具灰白色或淡皮黄色羽缘；尾羽深褐色，下体亮黄色。雄鸟非繁殖羽头部的红褐色较为斑驳。雌鸟上体淡橄榄褐色，具黑褐色纵纹，腰部、下体浅黄色。

分布与生境　主要分布于中国新疆地区的无树、稀树或具稀疏灌丛的干旱平原，为当地夏候鸟。1923 年 5 月一只雄鸟曾见于北京，被视为北京甚罕见的迷鸟。

习性与生态　非繁殖期集小群。繁殖期营巢于低矮的灌丛或草丛中；窝卵数 4~5 枚。

Morphological characteristics A medium-sized passerine (16-18 cm). Triangle, grey bill. Breeding male has reddish-brown head, chin, throat and upper breast; olive-green back with blackish streaks, yellow rump, blackish-brown wings with pale greyish-white or buff-white fringes; tail dark brown; bright yellow underparts. Non-breeding male has mottled red-brown head. Female has pale olive-brown finely streaked upperparts, pale yellow rump and underparts.

Distribution and habitat In China, a summer breeder in Xinjiang, where it inhabits arid plains with scattered trees or bushes. Very rare in Beijing, only one record of a male in May 1923. Residential status: vagrant.

Behaviour and ecology In small flocks in non-breeding season. In breeding season, nests on ground in grass or in low shrubs. Clutch size: 4-5.

灰头鹀	*Emberiza spodocephala*	Black-faced Bunting	鹀科 Emberizidae

形态特征 小型鸣禽，体长 14~15 厘米。喙淡粉色，喙峰灰色。雄鸟头至胸深灰色，眼先、颏黑色，上体褐色或橄榄褐色，具近黑色纵纹；尾羽黑褐色，最外侧尾羽具白斑；腹部黄色或淡黄色，两胁具褐色纵纹，尾下覆羽白色。雌鸟头和上体皆为褐色，具深色纵纹，眉纹皮黄色，下体污白色或浅黄色，胸和两胁具显著的黑色纵纹。

分布与生境 中国除西部地区外广泛分布。区域性常见于北京较低海拔的林地及湿地草丛、苇丛等生境，居留型主要为旅鸟，亦有少量越冬个体。

习性与生态 非繁殖期集群活动。鸣叫声为短促的"嗞"声。营巢在低矮的灌丛或草丛中；窝卵数 4~6 枚。

Morphological characteristics A small passerine (14-15 cm). Bill pale pink with grey culmen. Male is dark grey from head to breast; lores and chin black; upperparts brown or olive-brown with blackish streaks; tail blackish-brown, with white on the outermost tail-feathers; belly yellow or pale yellow; flanks streaked brown; undertail-coverts white. Female has brown head and upperparts with dark streaks, buff supercilium, off-white or pale yellow underparts, with dark streaks on breast and flanks.

Distribution and habitat Widespread in China except for western China. In Beijing, locally common in lowland woodlands, wet grassland and reedbeds at wetlands. Residential status: passage migrant, with a few overwintering.

Behaviour and ecology In flocks in non-breeding season. Calls short "zi". Nests in low shrubs or grasses. Clutch size: 4-6.

IUCN 红色名录 无危 (LC) 2016.10.1

雄鸟 Male

雄鸟 Male

雌鸟 Female

苇鹀	*Emberiza pallasi*	Pallas's Bunting	鹀科 Emberizidae

繁殖羽 Breeding

非繁殖羽 Non-breeding

非繁殖羽 Non-breeding

换羽中 Moulting

IUCN 红色名录　无危 (LC) 2018.8.9

形态特征　小型鸣禽，体长 13~14 厘米。喙呈锥状，喙峰较直；上喙灰色，下喙粉色。雄鸟繁殖羽上下喙皆为黑色；头顶至头两侧黑色，颚纹白色，甚宽，颈部白色；背部褐色，具黑白相间之纵纹；腰和尾上覆羽皮黄色或灰褐色；翼上小覆羽灰色，其余各羽大致为黑褐色，具淡色羽缘；尾羽黑褐色，最外侧尾羽具白色区域；颏、喉黑色，下体余部白色。雄鸟非繁殖羽和雌鸟头部褐色，眉纹皮黄色或淡黄白色；颏、喉白色。

分布与生境　广泛分布于中国北部、东部及华南地区。见于北京低山、平原的湿地苇丛和草丛，为常见冬候鸟。

习性与生态　非繁殖期集群活动，性活跃。鸣叫声为清脆的“嗞喳”声或鹀类典型的短促“嗞”声。营巢在地面草丛或灌木低枝上；窝卵数 4~5 枚。

Morphological characteristics A small passerine (13-14 cm). Bill triangular with strait culmen. Breeding male has all-black bill, black crown and head-sides; malar-stripe broad, white; neck white; back brown, with black and white streaks; rump and uppertail-coverts buff or greyish-brown; wings largely blackish-brown with pale fringes; lesser coverts grey; tail blackish-brown; outermost rectrices partly white; chin and throat black; remainder of underparts white. Non-breeding male and female have grey upper mandible and pink lower mandible; brown head with buff or pale yellowish-white supercilium; chin and throat white.

Distribution and habitat Northern, eastern and south China. In Beijing, common in reedbeds, grassland and scrub at lowland wetlands. Residential status: winter visitor.

Behaviour and ecology Forms flocks in non-breeding season. Active. Calls clipped "zi-chur" or short typical bunting-like "zi". Nests on ground in grasses or low shrubs. Clutch size: 4-5.

VI

鸣禽

红颈苇鹀	*Emberiza yessoensis*	Ochre-rumped Bunting	鹀科 Emberizidae

形态特征 小型鸣禽，体长 14~15 厘米。喙呈锥状，喙峰较直。雄鸟繁殖羽喙黑色，头、颏、喉至上胸部皆为黑色；后颈、背、腰及尾上覆羽红褐色，背部具黑白相间之纵纹；两翼黑褐色，羽缘棕红色，小覆羽蓝灰色；尾羽黑褐色，最外侧尾羽具白斑；下体白色。雄鸟非繁殖羽及雌鸟喙峰灰色，喙余部粉色；头顶和头两侧黑褐色，具宽阔的皮黄色眉纹；下体污白色或淡黄白色。相似种苇鹀和芦鹀的雄鸟繁殖羽后颈都有白色环及颚纹，而红颈苇鹀不具上述特征。

分布与生境 中国主要见于东部地区，于内陆地区有零星记载。见于北京平原湿地苇丛、灌丛和草丛，为罕见旅鸟和冬候鸟。

习性与生态 非繁殖期一般单独活动、集小群或与苇鹀混群。性活跃。鸣声为短促的“嗞”声。营巢于湿地草丛或灌丛中；窝卵数 5~6 枚。

Morphological characteristics A small passerine (14-15 cm). Triangular bill with strait culmen. Breeding male has black bill; all-black head, chin, throat and upper breast; reddish-brown hindneck, back, rump and uppertail-coverts, with black and white streaks on back; blackish-brown wings with rufous fringes; lesser coverts blue-grey; blackish-brown tail with white on the outermost rectrices; white underparts. Non-breeding male and female have pink bill with grey culmen; blackish-brown crown and head-sides; broad, buff supercilium; off-white or creamy underparts. Similar to Pallas's Bunting and Reed Bunting, but lacking white neck-ring (especially hindneck) and malar.

Distribution and habitat Mainly eastern China, scarce in central China. In Beijing, rare in reedbeds and scrubby grassland during migration and in winter. Residential status: passage migrant and winter visitor.

Behaviour and ecology Solitary, in small flocks or joins mixed flocks with Pallas's Buntings in non-breeding season. Active. Gives short, clipped calls "zi". Nests in grasses or shrubs. Clutch size: 5-6.

IUCN 红色名录 近危 (NT) 2020.8.6

雌鸟 Female

幼鸟 Juvenile

芦鹀	*Emberiza schoeniclus*	Reed Bunting	鹀科 Emberizidae

繁殖羽 Breeding

非繁殖羽 Non-breeding

非繁殖羽 Non-breeding

IUCN 红色名录 无危 (LC) 2018.8.9

形态特征 小型鸣禽，体长 14~17 厘米。喙深灰色，喙峰呈弧形。雄鸟繁殖羽头部黑色，颚纹白色；后颈及颈侧白色；背部红褐色，具清晰的黑色纵纹，腰及尾上覆羽灰色；两翼大致为黑褐色，各羽多具棕红色羽缘，小覆羽棕红色。尾羽黑褐色，最外侧尾羽部分白色；颏、喉至胸中央黑色；其余下体白色。雄鸟非繁殖羽及雌鸟头部褐色，眉纹皮黄色，下体污白色或淡皮黄色。甚似苇鹀，但本种个体稍大，且具略呈弧形的喙峰、灰色的下喙及棕红色的小覆羽。

分布与生境 分布于中国东部和北部地区。见于北京较低海拔的近水苇丛、草丛中，是不常见旅鸟和冬候鸟。

习性与生态 在北京常与苇鹀混群活动。性活跃。鸣叫声为清脆而尖细的“嗞”声。营巢于湿地苇丛、灌丛或草丛中；窝卵数 4~6 枚。

Morphological characteristics A small passerine (14-17 cm). Dark grey bill with arched culmen. Breeding male has black head with white malar; hindneck and neck-sides white; reddish-brown back with distinctive black streaks; grey rump and uppertail-coverts; wings largely blackish-brown, with rufous-brown fringes; lesser coverts rufous-brown; tail blackish-brown; outermost rectrices partly white; chin, throat to central breast black; remainder of underparts white. Non-breeding male and female have brown head, buff supercilium, off-white or pale buff underparts. Larger than Pallas's Bunting, identified by slightly arched culmen, grey lower mandible and rufous-brown lesser coverts.

Distribution and habitat Northern and eastern China. In Beijing, uncommon in reedbeds and grass at lowland wetlands. Residential status: passage migrant and winter visitor.

Behaviour and ecology In Beijing, often joins mixed flocks with Pallas's Bunting. Calls clear, high-pitched "zi". Nests in reedbeds, grasses or bushes at wetlands. Clutch size: 4-6.

主要参考文献
Key References

北京市园林绿化局，2018. 北京市园林绿化局公告 2007 年 第 1 号：北京市一级保护野生动物名录 [J]. 北京市人民政府公报 (1)：42-47.

北京市园林绿化局，2018. 北京市园林绿化局公告 2007 年 第 2 号：北京市二级保护野生动物名录 [J]. 北京市人民政府公报 (1)：49-59.

蔡其侃，1988. 北京鸟类志 [M]. 北京：北京出版社 .

陈服官，罗时有，等，1998. 中国动物志 . 鸟纲 . 第九卷，雀形目 (太平鸟科—岩鹨科)[M]. 北京：科学出版社 .

傅桐生，宋榆钧，高玮，等，1998. 中国动物志 . 鸟纲 . 第十四卷，雀形目 (文鸟科、雀科)[M]. 北京：科学出版社 .

高武，陈卫，付必谦，等，1994. 北京脊椎动物检索表 [M]. 北京：北京出版社 .

国家林业和草原局，农业农村部 . 国家重点保护野生动物名录 [EB/OL]. (2021-02-25) [2021-04-08]http://www.forestry.gov.cn/main/3957/20210205/153020834857061.html

李桂垣，郑宝赉，刘光佐，1982. 中国动物志 . 鸟纲 . 第十三卷，雀形目 (山雀科—绣眼鸟科)[M]. 北京：科学出版社 .

钱燕文，1995. 中国鸟类图鉴 [M]. 郑州：河南科学技术出版社 .

谭耀匡，关贯勋，2003. 中国动物志 . 鸟纲 . 第七卷，夜鹰目、雨燕目、咬鹃目、佛法僧目、鴷形目 [M]. 北京：科学出版社 .

王岐山，马鸣，高育仁，2006. 中国动物志 . 鸟纲 . 第五卷，鹤形目、鸻形目、鸥形目 [M]. 北京：科学出版社 .

张荣祖，2011. 中国动物地理 [M]. 北京：科学出版社 .

赵欣如，2014. 北京鸟类图鉴 [M]. 北京：北京师范大学出版社 .

赵欣如，2018. 中国鸟类图鉴 [M]. 北京：商务印书馆 .

赵正阶，2001. 中国鸟类志 [M]. 长春：吉林科学技术出版社 .

郑宝赉，等，1985. 中国动物志 . 鸟纲 . 第八卷，雀形目 (阔嘴鸟科—和平鸟科) [M]. 北京：科学出版社 .

郑光美，2017. 中国鸟类分类与分布名录 [M]. 3 版 . 北京：科学出版社 .

郑作新，等，1978. 中国动物志 . 鸟纲 . 第四卷，鸡形目 [M]. 北京：科学出版社 .

郑作新，等，1979. 中国动物志 . 鸟纲 . 第二卷，雁形目 [M]. 北京：科学出版社 .

郑作新，等，1997. 中国动物志 . 鸟纲 . 第一卷，第一部 中国鸟纲绪论；第二部 潜鸟目、鹱鹈目、鹱形目、鹈形目、鹳形目 [M]. 北京：科学出版社 .

郑作新，等，2010. 中国动物志 . 鸟纲 . 第十二卷，雀形目 (鹟科：Ⅲ . 莺亚科、鹟亚科) [M]. 北京：科学出版社 .

郑作新，龙泽虞，卢汰春，1995. 中国动物志 . 鸟纲 . 第十卷，雀形目 (鹟科：Ⅰ. 鸫亚科) [M]. 北京：

科学出版社 .

郑作新，龙泽虞，郑宝赉，1987. 中国动物志 . 鸟纲 . 第十一卷，雀形目 (鹟科：Ⅱ . 画眉亚科) [M]. 北京：科学出版社 .

郑作新，冼耀华，关贯勋，1991. 中国动物志 . 鸟纲 . 第六卷，鸽形目、鹰形目、鹃形目、鸮形目 [M]. 北京：科学出版社 .

Billerman S M, et al., 2021. Birds of the World[M]. Cornell Laboratory of Ornithology, Ithaca, NY, USA. https://birdsoftheworld.org/bow/home/.

Brazil M, 2009. Birds of East Asia[M]. Princeton: Princeton University Press.

Cofta T, 2021. Flight Identification of European Passerines and Select Landbirds: An Illustrated and Photographic Guide[M]. Princeton: Princeton University Press.

Gill F, et al., 2021. IOC World Bird List (Version 11.2)[DB/OL]. http://www. worldbirdnames. org/.

IUCN, 2021. The IUCN Red List of Threatened Species. Version 2021-1[DB/OL]. https:// www. iucnredlist.org/.

附　录
Appendix

中文名索引
Index to Chinese Names

J

K

L

学名索引
Index to Scientific Names

英文名索引
Index to English Names

附录

英文名索引

鸟名生僻字 GLOSSARY

（按第一字笔画数排序）

一至五画

勺［杓］sháo（芍）

兀鹫 wù jiù（误救）

刈 yì（意）

六至十画

凫 fú（服）

曳 yè（夜）

鸠 jiū（究）

鸢 yuān（冤）

鸨 bǎo（保）

䴗 jú（菊）

䴓 shī（师）

鹇 yán（炎）

鸫 dōng（冬）

鸱鸮 chī xiāo（吃消）

鸡 lì（立）

鸬鹚 lú cí（炉磁）

鸲 qú（渠）

隼 sǔn（损）

十一至十五画

笞 chī（吃）

鸸鹋 ér miáo（儿苗）

鸹 guā（瓜）

鸻 héng（恒）

鴷 liè（列）

鸺鹠 xiū liú（休留）

鵏 bǔ（补）

鹄 hú（湖）

鹡 jí（吉）

鵟 kuáng（狂）

鹂 lí（离）

椋 liáng（凉）

鹈鹕 tí hú（题胡）

鸵鸵 tuǒ kōng （妥空）

鹀 wú（吴）

鹇 xián（贤）

棕［椶］zōng（宗）

鹎 bēi（杯）

塍 chéng（成）

鹊 què（却）

鹗 è（饿）

鹛 méi（眉）

鹡鸰 jí líng（急灵）

鲣 jiān（坚）

鹟 wēng（翁）

鹞 yào（药）

十六至二十画

雕［鵰］diāo（刁）

鹨 liù（六）

薮 sǒu（叟）

鹧鸪 zhè gū（这估）

鵁鶄 jiāo liáo（交辽）

鹬 yù（玉）

鹱 hù（户）

鹮 huán（环）

鹭 lù（路）

鹲 méng（盟）

䴙䴘 pì tī（辟梯）

二十画以上

蠹 dù（杜）